Confronting the Climate Change and Health Nexus: Interactions, Impacts, and Adaptation Strategies

Confronting the Climate Change and Health Nexus: Interactions, Impacts, and Adaptation Strategies

Guest Editor

Ying Li

MDPI

Basel • Beijing • Wuhan • Barcelona • Belgrade • Novi Sad • Cluj • Manchester

Guest Editor
Ying Li
Department of Environmental
and Occupational Health and
Safety Sciences
East Tennessee State
University
Johnson City
USA

Editorial Office
MDPI AG
Grosspeteranlage 5
4052 Basel, Switzerland

This is a reprint of the Special Issue, published open access by the journal *Climate* (ISSN 2225-1154), freely accessible at: https://www.mdpi.com/journal/climate/special_issues/61O376Y658.

For citation purposes, cite each article independently as indicated on the article page online and as indicated below:

Lastname, A.A.; Lastname, B.B. Article Title. *Journal Name* **Year**, *Volume Number*, Page Range.

ISBN 978-3-7258-7508-5 (Hbk)
ISBN 978-3-7258-7509-2 (PDF)
https://doi.org/10.3390/books978-3-7258-7509-2

Contents

About the Editor

Ying Li

Ying Li is an Associate Professor at the Department of Environmental and Occupational Health and Safety Sciences, College of Public Health, East Tennessee State University in Johnson City, Tennessee, in the United States. Her main research and teaching interests are related to air pollution, climate change, risk assessment, and economic evaluation. Her doctoral dissertation evaluated the potential health benefits of traffic-related fine particular matter (PM2.5) control policies in the city of Bangkok, Thailand. She has conducted multiple large-scale projects related to population health risk assessments of outdoor/indoor air pollution and heat stress in various developed and developing countries, including the United States, China, and United Arab Emirates, and has published more than thirty journal articles, book chapters, and technical reports as a result of these projects.

climate

MDPI

Article

High-Resolution Temporal Variation of Thermal Discomfort Indices in the Eastern Mediterranean City of Athens, Greece, Since the Beginning of the 20th Century (1901–2024)

Basil E. Psiloglou *, Nikolas Gkinis and Christos Giannakopoulos

Institute for Environmental Research and Sustainable Development, National Observatory of Athens, P. Penteli, GR-15236 Athens, Greece; n.gkinis@noa.gr (N.G.); cgiannak@noa.gr (C.G.)

* Correspondence: bill@noa.gr

Abstract: This study analyzes more than a century of hourly meteorological data (1901–2024) from the Thissio station in central Athens, Greece, to assess the long-term changes in human thermal discomfort. Three simple and widely used bioclimatic indices, Thom's Discomfort Index (TDI), Humidex (HMDX), and Heat Index (HI), were calculated to capture the combined effects of air temperature and humidity. The results show a marked increase in the frequency, intensity, and duration of thermal discomfort since the 1980s, with a strong acceleration after 2000. The number of days with severe or dangerous heat stress has more than doubled compared with the mid-20th century, and periods of high discomfort now extend from June to September. The maximum values of HMDX and HI have exceeded critical health thresholds, highlighting increasing risks for the urban population. These findings demonstrate how rising temperature and humidity amplify heat stress in a Mediterranean city and emphasize the need for adaptation strategies in urban planning and public health to reduce vulnerability to extreme heat.

Keywords: climate change; thermal discomfort; discomfort indices; Thom's Discomfort Index; Humidex; Heat Index; Athens; Greece

1. Introduction

The issue of climate change, which is driven by global warming, is of the utmost importance in the twenty-first century, having a profound impact on various aspects of human life, including public health [1]. The climate crisis is triggering changes in climate parameters, engendering shifts in thermal comfort that necessitate urgent climate action [2]. Climate change is a complex and dynamic process associated with many uncertain externalities [3,4]. It has been demonstrated that global warming and extreme weather events, such as storms and heat waves, are associated with climate change [5–9]. According to different scenarios, the world's temperature is expected to increase between 1.4 and 5.8 °C by the end of the twenty-first century, depending on the level of emissions [10,11]. It is particularly probable that heat waves and extreme precipitation are likely to occur with greater intensity [12]. Since the time of Herodotus (430 BC) it has been known that climatic conditions exert a combined effect on the human body, thereby increasing thermal discomfort and potentially impacting health [13–16]. This makes it especially important to investigate changes in thermal comfort, particularly in outdoor applications.

Human comfort depends upon physiological, psychological, and behavioral conditions [17]; hence, it is challenging to define the term human comfort. The most extensively recognized definition of human comfort is "The condition of mind, which expresses satisfaction with the thermal environment" [18]. The human body experiences comfort when the environment is maintained in a manner that facilitates the body's ability to swiftly attain thermal balance with its surroundings. When the body's heat gain exceeds its thermoregulatory capacity, body temperature begins to rise, potentially leading to heat-related illnesses and disorders. The body may attempt to cool down by varying the rate and depth of blood circulation, by losing water through the skin and sweat glands, or as a last resort, by panting. Sweating plays a crucial role by cooling the body through the process of evaporation. However, high relative humidity can impede evaporation by diminishing the body's ability to cool itself effectively. The body's thermoregulatory ability varies considerably with gender, age, and size. Activity level and clothing also have a significant impact on thermal comfort.

Achieving thermal comfort from a climatological perspective hinges on factors like temperature, humidity, wind speed, solar radiation, metabolic heat, and clothing conditions [19] with temperature and humidity being the most influential for human comfort and health [20]. For this reason, several thermal comfort indices have been developed which translate meteorological data into scales that reflect people's responses to weather conditions, ranging from very comfortable to very uncomfortable [21–27]. These indices simplify the interpretation of atmospheric effects on human comfort and facilitate comparisons between different regions [28]. Alerts are typically issued when a heat stress index is forecasted to exceed a specific threshold. Examples of heat stress indices which are suitable for warm environmental conditions are Thom's Discomfort Index (TDI) [29], Canada's Atmospheric Environment Service Humidex (HMDX) [30], and NOAA's (U.S.A. National Oceanic and Atmospheric Administration) Heat Index (HI) [31–33].

TDI, the first physiological index of this kind presented in 1959, is based on the conditions of air temperature and relative humidity [34,35] and expresses their combined contribution to human thermal comfort [36] and is easier to be evaluated, especially at a comparative level. For these reasons, TDI has been adopted in various parts of the world, with different climatological conditions. Yousif and Tahir [34], using TDI, investigated the thermal effect on people in Khartoum, Sudan. Adegoke and Dombo [37] examined human thermal comfort in Akure, Nigeria. Talukdar et al. [38] applied TDI to assess the trend of outdoor thermal discomfort in Mymensingh City, Bangladesh, during 2006–2015. Ehsan et al. [39] applied TDI to evaluate thermal discomfort levels, building design concepts and some heat mitigation strategies in low-income neighborhoods of Faisalabad, Pakistan. Monforte et al. [40] used TDI to characterize from the prospective bioclimate the provinces that make up the island of Sicily in South Italy, a Mediterranean region defined as a hotspot of climate change. Maftei et al. [41] used TDI to establish the thermal comfort or thermal discomfort conditions for the coastal zone of the Black Sea, in the Dobrogea region (Romania). Azevedo et al. [42] used TDI to evaluate the thermal conditions of urban areas in Petrolina-PE, Brazil, for the year of 2012.

HMDX, developed at Canada's Atmospheric Environment Service (now the Meteorological Service of Canada) in 1979, is a relatively simple index that relies only on temperature and relative humidity, making it easier to calculate than more complex indices like the Universal Thermal Climate Index (UTCI) and the Wet-Bulb Globe Temperature (WBGT). By combining air temperature and humidity, the Humidex offers a more accurate representation of how weather conditions feel to individuals, particularly in humid environments. Due to its simplicity, HMDX has also been used in climatologically different areas worldwide. Diaconescu et al. [43] proposes an approach that enables the reliable estimation of the daily HMDX index, using daily temperature and humidity data in Canada. Orimoloye et al. [44] appraised the indications of climate variability and heat stress index on human health incorporating HMDX for the period of 1986 to 2016, over

East London, South Africa. Golbabaei et al. [45] and Fallah Ghalhari et al. [46] used HMDX to demonstrate a comprehensive view of the heat stress in Iran, analyzing meteorological data for two periods, 1965–2009 and 1985–2014, respectively. Scoccimarroet al. [47] applied HMDX to include the effect of humidity on the temperature perceived by the human body under moderate to extreme conditions, in exposed areas over Europe. Středová et al. [48] used HMDX to evaluate the impact of an urban heat island on the residents and visitors of Hradec Králové, Czech Republic. Infusino et al. [49] used HMDX to analyze the combined effects of temperature and relative humidity on people living in the Crati River valley (south Italy), a very interesting area for its territorial, climatic, and socioeconomic particularities.

HI, also known as the apparent temperature, described by Lans P. Rothfusz [33] in a 1990 U.S.A. National Weather Service Technical Attachment after analyzing original data tables of apparent temperature proposed in 1979 by Steadman [31], is what the temperature feels like to the human body when relative humidity is combined with the air temperature. This has important considerations for the human body's comfort, and because of this, it has been used as a measure of heat exposure in studies throughout the world. Dahl et al. [50] applied HI to evaluate the increase in frequency and exposure of the population to days with extreme heat in the United States during the 21st century. Michelozzi et al. [51] used HI to evaluate the impact of high environmental temperatures on hospital admissions during April to September in 12 European cities. Khalaj et al. [52] used this index to determine and characterize the health impacts of extreme heat events on the population in five regions of New South Wales, Australia. Kyselý et al. [53] also applied HI to examine heat-related mortality in the population of South Korea between 1991 and 2005 and examine the extent to which the 1994 heat wave was unusual in terms of mortality impacts and recurrence probability. Bell et al. [54] also used this index to investigate heat-related mortality for three Latin American cities (Mexico City, Mexico; São Paulo, Brazil; Santiago, Chile).

Several studies have also examined thermal discomfort in Greece, which possesses a Mediterranean climate, characterized by cool and wet winters and hot and dry summers. Angouridakis and Makrogiannis [55] determined the environmental thermal sensation for summer in Thessaloniki by analyzing air temperature and relative humidity data for the period 1950–1957. Tsitoura et al. [56] performed outdoor questionnaires about the thermal comfort effect on people in four cities of Crete. Matzarakis and Mayer [57] examined heat stress in Greece, for 12 selected synoptic stations of the Greek Weather Service, using the Predicted Mean Vote (PMV) thermal index. Kambezidis et al. [58] investigated the future thermal sensation of the Greek population, using TDI, for 33 locations in Greece (including Athens), incorporating input data from Typical Meteorological Years (TMYs) developed specifically for each location for meteorology–climatology applications [59]. Each TMY consists of a set of typical meteorological months (TMMs) selected from individual years covering a period of 30 years (1985–2014), integrated into a complete year, reflecting all the climatic information of the location for a period as long as the mean life of the system.

Specifically, for Athens there have been several studies about the effect of heat-related stress on the population. Giles et al. [60] examined the heat waves that affected Athens (and Thessaloniki) in July 1987 and July 1988 using TDI and the "relative strain" index. Paliatsos and Nastos [61] compared the thermal effect during air pollution episodes in Athens during the summers of 1993–1995 using TDI and found that it was related to the ozone concentration. Tselepidaki et al. [62] examined the summer thermal perception in Athens for cooling purposes using TDI. Bartzokas et al. [63] examined the characteristics of the inter-annual and intra-annual variability of thermal discomfort in Athens for a longer time period, utilizing the PMV discomfort index. Stathopoulou et al. [36] explored the use of thermal AVHRR satellite data to map bioclimatic comfort conditions by estimating

TDI, for the case of Athens. Matzarakis and Nastos [64] analyzed heat waves in Athens during 1955–2001 using the daily physiologically equivalent temperature (PET) index and the daily minimum air temperature, presenting an increasing trend in heat wave frequency, with heat waves being more pronounced since the mid-1970s. Nastos and Matzarakis [65] investigated whether there is any association between daily mortality and air temperature and thermal indices such as PET and Universal Thermal Climate Index (UTCI) within the wider region of Athens, Greece, for a 10-year period, 1992–2001. Pantavou et al. [66] used various thermal comfort indices in Athens (see Table 2 in [66]). Also, Pantavou et al. [67] estimated the thermal sensation of pedestrians in four locations of Athens using UTCI. From a time scale perspective, previous studies have mainly focused on a daily basis or certain hours of the day. Katavoutas and Founda [68] using hourly values of air temperature, relative humidity, and solar radiation for Athens, examined a smaller period of 1960–2017, and found an increasing risk of heat stress in Athens using two simple (HMDX, HI) and two rational (UTCI, PET) indices.

Based on the existing background knowledge, this research aims to explore the temporal variation in thermal sensation and discomfort in Athens, Greece. By using one of the oldest, largest, and most homogeneous meteorological datasets for an urban environment in the Mediterranean, the present study attempts to make a significant contribution to the research on thermal discomfort. The combination of the large temporal coverage, more than a century (1901–2024), and the high resolution of this dataset, 1 h time step, enables a reliable analysis of both long-term climate trends and short-term thermal extremes, offering an evidence-based perspective that complements and extends already-existing studies based on daily data or shorter time intervals. Three simple and widely known discomfort indices, TDI—the first physiological index of this kind introduced— and HMDX and HI—both used operationally on our days for excessive discomfort warnings—have been analyzed across the globe in areas with different climatology, thus adding an important dimension to traditional temperature-based climate studies, offering a differentiated perspective on how climatic factors interact and affect human comfort and environmental stress in the region. The revealed changes in thermal discomfort levels in the region provide valuable insights into potential impacts on public health. These elements reinforce the scientific value of the present research, positioning Athens as a representative case study for Mediterranean metropolitan areas experiencing intensified heat stress due to urbanization and climatic change.

2. Materials and Methods

2.1. Data Sources

The present analysis was based on meteorological data records from Thissio meteorological station, maintained by the National Observatory of Athens (NOA), Greece (latitude: 37°58′ North, longitude: 23°43′ East), for more than a century. The station is located on a small hill near the Acropolis at an elevation of 107 m above mean sea level. The distance to the coastline is about 5 km. Although the station lies near the center of Athens, it is isolated from heavy traffic and densely built areas.

Specific information on the observation methods and measuring instruments employed for the measurement of air temperature and humidity parameters at Thissio meteorological station, since its commencement of operation, is provided in Appendix A [69–73]. In May 2017, according to a list of criteria, Thissio station was recognized by the World Meteorological Organization (WMO) as a long-term observing station for its high-quality climate records (WMO Decision 8, EC-69, May 2017).

Although air temperature is usually used to express a component of thermal comfort, it is rarely the reason for heat stress by itself alone [74]. Sweating plays a key role in the body's thermoregulatory process by cooling the body through the process of evaporation. However, high relative humidity retards evaporation, reducing the body's ability to cool itself. Thus, both temperature and relative humidity parameters are necessary for the calculation of thermal discomfort indices, due to their dominant role in determining human heat perception.

For the needs of this study, measurements of air temperature and relative humidity at a 1 h time step were collected and analyzed, covering a period of 124 years, 1901–2024 (in Local Standard Time, LST = GMT + 2 h). Although the long-term air temperature time series has been checked for its homogeneity and accuracy [73], an extra extensive quality control was performed on temperature–humidity pairs, checking and correcting any erroneous values that may have passed automated procedures. Based on input parameter availability (1 h temperature and humidity pairs), a total of 1,079,938 calculated indices were analyzed, with missing values corresponding to less than 0.65%.

Athens, the capital and the largest city in Greece, is located in a region of complex topography within the Athens basin (~450 km^2), presenting an elongated shape spanning from the NE to the SW direction, surrounded by mountains ranging in height from 400 to 1500 m at the western (Aegaleo, 458 m), northern (Penteli, 1107 m), and eastern (Imitos, 1026 m) sides. The city is also open to the sea to the south (Saronikos Gulf) and is close to the port of Pireaus (Figure 1). To the east of the basin's axis, the city is less densely populated. To the west, the area is designated as 75% industrial and 25% residential. Athens enjoys a typical Mediterranean climate with mild winters (DJF) (December, January, February; mean seasonal air temperature of 10.1 °C) and warm summers (JJA) (June, July, August; mean seasonal air temperature of 26.1 °C). The annual rainfall is approximately 376 mm, with the majority occurring in early October and the winter months. However, during the summer months, occasional severe local thunderstorms of short duration may also contribute to significant precipitation [75]. The Athens region experiences an average of 2919 h of sunshine per year. The prevailing winds in the Athens basin blow from N and NE in late summer, fall, and winter and from SSW and SW in the spring and early summer. It is important to note that these NE and SW directions coincide with the major geographical axis of the basin. The basin's ventilation is inadequate during periods of local circulation systems, such as sea–land breezes along the major NE–SW geographical axis of the basin. Kallos et al. [76] and Kassomenos et al. [77] cited additional information on the flow regime and the structure of the boundary layer over Athens.

In 2001, the population of Athens' urban area was 3.7 million, accounting for more than one-third of the country's total population. The surface area of the site was measured at 433 Km^2, and it continues to expand at an accelerating rate. According to Eurostat, the Athens metropolitan area is the eighth most populous in the European Union, with a population of approximately 4 million inhabitants in 2004. However, this number is expected to increase due to the presence of a significant number of undocumented migrants in the city. Despite some attempts toward decentralization between 1980 and 1990, Athens remains the center of the country's commercial, economic, industrial, political, and cultural activities. Figure 2 presents the population growth of Athens since the mid-19th century (data are provided in Appendix A) [78–82]. The city's rapid growth, particularly after World War II, was accompanied by a large number of automobiles, leading to poor air quality and traffic problems.

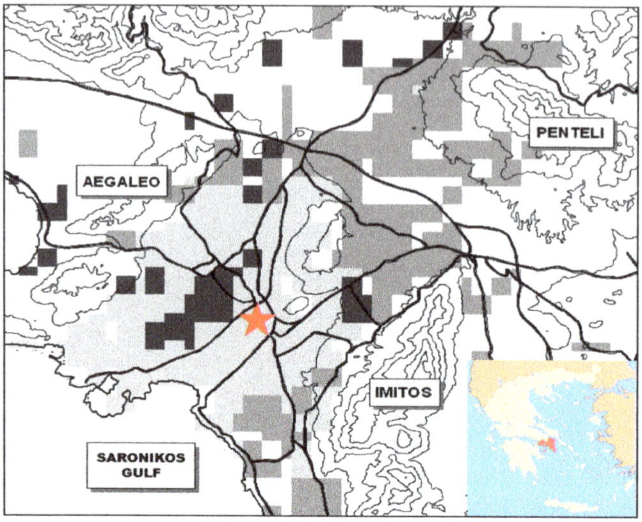

Figure 1. Map of the greater Athens area, where the location of the historical station of the National Observatory of Athens (NOA) in the Thissio area is marked (red star). Urban areas are marked with light gray, semi-urban areas with medium gray, and finally, industrial areas are marked with dark gray.

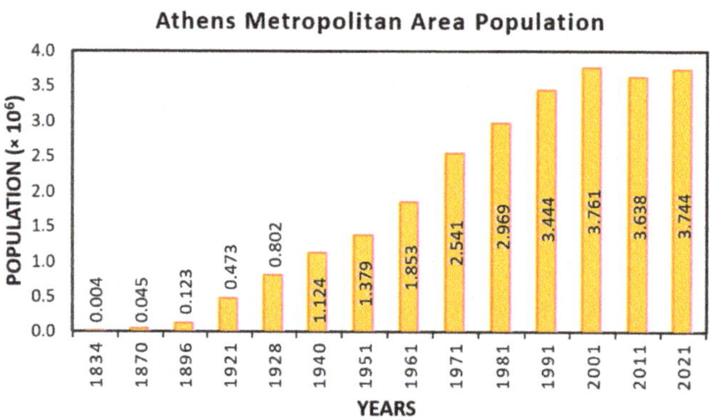

Figure 2. Population of the Greater Athens metropolitan area since the middle of the 19th century (sources: [78–82]).

Over the past few decades, Athens has experienced regional climate change along with significant urbanization. However, a delayed summer temperature increase in the eastern Mediterranean relative to the western Mediterranean has been observed, attributed to the increased frequency of northerly winds from the mid-1960s to mid-1980s [83]. This feature is a notable aspect of the observed climate change in this region. The observed increase in Athens' mean temperature began in the mid-1980s [84]. Moreover, the ongoing process of urbanization and its associated urban heat island effect serve as a secondary contributor to the local warming and the modification of climate characteristics, particularly during the summer months. A potential factor that could affect the climate in Athens is the recent destruction of a significant portion of the green areas located on the northern outskirts of the city, which was caused by the recent forest fires.

2.2. Thom's Discomfort Index (TDI)

The subjective sensations of comfort and discomfort are influenced by individual perceptions and external weather conditions. The American Society of Heating, Refrigeration and Air Conditioning Engineers (ASHRAE) has developed the concept of "effective temperature" over the course of several years. This concept is thoroughly discussed in ASHRAE [85], where it is defined as follows: "An empirically determined index of the degree of warmth perceived when exposed to different combinations of temperature, humidity and air movement". Research by Bosen and Thorn of the U.S. Weather Bureau, as described by Thom in 1957 [86], demonstrated that a simple linear equation based on dry- and wet-bulb readings yielded values that closely approximated the effective temperature over a range of temperatures and relative humidities commonly experienced in the U.S. during summer. The original equation

$$\mathrm{TDI} = [(\mathrm{Ta} + \mathrm{Tw})/2] + 0.1 \times [-150 - (\mathrm{Ta} + \mathrm{Tw})] \quad (1)$$

is usually simplified to

$$\mathrm{TDI} = 0.4 \times (\mathrm{Ta} + \mathrm{Tw}) + 15 \quad (2)$$

where Ta is the dry-bulb temperature in °F and Tw the wet-bulb temperature in °F.

Alternative versions have been proposed for use with air temperatures measured in °C and/or humidities described in terms of relative humidity expressed in %. The following mathematical expression was used:

$$\mathrm{TDI} = \mathrm{Ta} - 0.55 \times (1 - \mathrm{RH}/100) \times (\mathrm{Ta} - 14.5) \quad (3)$$

where Ta is the air temperature in °C and RH is relative humidity in %.

In a subsequent article, Thom [29] provided a list of boundary values that indicated degrees of discomfort. These are presented in Table 1, where the discomfort index is based on the Celsius temperature scales. Since these values were formulated for the U.S., they may also be considered valid for countries at a similar latitude, such as Greece.

Table 1. Thom's Discomfort Index classes, ranges, and description.

Range (°C)	Discomfort Level
TDI < 21	No discomfort
21 ≤ TDI < 24	Under 50% of the population feels discomfort
24 ≤ TDI < 27	Over 50% of the population feels discomfort
27 ≤ TDI < 29	Most of the population suffers discomfort
29 ≤ TDI < 32	Everyone feels severe stress
TDI ≥ 32	State of medical emergency

2.3. Humidex Index (HMDX)

Humidex, created by Canadian meteorologists, was initially used to describe how humidity affects human comfort. On a warm day, high relative humidity can make it feel even hotter because perspiration does not evaporate easily. When the Humidex value exceeds 38 °C, it is considered an extreme weather event. The threshold of HMDX ≥ 38 °C was used by Giannakopoulos et al. in 2011 [87] to identify days with significant thermal discomfort, as it marks a level where human health may be adversely affected, especially during prolonged outdoor exposure. This threshold aligns with guidelines from Environment Canada [88], which is used operationally for its excessive discomfort warnings, and is widely adopted in climate–health impact assessments and extreme heat studies. At a Humidex value of 30 °C, outdoor activities should be carefully managed, paying particular attention to the elderly and children and the overall health of adults [88]. The Humidex value represents an equivalent temperature, indicating how hot it feels to the human body by factoring in both temperature and relative humidity. The equation for calculating the Humidex was proposed by Masterson and Richardson [30]. The HMDX index is calculated using the following Equation (4), which is valid for air temperatures above 21 °C, and the interpretation of its range is detailed in Table 2.

$$\text{HMDX} = \text{Ta} + (5/9) \times (\text{e} - 10) \quad (4)$$

where Ta is the air temperature in °C and RH is relative humidity in %, and e is the vapor pressure given by the expression

$$\text{e} = 6.112 \times 10^{(7.5 \times \text{Ta}/(237.7 + \text{Ta}))} \times \text{RH}/100 \quad (5)$$

Table 2. HUMIDEX classes, ranges, and description.

Range (°C)	Discomfort Level
HMDX < 20	No discomfort
20 ≤ HMDX < 30	Little to no discomfort
30 ≤ HMDX < 38	Some discomfort
38 ≤ HMDX < 45	Great discomfort, avoid exertion
HMDX ≥ 45	Dangerous, heat stroke quite possible

2.4. *Heat Index (HI)*

The Heat Index (also known as Steadman's Apparent Temperature) is one of the most popular indices used in environmental health research [31,32,89], a variant of which serves as the foundation for heat advisories in numerous communities across the United States [90]. HI translates current weather conditions (air temperature and air moisture in the most basic formulations) into the air temperature that would "feel" the same to humans if the dew point temperature was 14 °C [31,33]. Steadman's approach involved expressing weather conditions in terms of the equivalent temperature if the dew point temperature was 14 °C. This method enabled Steadman to translate combinations of air moisture and temperature, as well as other factors such as wind speed and sun radiation, into a single scale [31,32], measured in the same units as air temperature. This index is a simplified version that relies exclusively on air temperature and moisture [31].

The formula for the HI was developed based on work conducted in the late 1970s. The R. G. Steadman paper entitled "The assessment of sultriness" employed a list of twenty factors to determine a person's perceived temperature on a given day. These factors included the rate of sweat production, the type of clothes worn, the surface area of the body, and the activity being performed. Meteorologists within the U.S. National Weather Service used Steadman's table to derive a simpler formula for HI. They did so by creating a mathematical formula that approximates its values, within ±0.7 °C (±1.3 °F), using only two variables—air temperature and percent relative humidity—being valid for air temperatures above 20 °C:

$$\begin{aligned} \text{HI} = c_0 + (c_1 \times \text{Ta}) + (c_2 \times \text{RH}) + (c_3 \times \text{Ta} \times \text{RH}) + (c_4 \times \text{Ta}^2) + (c_5 \times \text{RH}^2) + \\ (c_6 \times \text{Ta}^2 \times \text{RH}) + (c_7 \times \text{Ta} \times \text{RH}^2) + (c_8 \times \text{Ta}^2 \times \text{RH}^2) \end{aligned} \quad (6)$$

where Ta is the air temperature in °C and RH is relative humidity in %, and constants c_i take the values of

$c_0 = -8.784695$, $c_1 = 1.61139411$, $c_2 = 2.338549$, $c_3 = -0.14611605$, $c_4 = -0.012308094$, $c_5 = -0.016424828$, $c_6 = 0.002211732$, $c_7 = 0.00072546$ and $c_8 = -0.000003582$.

The development of the HI was not intended to study human health, but rather to measure thermal comfort [91]. However, it has gained traction as a prevalent exposure metric in the field of environmental health. The U.S. National Weather Service (NWS) has established a correlation between distinct Heat Index values and environmental health hazards. For instance, a Heat Index of 41 °C indicates a "danger" level associated with heat-related medical complications [92]. The NWS utilizes HI for its operational excessive heat warnings [93]. The interpretation of its range, with the assumption that the individual is healthy and has easy access to water and shade, is outlined in Table 3. Accidental exposure to full sunshine could increase HI values by up to 8 °C [94].

Table 3. Heat Index classes and description, with shaded caution/danger ranges.

Range (°C)	Discomfort Level
HI < 27	No effect—temperature and humidity should not have any effect.
27 ≤ HI < 32	Caution: fatigue is possible with prolonged exposure and activity. Continuing activity could result in heat cramps.
32 ≤ HI <41	Extreme caution: heat cramps and heat exhaustion are possible. Continuing activity could result in heat stroke.
41 ≤ HI < 54	Danger: heat cramps and heat exhaustion are likely; heat stroke is probable with continued activity.
HI ≥ 54	Extreme danger: heat stroke is imminent.

2.5. Methodology

This study applies a detailed and reproducible methodology to examine thermal discomfort over Athens, Greece, over a historical period spanning more than a century (1901–2024), incorporating high-resolution meteorological records. The three widely recognized bio-meteorological indices (TDI, HMDX, HI) were calculated for the entire study period, whenever the Ta-RH pair was available, in 1 h intervals, using FORTRAN code especially developed for this purpose. All index results were kept with one decimal place, maintaining the same decimal precision as the air temperature values in the historical archive. The relative humidity values were provided as integers.

The main differences between the three indices (TDI, HMDX, and HI) lie in their composition, limits, and functional use. The TDI is the oldest and simplest index proposed in 1959 by Thom [29], primarily designed to describe discomfort at the population level (e.g., what percentage of people feel stressed) and historically used in many regions due to its ease of calculation and interpretation. On the other hand, HMDX, developed with clearly defined health-related thresholds [30], tends to be more sensitive to high humidity than TDI, thus offering the enhanced capture of discomfort in humid heat conditions. Finally, the HI, employing a more sophisticated expression in order to approximate human thermal equilibrium, also based on air temperature and relative humidity [31,33], produces thresholds that are directly linked to health advisories. Its extensive utilization in global environmental health research has led to its adoption as a policy-oriented and health-focused tool.

With regard to the question of representativeness, it is not possible to consider any index to be universally superior, as each index has its own advantages, depending on the application. In our framework (Athens metropolitan city with a Mediterranean climate) the TDI is useful for historical comparability and evaluating relevant trends over long periods of time. HMDX and HI, incorporating an enhanced weighting of humidity and utilized directly in operational health warnings, may prove more representative of actual perceived heat stress and health risks. The HI is more widely used in epidemiological and public health studies, while HMDX is simpler to calculate and more common in climatological studies outside the U.S. Therefore, in this study, our approach of analyzing all three indices in parallel provides a more comprehensive and comparative perspective: TDI ensures long-term comparability, HMDX highlights humidity-driven thermal stress, and HI directly links to health advisories and epidemiological evidence.

Having available Ta and RH as input parameters and three index (TDI, HMDX, HI) values in 1 h intervals, the average, maximum, and minimum values per day and parameters were calculated, but only if hourly values per day exceeded 80% availability (i.e., at least 19 out of 24 values were available). In instances where the availability was less than 100%, we also verified whether the missing air temperature values were during midday, as this would affect the daily maximum value and our subsequent analysis.

To characterize both short-term variability and long-term climate trends, the indices were analyzed on a daily, monthly, seasonal, annual, decadal, and twenty-year basis. The daily scale identifies daily fluctuations, extreme events, and threshold exceedances related to human health. The monthly scale highlights intra-annual variability, identifying the months with the highest risk of thermal discomfort. The seasonal scale focuses on climatologically critical periods (e.g., July and August), assessing changes in the intensity of seasonal heat stress. The annual and decadal scales smooth out variability over time, providing a reliable long-term picture of climate change. For the most recent decade, the results for the four-year period 2021–2024 have been included into the decade 2011–2020. The multi-climate approach allows for a comprehensive examination of how urbanization and climate change have influenced discomfort trends over more than a century.

To quantify extreme heat or discomfort events, the 90th or the 95th percentile of temperature (or daily maximum temperature) is widely adopted in climatological and bio-meteorological studies, guidelines, or practical applications as a threshold, representing conditions that are statistically extreme but frequent enough to be important for adaptation and planning. This approach standardizes thresholds for extreme heat events, aligns with health impacts, and accounts for regional variability [95,96]. The 90th percentile of temperature is generally better for ensuring consistency across heat stress indices like Heat Index and Humidex in most operational contexts, as it captures a broader range of extreme events, aligns with public health warnings, and is widely supported by studies [96–102]. The 95th percentile is preferable for rare, high-severity events, particularly in hot climates or for climate change projections [103,104].

By clarifying the analytical structure and decision-making process, this expanded methodology enhances the relationship between the detailed results presented in the following sections and a transparent, reproducible framework.

3. Results and Analysis

3.1. Distribution of Thermal Discomfort Indices per Day—Thermal Stress Calendars

In order to emphasize the identification of seasonal variations and periods with the highest risks of thermal discomfort per index, a classification of each day in the examined period is performed, based on its daily maximum discomfort value. This classification follows the categories presented in Tables 1–3 for each discomfort index. This approach facilitates the construction of a comparative picture of the evolution of human thermal discomfort over a period exceeding a century, under changing climatic and urban conditions.

Thermal stress calendars were developed for Athens for the period 1901–2024, for each discomfort index (TDI, HMDX, HI), which are presented in Figure 3a–c. The horizontal axis represents calendar years and the vertical axis represents Julian days (1–365 or 366). Each cell corresponds to one day, while its color represents the level of thermal discomfort based on five defined categories: *green*: no discomfort (TDI < 21 °C, HMDX < 20 °C, HI < 27 °C); *blue*: little discomfort (21 ≤ TDI < 24 °C, 20 ≤ HMDX < 30 °C, 27 ≤ HI < 32 °C); *orange*: over 50% of the population feels discomfort (24 ≤ TDI < 27 °C, 30 ≤ HMDX < 38 °C,

32 ≤ HI < 41 °C); *red*: everyone feels severe stress: (27 ≤ TDI < 29 °C, 38 ≤ HMDX < 45 °C, 41 ≤ HI < 54 °C); and *purple*: extreme danger (TDI ≥ 29 °C, HMDX ≥ 45 °C, HI ≥ 54 °C). White boxes correspond to days for which no data was available (i.e., one or both of the input variables Ta or RH are missing), which did not affect the research, corresponding to less than 0.65% of input parameter availability for a period of 124 years.

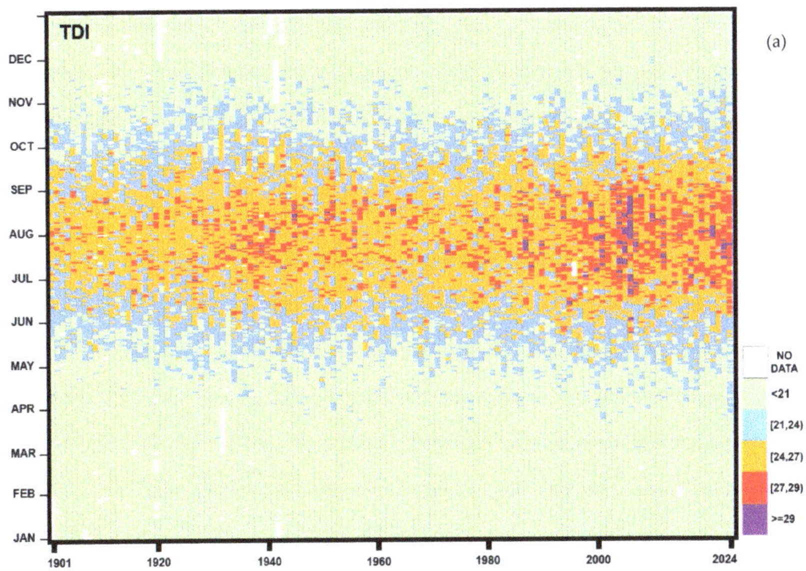

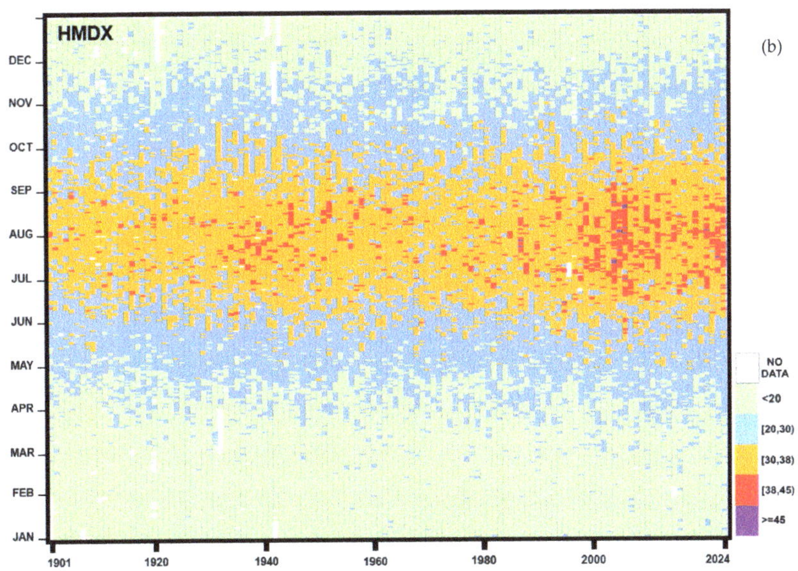

Figure 3. *Cont.*

(c)

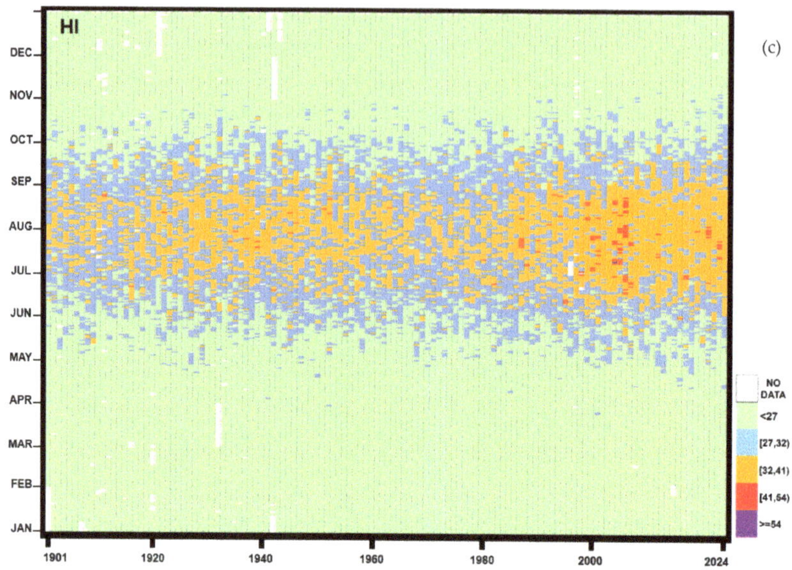

Figure 3. Calendar evolution of (**a**) Thom's Discomfort Index (TDI), (**b**) Humidex Index (HMDX) based on adjusted discomfort thresholds, and (**c**) Heat Index (HI), for Athens, Greece, for the examined period 1901–2024.

All three indices exhibit steady deterioration since 1980. The TDI (Figure 3a) shows an expanding discomfort core during July–August, with a rapid increase in days exceeding most of the population suffering discomfort and everyone feeling severe stress (red and purple). In the HMDX graph (Figure 3b), the orange zone (HMDX $\geq$ 30 °C) expands significantly after 1960 and, after 1990, covers almost the entire July–August period. The red cells (HMDX $\geq$ 38 °C) have gone from being rare to happening every year, while the first purple days ($\geq$45 °C) showed up in the mid-2000s, pointing to a new level of danger. Regarding the HI (Figure 3c), despite the higher "base" (discomfort when TDI $\geq$ 27 °C), the orange cells only become widespread after 1980. The HI $\geq$ 41 °C (red) risk category entered the climate "norm" after 2000. Despite threshold differences, all indices reveal a synchronized acceleration post-1980: the duration of thermal stress expanded and new "health emergency" classes emerged. Indices incorporating humidity (HMDX, HI) show stronger upward trends, confirming increases in relative humidity alongside warming.

The qualitative trend of the expansion of the period of great thermal discomfort, which is reflected in Figure 3a–c, is quantitatively confirmed in Figure 4. Figure 4b–d presents the day of the first and last appearances of the hazardous values of all three indices, i.e., TDI $\geq$ 27 °C, HMDX $\geq$ 38 °C, and HI $\geq$ 32 °C. Also, in the upper part of each sub-figure of Figure 4, the duration of the period between the first and last day is given as well, for each year.

Moreover, in order to quantify extreme heat events in Athens during the specified period (1901–2024), not necessarily during consecutives days, the 90th percentile of maximum temperature—a widely adopted metric in climatological studies—was implemented. The 90th percentile of temperature is more consistent across heat stress indices such as HI and HMDX in most operational contexts, as it encompasses a broader range of extreme events, aligning with public health warnings [96–102]. For each year between 1901 and 2024, the 90th percentile of daily maximum temperatures (Tmax) was determined. The mean of these values was then calculated, resulting in a critical value of 32.6 °C for the entire period under review. In Figure 4a the day of the first and last appearance of the critical value Tmax ≥ 32.6 °C is presented.

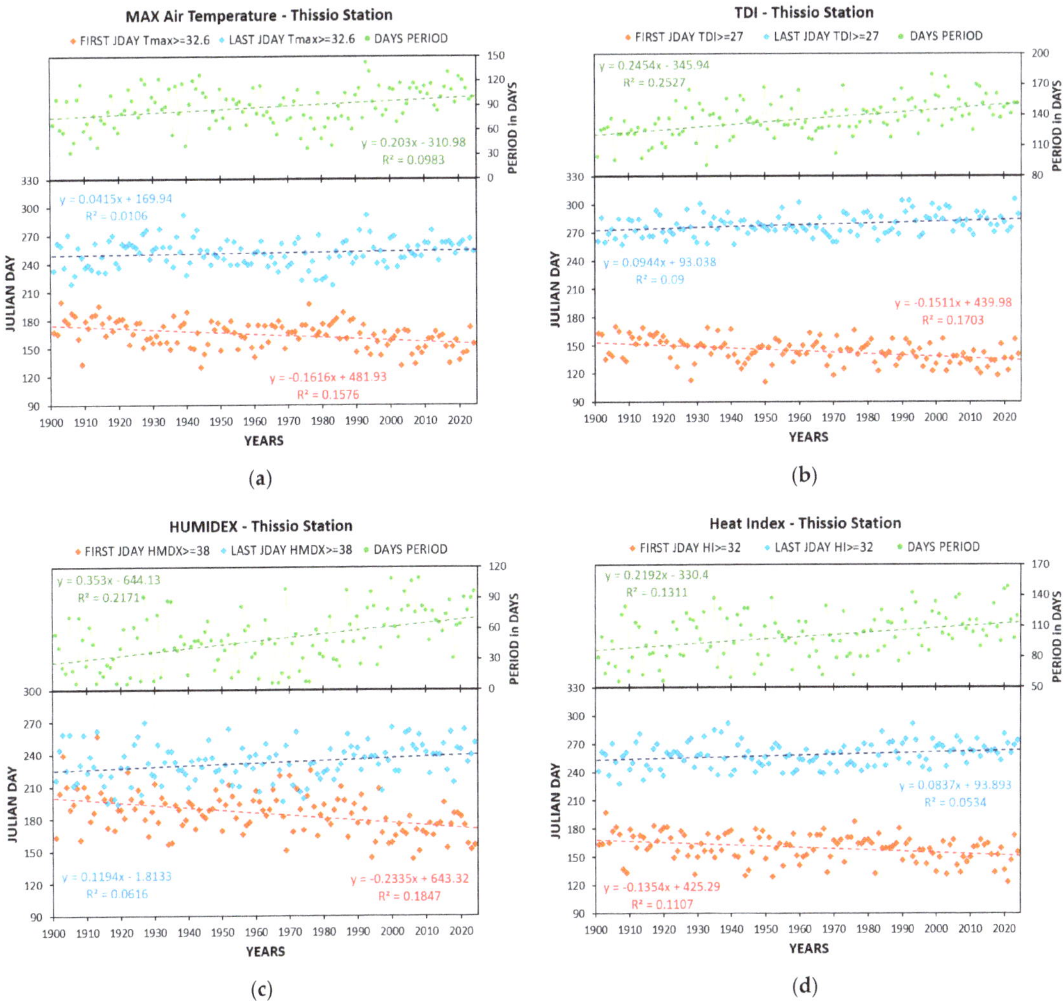

Figure 4. Temporal trends in the timing and duration of hazardous thermal conditions at the Thissio station in Athens, based on (**a**) Tmax ≥ 32.6 °C, (**b**) TDI ≥ 27 °C, (**c**) HMDX ≥ 38 °C, and (**d**) HI ≥ 32 °C. For each index, the lower panel displays the dates (Julian day) of the first (orange dots) and last (blue dots) occurrence of hazardous values, while the upper panel (green dots and lines) shows the annual length of the period (in days) between the first and last occurrence. Linear regression lines with R^2 values are also included (red, blue and green dashed lines).

According to the regression results, based on Tmax (Figure 4a), the first occurrence of such hot days within the year has been shifting progressively earlier, at a rate of 1.6 days per decade. This trend indicates that extreme temperature episodes tend to emerge earlier in the warm season, particularly since the late 20th century. On the other hand, the last annual occurrence is shifted later over the years by only 0.4 days per decade, a trend that is statistically weak. The overall duration of the period spanning from the first to the last appearance of such events increase over time, with a slightly increasing trend of 2 days per decade, suggesting that extreme heat events, in addition to starting earlier, are steadily extending later in the year.

The corresponding shifts for the TDI (Figure 4b) are similar to those for Tmax regarding the first day and the last day within the year, where TDI's specific threshold (TDI $\geq$ 27 °C) is observed to be exceeded. The first occurrence of a day on which the majority of the population suffers discomfort shifts earlier by 1.5 days per decade, while the last occurrence shifts later by approximately 1 day per decade over the years, a trend which is not statistically significant. However, the total duration of the period from the first to the last occurrence of such events increases significantly over time, by approximately 2.5 days per decade. This increase is evident especially after the 1990s, when this has been observed systematically from June to September, while in some cases after 2000 it extends to October.

The results based on HMDX (Figure 4c) indicate that the shifts in the first and the last appearance of very strong heat stress (HMDX $\geq$ 38 °C) are more prominent than the shifts in the TDI. The first appearance tends to occur earlier, at a rate of 2.335 days per decade. The last appearance occurs later, at a rate of approximately 1.2 days per decade. The total period from the first to the last occurrence of such events lasts 3.53 days per decade. The progressive lengthening of the period of dangerous HMDX's values has been more intense in the last three decades (1991–2024), a period during which both the frequency and duration of these days have increased. The duration of this period is over 70 days per year on average, compared with around 40 days per year in the first half of the 20th century. Furthermore, after 2000, the period has exceeded 90 days in several years, indicating an accelerating trend.

The corresponding shifts for the HI (Figure 4d) are not as pronounced as those for HMDX and the slopes of the trend lines are not as steep as HMDX's. The first day with extreme caution (HI $\geq$ 32 °C) is shifted earlier by 1.4 days per decade, while the last occurrence is extended by 0.8 days per decade, a trend which is not statistically significant. However, although the rate of increase in duration is smaller compared with HMDX, a systematic extension of the period of intense discomfort is observed, with its duration increasing at a rate of 2.2 days per decade. This increase, as in the case of TDI, is particularly evident after the 1990s, when it is systematically observed from June to September, while in some cases after 2000 it extends until October.

3.2. *Monthly Frequency of High Thermal Discomfort Days*

For the purpose of investigating the appearance of outlier cases for TDI, HMDX, and HI indices, incorporating estimated values with a 1 h interval, the total number of days per month (not necessarily consecutives) where at least one hour of great to extreme discomfort (TDI $\geq$ 27 °C, HMDX $\geq$ 38 °C, HI $\geq$ 32 °C) appeared are introduced in Figure 5, for each year of the examined period 1901–2024. The results of our analysis are given in table format in Appendix C.

In the early decades of the 20th century, exceedances of TDI $\geq$ 27 °C and HMDX $\geq$ 38 °C were isolated and generally confined to the core summer months (July and August). Over time, and especially after the 1980s, the frequency and temporal distribution of threshold exceedances expand significantly. Especially after 2000, Figure 5 shows a notable increase in both the frequency and duration of conditions of increased TDI and HMDX, as the boxes show an increasingly intense red hue, indicating an increase in the number of days with intense discomfort. The increase in months outlined in black reflects the continuous presence of days with severe thermal discomfort. Almost every year in the 2010s and 2020s during summer show consistently more intense red shading and numerous black boxes, highlighting the existence of too many days, even entire months, with episodes of severe thermal discomfort. It is worth noting that extreme exceedances of the thresholds (TDI ≥ 32 °C, HMDX ≥ 45 °C) were almost non-existent in the first half of the century, but have been occurring since 1980, even at the beginning of summer. Regarding the HI, by the 1990s, virtually every summer displayed up to 25 exceedance days during July and August, while in some years there were several days with HI ≥ 32 °C in June and September. The escalation in severity is emphasized by the increasing prevalence of months with days with HI ≥ 41 °C since the late 1980s, as well as by the first appearance of months with HI ≥ 54 °C. These findings demonstrate the increased heat stress attributed to both rising temperatures and increasing humidity, requiring targeted adaptations and public health interventions for Mediterranean urban areas.

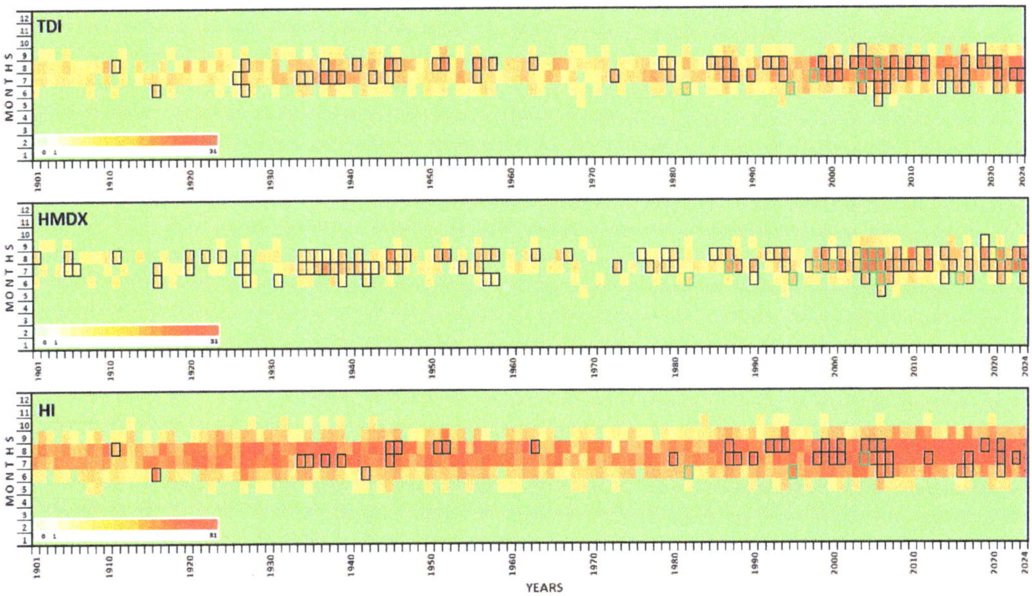

Figure 5. Number of days per month for each examined year where TDI values ≥ 27 °C (**up**), HMDX values ≥ 38 °C (**middle**), and HI values ≥ 32 °C (**down**) are observed. In black boxes, months also including days where values of TDI ≥ 29 °C (**up**), HMDX ≥ 40 °C (**middle**), and HI values ≥ 41 °C (**down**) are marked. In bright green boxes, months including days where extreme values of TDI ≥ 32 °C (up), HMDX ≥ 45 °C (**middle**), and HI ≥ 54 °C (**down**) are also marked. Months with zero days of occurrence are given with light green color.

3.3. Seasonal Distribution of Thermal Discomfort

In order to investigate in detail the hot season from June to August, including also the neighboring months, one before (May) and two after (September and October), available 1 h indices calculations are classified according to the categories in Tables 1–3, for 20-year intervals. The four-year period 2021–2024 has been included in the 2001–2020 period. The calculated percentage frequency per month and period was then normalized using the number of days per month (30 or 31) in order for the results to be expressed as number of days per class (not as %), for each index separately. The results for each index are presented in Figure 6a–c. Furthermore, Table 4 presents per index the total number of hours with high discomfort (TDI ≥ 27 °C, HMDX ≥ 38 °C, HI ≥ 32 °C), only from June to September, as most cases of thermal discomfort occur during these months. As can be easily seen, discomfort indices increase from June, peak in July and August, and decrease until October.

The analysis clearly shows an increase in thermal discomfort over a period of 20 years, especially after 1980, with the peak occurring after 2000. In the first decades of the 20th century, the overwhelming majority of summer days were recorded in categories of low (TDI < 21 °C, HMDX < 20 °C, HI < 27 °C) or moderate discomfort (21 ≤ TDI < 24 °C, 20 ≤ HMDX < 30 °C, 27 ≤ HI < 32 °C), while high discomfort categories were almost never observed. From 1941 to 1980, there was an increase of up to 34% of these days compared with previous years, with this growth rising to 38% in July and August alone, which are the most dangerous months. After the 1980s, the number of these days continued to increase gradually.

During the period 1981–2000, the days of the peak of summer, when more than half of the population felt discomfort, were almost as numerous as those days during the previous forty years. In the last two decades, these days have become more numerous than low-risk ones and have increased by more than 100% compared with the previous two decades. More specifically, the three discomfort indices show an increase in these days of 175%, 219%, and 112%, respectively, compared with the period 1981–2000. Also, looking at the TDI and HMDX charts (Figure 6a,b), it is clear that there are now more "dangerous" days, when everyone feels intense discomfort, indicating clear climate deterioration and increased heat stress conditions. This significant change reflects the growing impact of climate change, which has intensified the frequency, duration, and severity of heat waves in recent decades. The sharp increase in thermal discomfort indices corresponds to ob-served global warmings trends, highlighting the direct impact of climate change on ther-mal stress.

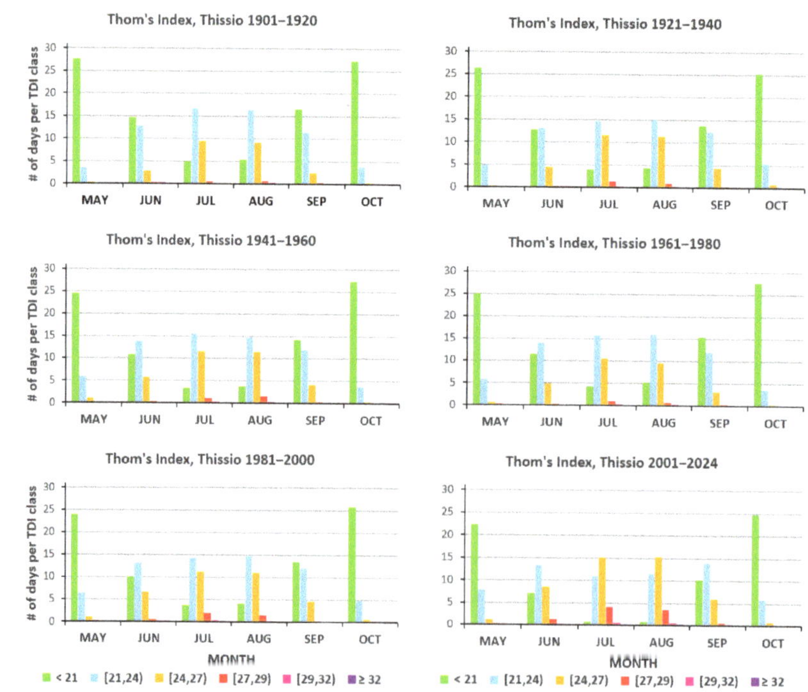

(a)

Figure 6. *Cont.*

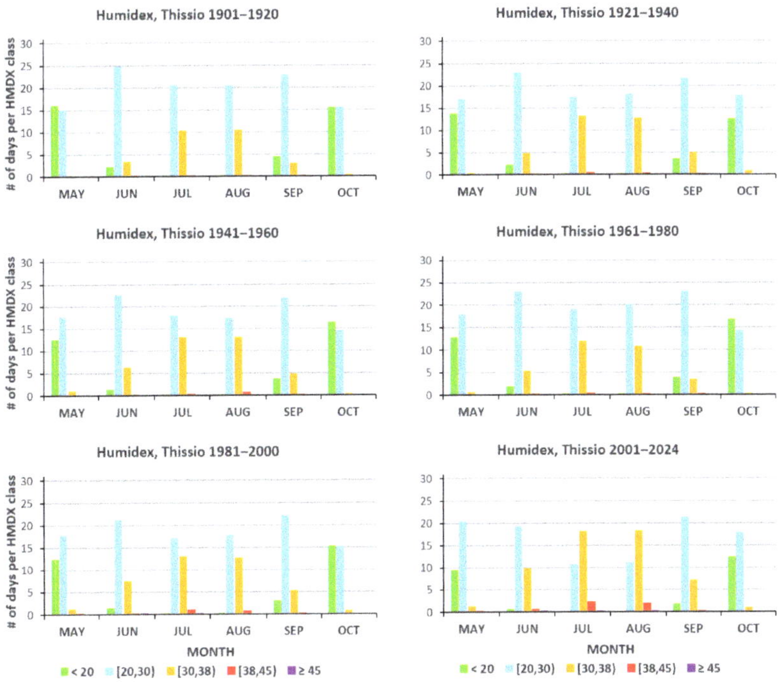

(b)

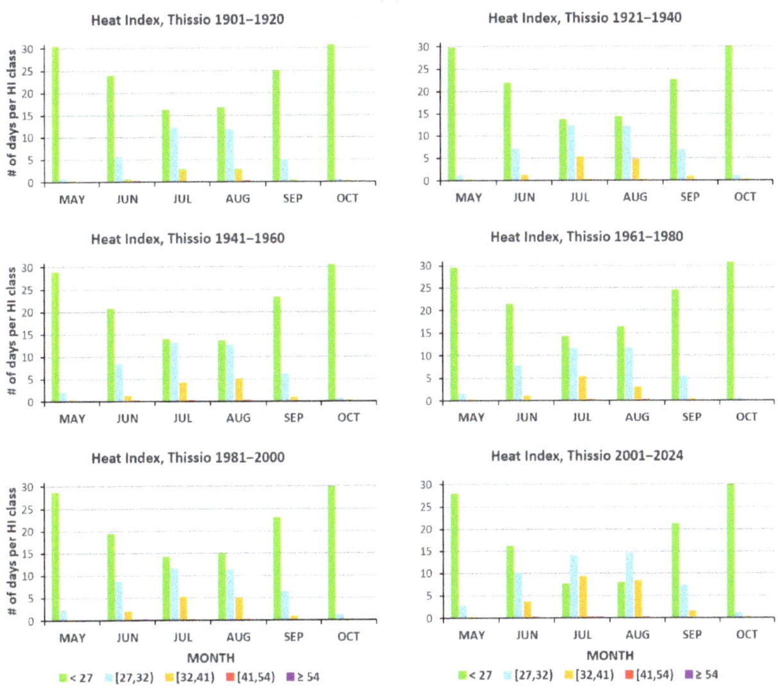

(c)

Figure 6. Monthly distribution of (**a**) TDI, (**b**) HMDX and (**c**) HI classes from May to October, in Athens, Greece, per 20-year intervals for the examined period 1901–2024. The four-year period 2021–2024 has been included into the 2001–2020 period.

Table 4. Total number of great discomfort hours from June to September in Athens, Greece, during the examined period 1901–2024, per 20-year intervals. The four-year period 2021–2024 has been included into the 2001–2020 period.

	Month	1901–1920	1921–1940	1941–1960	1961–1980	1981–2000	2001–2024
Number of hours where TDI ≥ 27 °C	JUN	54	82	91	41	242	746
	JUL	166	551	441	386	991	2534
	AUG	260	377	741	284	742	2224
	SEP	16	29	64	10	60	204
	Total	*496*	*1039*	*1337*	*721*	*2035*	*5708*
Number of hours where HMDX ≥ 38 °C	JUN	25	22	26	4	87	344
	JUL	54	198	145	117	431	1297
	AUG	95	119	305	75	314	1077
	SEP	3	5	15	2	8	53
	Total	*177*	*344*	*491*	*198*	*840*	*2771*
Number of hours where HI ≥ 32 °C	JUN	278	555	543	475	966	2138
	JUL	1267	2497	1988	1659	2520	5421
	AUG	1332	2212	2356	1441	2361	4951
	SEP	160	368	407	138	390	858
	Total	*3037*	*5632*	*5294*	*3713*	*6237*	*13,368*

3.4. Statistical Distribution of Meteorological and Discomfort Variables

In Figure 7, box and whisker plots incorporating all available 1 h values of the input parameters (Ta, RH) and calculated indices (TDI, HMDX, and HI) values are presented per decade and for the whole period 1901–2024. The four-year period 2021–2024 has been included in the 2011–2020 period. In Table 5, the main statistical values of the above parameters and indices but for the whole examined period 1901–2024 as a total are also given.

The interquartile range (IQR) of air temperature presents a gradual increase over the decades, starting from 11.4 in the first decade and rising to 12.1–12.5 in the last. This corresponds to an overall increase from 6.14% to 9.65%, suggesting that the intermediate temperatures (between 25th and 75th percentiles) are increasingly different. The distance between the adjusted trend lines shows a steady mean increase of 0.78% per decade, reflecting a gradual but steady increase in variability.

The IQR variation over time for relative humidity shows a significant decrease over the examined period. From 29.0 in the first decade, it falls to 24.0 in the last, recording a decrease of 17.24%. The linear regression confirms a steady downwards trend at a rate of −1.25% per decade. This trend reveals that the intermediate values of relative humidity converge to a smaller range, which may reflect the prevalence of drier and more stable conditions, possibly due to the impact of climatic change on the hydrological cycle. The IQR of the TDI starts at 8.7, reaches 9.7 during the decade of 1991–2000, and drops to 9.0 during the last decade. This increase shows that TDI values have a greater dispersion over the decades. This implies that the range of discomfort is widening, and possibly more days are reaching values closer to thresholds of severe discomfort. The regression values demonstrate a steady mean increase in IQR at a rate of +0.74% per decade, confirming the increasing trend. The IQR variation over time for HMDX reveals a clear trend of increasing intra-decadal variance. HMDX shows an overall increase in IQR as high as 14.94% from 1991 to 2010, with a constant rate of +0.96% per decade. These rises suggest that thermal discomfort conditions are becoming increasingly unstable. The HI exhibits the most significant rise in IQR, increasing from 11.4 during the first decade to 12.8 in the last, an increase of 12.8%. The linear regression confirms an upward trend at a rate of +1.03%

per decade. This marked change indicates that the variability of thermal stress is rapidly increasing, causing increasingly unpredictable discomfort conditions. The widening of the mean values reinforces the need to adopt adaptation strategies to ensure the protection of public health.

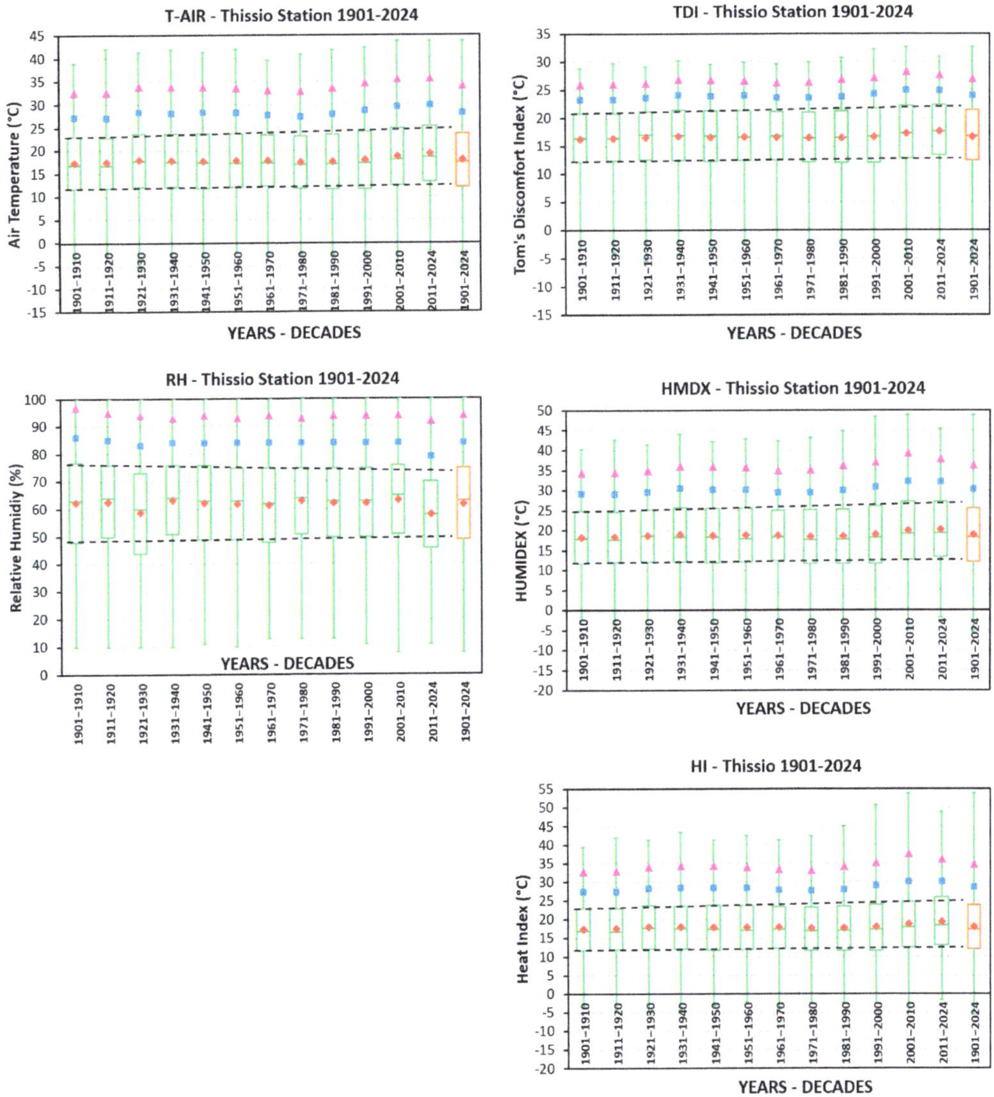

Figure 7. Box and whisker plots of air temperature, relative humidity, TDI, HMDX, and HI hourly values for the period 1901–2024, in 10-year intervals (green boxes) and for the whole period (red boxes), presenting 25% (Q1) and 75% (Q3) percentiles, median (red diamonds), and average (green lines) values, and max–min outliers. Also, both 90th (blue squares) and 99th (magenta tringles) percentiles are presented. Linear trends (dash lines) at each quartile (25% and 75%) of all parameters examined are also included.

Table 5. Main statistical values for each parameter, for the examined period 1901–2024, for Athens.

Statistical Parameter	Ta	RH	TDI	HMDX	HI
Number of values	1,083,876	1,082,598	1,079,938	1,079,938	1,079,938
Average	17.97	61.6	16.76	19.00	18.09
Standard deviation	7.57	17.1	5.69	8.35	7.70
Minimum	−5.4	8	−5.4	−5.4	−5.4
1st Quartile (25%)	12.1	49	12.5	12.3	12.2
Median (50%)	17.5	63	16.9	18.4	17.5
3rd Quartile (75%)	23.8	75	21.5	25.7	23.9
Maximum	43.8	100	32.7	49.0	54.0
90th Percentile	28.2	84	24.0	30.3	28.6
99th Percentile	34.1	94	27.0	36.4	34.8

A progressive and statistically meaningful increase in both the median and maximum values is clearly evident from the 1980s, reflecting a marked intensification of thermal discomfort over time. Specifically, while the median TDI values remained relatively stable, between 16.5 °C and 17 °C, from 1901 to 1980, a significant upward trend occurred in the last decade, when the median rose to ≈18 °C. This shift indicates that typical summer thermal conditions are no longer confined to moderate discomfort levels but are transitioning towards more intense bioclimatic stress for the general population. The observed rise in TDI median values is mirrored by a concurrent increase in maximum values, which reaches nearly 35 °C in the decades of 1981–1990 and 2001–2010, which is significantly higher than those recorded in earlier periods. These elevated maximum values are accompanied by the higher interquartile ranges and upper whisker extensions, indicating a growing occurrence of high-discomfort events. Even more pronounced trends are identified for the HMDX and HI statistics. While the median HMDX and HI values increased from ≈18 °C and ≈17 °C (1901–1930) to 19.3 °C and 18.4 °C accordingly in the 2011–2024 period, the mean values rose from 18.3 °C and 17.5 °C to over 20 °C and 19 °C in the same time frame, highlighting a systemic shift towards more oppressive summer conditions across the entire distribution. Most critically, the maximum values of HMDX and HI rose dramatically, peaking at 49 °C and 54 °C during the 2001–2010 decade. These values are not only exceptionally high, but surpass the well-established danger thresholds, above which heat-related health effects such as heat exhaustion or even heat stroke become a serious concern. This finding underscores the growing public health implications of climate change, especially for vulnerable populations.

The evolution of the TDI, HMDX, and HI over the 20th and early 21st centuries reveals not only a clear warming trend but also a transition towards more frequent and intense thermal discomfort conditions. The need for adaptive urban planning and targeted public health strategies is underscored by these trends in response to rising heat stress, especially in densely populated urban environments like Athens.

3.5. Decadal Classification of Thermal Discomfort Conditions

In order to illustrate the decadal distribution of the number of days per year falling into different thermal discomfort categories, the daily maximum values of each index were employed once again for the days' characterization. The four-year period 2021–2024 has been included in the 2011–2020 period. In Figure 8a–c (left column), each vertical bar represents the average number of days per decade in specific discomfort ranges, expressed as the percentage (%) of the total days available in each decade. In order for more details to be revealed in Figure 8a–c, a magnification of the top 2–3% was provided for each index (extreme classes). These three figures provide information on the temporal evolution of thermal stress levels observed in Athens from 1901 to 2024, reflecting both the frequency and severity of thermal stress over time.

Figure 8. Decadal distribution of thermal discomfort conditions in Athens during the 1901–2024 period, based on the daily maximum values of the TDI (**a**,**d**), HMDX (**b**,**e**), and HI (**c**,**f**). In (**a**–**c**), each bar represents the total number of days per decade that fall within specific discomfort classes, expressed as percentage of days available in each decade.

For the TDI (Figure 8a) and from 1951 to 1980, the frequency of days on which more than half of the population felt discomfort ($24 \leq \text{TDI} < 27$ °C) and the majority of the population suffered discomfort ($27 \leq \text{TDI} < 29$ °C) began to increase, with the majority of the population beginning to experience significant thermal discomfort compared with the early decades. An even more pronounced change occurred after 1980, especially during the decades 2001–2010 and 2011–2020, when the number of days on which most of the population suffered discomfort more than doubled, and for the first time there were days when emergency medical assistance was required (TDI ≥ 32 °C).

A similar trend is observed in the distribution of the HMDX (Figure 8b) and HI (Figure 8c) indices. During the first half of the 20th century, the majority of summer days have taken values of HMDX < 30 °C and HI < 32 °C, and consequently, especially in the categories for HMDX < 20 °C and HI < 27 °C, days are not characterized as discomfort days. Since the 1950s, there has been a decrease in the number of discomfort-free days, according to the HMDX and HI indices, and an increase in the number of days classified as "some discomfort", which represents a certain level of discomfort that makes it necessary to limit physical activity. HMDX analysis shows that days indicating dangerous levels of discomfort have been steadily increasing since 1980 and have doubled in recent decades. The most extreme category of HMDX—and the days when, according to HI, heatstroke is likely—first appeared after 1980. This development highlights the transition from thermally safe to increasingly dangerous conditions, reinforcing the role of humidity in amplifying the risk from heat.

Complementary to Figure 8a–c, Figure 8d–f (right column) present more details on the distribution of great discomfort days in Athens during the period under study. More specifically, it depicts the total number of days whose daily maximum falls within the highest discomfort classes of each index per decade (TDI ≥ 27 °C, HMDX ≥ 38 °C, HI ≥ 32 °C). The first notable increase in discomfort days is observed in the decade 1931–1940, when days of high discomfort almost doubled compared with previous decades, adding on average one more day per decade of severe discomfort observed (see Figure A1 given in Appendix D, where the % increase is expressed in equivalent number of days per decade).

A rapid change occurs after the 1980s. The peak is at the beginning of the 21st century, when the number of days on which the majority of the population suffers from discomfort almost sextupled compared with the first decades of the 20th century. Based on the average of all three indices, during the decade 2001–2010, around 4 days per decade of severe thermal discomfort are identified, followed by the next period 2011–2024 where the number of these days found to be 3.8 per decade. This increase is noteworthy because, until the 1971–1980 decade, no more than two such days per decade were observed.

In addition, days with health hazards are becoming more frequent: from sporadic occurrences to more than 100 days in the decade 2001–2010, marking a shift from occasional dangerous episodes of thermal discomfort to recurring phenomena with implications for public health. The three graphs show (Figure 8d–f) not only the clear transition to more severe conditions of thermal discomfort, but also the increase in the number of days that exceed dangerous levels of thermal stress, emphasizing the intensification of thermal risk in recent decades.

4. Discussion

The present analysis, based on more than a century (1901–2024) of hourly meteorological observations of Athens, underlines a marked and accelerating increase in thermal discomfort, as expressed by three simple but widely used indices (TDI, HMDX, HI). The results indicate that both the frequency and intensity of heat stress conditions have increased significantly over the last four decades, with the most dramatic increase occurring after 2000. This pattern is consistent with the broader phenomenon of warming in the eastern Mediterranean but is further amplified by local urban dynamics. The intensification is clearly reflected in the graphical analyses, which provide complementary perspectives on daily, monthly, seasonal, and decadal patterns.

Daily calendar analyses emphasize the progressive expansion of the core of discomfort since the mid-20th century. While in the early decades, days of high discomfort were sporadic and limited to July and August, after 1980 they became systematic and extended to June through September, and occasionally into October. Extreme categories (TDI $\geq$ 32 °C, HMDX $\geq$ 45 °C, HI $\geq$ 54 °C), which were almost non-existent before 1980, have appeared in the last two decades. This extension of the dangerous period is consistent with earlier reports of a delayed summer temperature increase in the eastern Mediterranean until the 1980s [83] and the subsequent sharp increase in average summer temperatures in Athens [84].

The total duration of the hazardous period now exceeds 90 days in some years, compared with less than 40 days in the early 20th century. These results corroborate Katavoutas and Founda [68], who found an increasing risk of heat stress in Athens during 1960–2017, and extend their findings by demonstrating that the acceleration is strongest after 2000.

The monthly and seasonal decadal averages reveal a structural upward shift of $\approx$3 °C across all indices between 1901 and 1940 and 2001–2024. Particularly noteworthy is the rise in September values, confirming that the thermal stress season now extends into early autumn. These findings support earlier work by Bartzokas et al. [63], who reported an intensification of intra-annual discomfort variability, and by Pantavou et al. [66,67], who identified prolonged periods of elevated heat stress in Athens.

The classification of discomfort categories by month further emphasizes this expansion. While in 1901–1940, almost all summer days fell within the low - to moderate - discomfort classes, after 2000 more than 100% increases are observed in severe categories compared with 1981–2000. The cumulative hours of high discomfort rose from $\approx$500–700 h in 1901–1920 to over 13,000 h in 2001–2024 for HI $\geq$ 32 °C, a nearly 20-fold increase. These results extend the findings by Paliatsos and Nastos [61], who linked high TDI values to pollution episodes in the summers of 1993–1995 by showing that such discomfort is no longer tied to episodic conditions but dominates the warm season climatology.

Statistical distributions demonstrate both rising central tendencies and widening variability. These findings are in agreement with Stathopoulou et al. [36], who mapped high TDI values from satellite observations in the 1990s, but the present analysis shows that both the frequency and variability of dangerous conditions have sharply escalated since then.

Finally, the decadal classification of daily maxima shows that the number of severe discomfort days sextupled after the 1980s and, in the decade 2001–2010, over 100 days per decade reached hazardous categories. This confirms the findings of Nastos and Matzarakis [64], who linked high PET/UTCI values to increased mortality in Athens in the 1990s by showing that the underlying climate has shifted into a regime of the persistent annual recurrence of such dangerous days.

Taken together, the graphical evidence demonstrates that Athens has undergone a transition from sporadic extreme heat events to a chronic regime of extended, frequent, and severe thermal discomfort. This transition is amplified by urbanization and the urban heat island effect, as noted in earlier works [63,68], and is consistent with broader Mediterranean and global trends [10,47,65]. The implications for public health, urban sustainability, and economic activity are severe. The observed trends highlight the need for integrated adaptation strategies, combining urban greening, heat warning systems, and the targeted protection of vulnerable populations.

5. Conclusions

The present study, which is based on 124 years (1901–2024) of high-resolution hourly meteorological records from the Thissio station in central Athens, Greece, demonstrates a significant and accelerating increase in thermal discomfort conditions during the 20th and early 21st centuries. A comprehensive analysis of the widely used bioclimatic indices, TDI, HMDX, and HI, reveals a pronounced trend toward more intense and protracted heat stress episodes since the 1980s, with the strongest acceleration occurring after 2000. Discomfort values that were once considered extreme are now common, occurring earlier in the summer and later in the autumn, extending the period of heat stress from June to September. The mean values, interquartile ranges, and maximum values of all indices exceed the thresholds associated with serious health risks, underscoring the combined impact of rising air temperatures, increasing humidity, and rapid urbanization on heat risk. However, this study is not without its limitations. The analysis is based on a single historical station, which is highly reliable with a long measurement record. However, it may not fully represent microclimatic variability across different urban landscapes in Athens. Furthermore, the selection of relatively simple indices facilitates long-term comparability; however, it does not capture the complete physiological complexity provided by more sophisticated indices. Future research should expand the spatial coverage by incorporating data from multiple stations or satellites, as well as information on land use and socioeconomic conditions. Moreover, it should also link meteorological trends directly to health records in order to better quantify heat-related risks. Such measures would enhance the process of evidence-based decision-making for sustainable and climate-resilient Mediterranean cities.

Author Contributions: Conceptualization, B.E.P.; methodology, B.E.P. and C.G.; data collection and quality control, B.E.P.; necessary software code development, B.E.P. and N.G.; analysis, investigation, and visualization, B.E.P. and N.G.; writing—original draft preparation, B.E.P. and N.G.; writing—review and editing, B.E.P., N.G. and C.G.; supervision, B.E.P.; funding acquisition, B.E.P. All authors have read and agreed to the published version of the manuscript.

Funding: This research was funded by the project "CLIMPACT II: Support for the upgrading of the National Research Network for Climate Change and Its Effects" (2023NA11900001, OPS 5201588), financed by the National Component of the Public Investment Program, National Development Program 2021–2025, Ministry of Development, General Secretariat for Research and Innovation.

Institutional Review Board Statement: Not applicable.

Informed Consent Statement: Not applicable.

Data Availability Statement: The datasets generated during the current study are available from the corresponding author upon reasonable request. Part of the analysis results are given freely in table form in Appendix C.

Acknowledgments: The authors of this article would like to thank all the anonymous observers who served at NOA's meteorological station in Thissio for their dedication in the day-by-day collection of all necessary measurements/observations and their careful processing/archiving, especially under very adverse conditions, often with the risk of their lives during times of war (World War II, Greek Civil War, etc.). The first author of this article wishes to express his personal gratitude to S. Lykoudis, D. Papagiannis, and V. Mitsopoulos for their invaluable assistance during the inaugural automation attempt of the Thissio meteorological station in 1995. Finally, the authors of this article would also like to thank all three anonymous reviewers for the valuable time they devoted to reading and commenting on this manuscript, thereby resulting in its improvement.

Conflicts of Interest: The authors declare no conflicts of interest.

Abbreviations

The following abbreviations are used in this manuscript:

TDI	Thom's Discomfort Index
HMDX	Humidex Discomfort Index
HI	Heat Index
UTCI	Universal Thermal Climate Index
WBGT	Wet-Bulb Globe Temperature
PET	Physiologically Equivalent Temperature
PMV	Predicted Mean Vote
IRQ	Interquartile Range
Ta	Air Temperature in °C
Tmax	Daily Maximum Air Temperature
RH	Air Relative Humidity in %
e	Vapor Pressure in hPa
TMY	Typical Meteorological Year
TMM	Typical Meteorological Month

Appendix A

The National Observatory of Athens (NOA) is the first research institution created in Greece (on 26 June 1842, during a solar eclipse) after its liberation from the Ottoman Empire (1828), the arrival of King Otto as the head of the modern Greek Kingdom (1833), and the establishment of Athens as the capital of the modern Greek state (1834). The hill of the Nymphs, selected as the place to build the Observatory, is one of the seven hills of Athens, a sanctuary of the Nymphs in antiquity, next to the Pnyka hill.

NOA's meteorological activities began in early 1858 with the systematic collection of daily observations in central Athens, but it was not until 1890 that a first-class meteorological station was permanently installed in Thissio, being in continuous operation ever since. According to the World Meteorological Organization's guidelines [69–71], the most important requirements for meteorological instruments to equip an observation station are reliability, accuracy, the simplicity of design, the convenience of operation and maintenance, and the strength of construction.

The psychrometric method was adopted at Thissio station to record the two basic parameters, temperature and humidity, that characterize atmospheric air, incorporating a pair of identical high-precision liquid-in-glass mercury thermometers, the dry- and the wet-bulb thermometers, with preferable scale intervals of 0.2 °C, having approximately the same lag coefficient. It is considered that the lag coefficient, defined as the time required by the thermometer to respond to 63% of a sudden change in temperature, should be between 30 and 60 s in a wind speed of 5 m/s. Thermometers for psychrometry should be accompanied with a factory correction certificate providing corrections, with an accuracy of ±0.1 °C, for at least six points spaced evenly across the range.

In addition, the set of psychrometric thermometers is accompanied by high-precision maximum and minimum thermometers, with scale intervals of 0.2 °C, for obtaining the required daily extreme observations. The maximum or Negretti thermometer is a mercury-in-glass thermometer, with a constriction in the bore below the lowest graduation, permitting its resetting without undue effort; however this prevents the mercury column from receding with falling temperature. The minimum or Rutherford thermometer is a spirit-in-glass (ethyl alcohol) thermometer, with a dark glass index (about 2 cm long) immersed in the spirit. Both maximum and minimum thermometers are supported nearly horizontally.

A wooden louvered Stevenson screen, with a volume of one cubic meter and natural ventilation, is used for all thermometers to be installed inside, protected from radiation from the Sun, sky, Earth, and any surrounding objects (i.e., large buildings and expanses of concrete or tarmac), but at the same time be adequately ventilated for measurements to be representative of the free air circulating in the locality. The screen should be mounted at such a height that thermometer bulbs are between 1.25 and 2.0 m above the ground. Observations for synoptic purposes are taken on a daily basis at 08:00, 14:00, and 20:00 LST (Local Standard Time, LST = GMT + 2 h), with an allowed tolerance of 10 min before or after the synoptic hour. Thermometers should be read to the nearest 0.1 degree.

In order to cover the rest of the day with measurements in 1 h intervals, following standard practice, a pair of self-recording instruments was placed inside the meteorological screen: a bimetallic thermograph and a hair hygrograph. Both of them incorporate a sensing element and a form of rotating chart mechanism. A typical thermograph in good condition has a small lag coefficient of about 25–30 s in wind of 5m/sec, with a maximum error not exceeding 1 °C at any point of the range. On the other hand, for air temperatures between 0 and 30 °C and relative humidity between 20 and 80%, an hygrograph in good condition subjected to a sudden change in relative humidity indicates 90% of the change within about 3 min, to an accuracy within ±3%.

Paper thermo- and hygro-charts are subject to change once a week, during the observational time of 08:00 LST on Monday morning, and are read by station staff at 1 h intervals. Using the three synoptic observations available per day (08:00, 14:00, 20:00 LST), errors in the chart readings are evenly distributed using a linear interpolation technique and temperature and humidity measurements are finally corrected and stored in NOA's archives. These are the "true" 24-hourly values produced according to Professor Eginitis [72], who served as NOA's director for the period 1890–1934, organizing meteorological observations in NOA and establishing the first network of meteorological stations throughout the country.

In June 1995, the first attempt for automating the Thissio historical station was made by introducing inside the meteorological screen two high-precision platinum Pt100 thermometers (electrical thermometers with resistance close to 100 Ohm at 0 °C), in a shiny stainless steel outer case with a small diameter, with a nominal accuracy of less than ±0.2 °C and connected to a digital recorder (four-wire connection on a CR-21X datalogger, Campbell Scientific Ltd., Logan, UT, U.S.A.) to measure dry- and wet-bulb temperatures alongside with the traditional liquid-in-glass mercury thermometers.

The station's new digital files consist of one-minute averages of 15 s samples of both dry and wet temperatures, allowing the psychrometric method to be applied, estimating a number of humidity parameters at the same time step (relative humidity included). After an extensive comparison from April 1999 to February 2004, the bias between conventional and new instruments was estimated and monthly coefficients were developed and published [73]. The homogeneity at NOA's temperature time series was restored by eliminating the bias caused from the 1995 instrument change.

The availability of digital files with high temporal resolution (1 min time step) renders the process of reading and correcting paper charts redundant. The extraction of instantaneous measurements at specific times (e.g., 10:00, 11:00, or 12:00 LST) from digital files is now a straightforward process. Even so, the two autographic instruments, which were replaced in 2005 by a new combined thermo–hygrograph, continue to be located within the meteorological screen in order to serve as a backup in emergency situations involving extensive power failures.

In May 2017, according to a list of criteria, Thissio station was recognized by the World Meteorological Organization (WMO) as a long-term observing station for its high-quality climate records (WMO Decision 8, EC-69, May 2017).

Appendix B

Table A1. Population of the Greater Athens metropolitan area since the middle of the 19th century.

Year	Population	Notes
1834	~4000	Capital moved to Athens
1870	~44,500	Beginning of urban development
1896	~123,000	Year of the first modern Olympic Games
1921	~473,000	Before the Asia Minor Catastrophe
1928	~801,622	Large influx due to refugee crisis
1940	~1,124,000	Pre-war growth
1951	~1,379,000	Post-war reconstruction
1961	~1,853,000	Industrial development
1971	~2,541,000	Urban peak
1981	~2,969,000	Expansion of suburbs
1991	~3,444,000	Gradual stabilization
2001	~3,761,000	Post-Olympic infrastructure era
2011	~3,638,000	Start of demographic decline
2021	~3,744,059	Slight recovery, stabilization

Data sources: [78–82].

Appendix C

Table A2. Number of high discomfort hours and days (not necessarily consecutives), from May to October, for Athens, Greece, during the examined period 1901–2024 according to TDI, HMDX and HI indices.

Year	Month	TDI [27,29)		TDI [29,32)		TDI ≥ 32		HMDX [38,40)		HMDX [40,45)		HMDX ≥ 45		HI [32,44)		HI [41,54)		HI ≥ 54	
		Total Hours	Days	Total Hours	Days	Total Hours	Days	Total Hours	Days	Total Hours	Days	Total Hours	Days	Total Hours	Days	Total Hours	Days	Total Hours	Days
1901	MAY																		
	JUN	1	1					1	1					5	2				
	JUL	5	3											73	15				
	AUG	34	4					20	3	1	1			56	10				
	SEPT																		
	OCT																		
1902	MAY																		
	JUN													2	2				
	JUL	12	4					4	2					67	11				
	AUG	6	5											95	17				
	SEPT	5	1					1	1					27	8				
	OCT																		
1903	MAY																		
	JUN																		
	JUL	1	1											38	7				
	AUG	5	3											65	11				
	SEPT	1	1					1	1					13					
	OCT																		
1904	MAY																		
	JUN													6	2				
	JUL	8	3					2	2					71	15				
	AUG	5	3											80	13				
	SEPT																		
	OCT																		

Table A2. *Cont.*

Year	Month	TDI [27,29)		TDI [29,32)		TDI ≥ 32		HMDX [38,40)		HMDX [40,45)		HMDX ≥ 45		HI [32,44)		HI [41,54)		HI ≥ 54	
		Total Hours	Days	Total Hours	Days	Total Hours	Days	Total Hours	Days	Total Hours	Days	Total Hours	Days	Total Hours	Days	Total Hours	Days	Total Hours	Days
1905	MAY																		
	JUN													1	1				
	JUL	7	3					2	1	1	1			60	14				
	AUG	36	6					15	4	1	1			116	15				
	SEPT	8	2					1	1					21	4				
	OCT																		
1906	MAY																		
	JUN																		
	JUL	3	2							1	1			44	13				
	AUG	3	2											25	7				
	SEPT													1	1				
	OCT																		
1907	MAY																		
	JUN													7	4				
	JUL	2	2											62	14				
	AUG	2	1					1	1					51	8				
	SEPT																		
	OCT																		
1908	MAY													3	2				
	JUN	7	2					1	1					34	6				
	JUL	11	4											47	12				
	AUG	9	2					2	1					76	15				
	SEPT													5	1				
	OCT																		
1909	MAY													9	4				
	JUN													11	4				
	JUL													91	12				
	AUG	5	2					1	1					49	12				
	SEPT													15	7				
	OCT																		
1910	MAY																		
	JUN													7	1				
	JUL	21	5					2	1					70	14				
	AUG	24	9					3	3					97	20				
	SEPT																		
	OCT																		
1911	MAY							4	2										
	JUN	9	2											29	4				
	JUL							9	2	11	2			18	6				
	AUG	31	4	4	1									84	8	2	2		
	SEPT													4	2				
	OCT																		
1912	MAY																		
	JUN							2	2					5	2				
	JUL	7	2											42	7				
	AUG	11	3											62	15				
	SEPT	1	1											5	2				
	OCT																		
1913	MAY																		
	JUN													1	1				
	JUL													12	4				
	AUG													14	5				
	SEPT													29	7				
	OCT													1	1				

Table A2. *Cont.*

Year	Month	TDI [27,29)		TDI [29,32)		TDI ≥ 32		HMDX [38,40)		HMDX [40,45)		HMDX ≥ 45		HI [32,44)		HI [41,54)		HI ≥ 54	
		Total Hours	Days	Total Hours	Days	Total Hours	Days	Total Hours	Days	Total Hours	Days	Total Hours	Days	Total Hours	Days	Total Hours	Days	Total Hours	Days
1914	MAY																		
	JUN													1	1				
	JUL	1	1					6	3					32	10				
	AUG	20	5											68	11				
	SEPT																		
	OCT																		
1915	MAY																		
	JUN													5	2				
	JUL	6	4											80	18				
	AUG													34	7				
	SEPT													7	1				
	OCT																		
1916	MAY																		
	JUN	22	5	1	1			7	1	7	2			112	17	2	1		
	JUL	44	5					26	4	1	1			145	18				
	AUG													26	8				
	SEPT																		
	OCT																		
1917	MAY																		
	JUN																		
	JUL	5	2					2	1					82	14				
	AUG	13	5					4	2					117	21				
	SEPT													10	3				
	OCT																		
1918	MAY																		
	JUN	14	2					5	1					29	5				
	JUL	13	6					1	1					96	17				
	AUG													61	12				
	SEPT	1	1											23	6				
	OCT																		
1919	MAY																		
	JUN													10	2				
	JUL	6	4					2	1					51	10				
	AUG	1	1											21	8				
	SEPT																		
	OCT																		
1920	MAY																		
	JUN													11	2				
	JUL	14	3					5	1	3	1			86	14				
	AUG	51	10					20	5	1	1			133	19				
	SEPT																		
	OCT																		
1921	MAY																		
	JUN																		
	JUL	2	1											118	22				
	AUG	2	1											116	21				
	SEPT													8	2				
	OCT																		
1922	MAY																		
	JUN													10	6				
	JUL	3	2											120	22	1	1		
	AUG	31	8					12	3	4	1			143	18	2	1		
	SEPT													7	4				
	OCT																		

Table A2. *Cont.*

Year	Month	TDI [27,29)		TDI [29,32)		TDI ≥ 32		HMDX [38,40)		HMDX [40,45)		HMDX ≥ 45		HI [32,44)		HI [41,54)		HI ≥ 54	
		Total Hours	Days	Total Hours	Days	Total Hours	Days	Total Hours	Days	Total Hours	Days	Total Hours	Days	Total Hours	Days	Total Hours	Days	Total Hours	Days
1923	MAY													2	2				
	JUN													22	5				
	JUL	15	4					2	1					77	13				
	AUG	4	1											59	11				
	SEPT													24	7				
	OCT																		
1924	MAY																		
	JUN	1	1											27	5				
	JUL	4	2					1	1					121	22				
	AUG	33	7					10	3	3	1			101	16				
	SEPT	8	3					2	1					64	13				
	OCT																		
1925	MAY																		
	JUN													4	1				
	JUL	3	1											99	18				
	AUG	2	2											92	18				
	SEPT	1	1											26	6				
	OCT																		
1926	MAY																		
	JUN	2	1											14	4				
	JUL	4	2	3	1					3	1			47	10				
	AUG	1	1											44	9				
	SEPT													10	3				
	OCT																		
1927	MAY													6	1				
	JUN	11	4	1	1			1	1	1	1			68	16				
	JUL	31	10					4	2	2	1			165	25				
	AUG	25	8	1	1			11	3	1	1			168	23				
	SEPT	4	2					2	1					37	9				
	OCT																		
1928	MAY																		
	JUN	5	3					1	1					65	10				
	JUL	5	3											174	22				
	AUG	7	5											174	26				
	SEPT													26	8				
	OCT													7	2				
1929	MAY													2	1				
	JUN													16	5				
	JUL	10	4											99	16				
	AUG	19	7					3	3					157	25				
	SEPT													6	2				
	OCT																		
1930	MAY													2	1				
	JUN													22	5				
	JUL	5	2											91	18				
	AUG	9	2					4	2					109	15				
	SEPT													16	8				
	OCT																		
1931	MAY																		
	JUN	7	1					3	1	2	1			56	11				
	JUL	21	9					4	1					169	25				
	AUG	31	8					6	3					155	23				
	SEPT	7	4					1	1					35	6				
	OCT																		

Table A2. *Cont.*

Year	Month	TDI [27,29)		TDI [29,32)		TDI ≥ 32		HMDX [38,40)		HMDX [40,45)		HMDX ≥ 45		HI [32,44)		HI [41,54)		HI ≥ 54	
		Total Hours	Days	Total Hours	Days	Total Hours	Days	Total Hours	Days	Total Hours	Days	Total Hours	Days	Total Hours	Days	Total Hours	Days	Total Hours	Days
1932	MAY																		
	JUN													22	7				
	JUL	15	8					5	2					101	17				
	AUG													40	13				
	SEPT	1	1											22	7				
	OCT													9	3				
1933	MAY																		
	JUN													8	3				
	JUL	29	5					7	3					51	8				
	AUG	12	7											82	15				
	SEPT	1	1											4	1				
	OCT																		
1934	MAY																		
	JUN	11	4					3	2					45	10				
	JUL	44	5	3	2			18	2	8	2			120	17	3	1		
	AUG	31	8					6	2	2	1			116	24				
	SEPT													3	1				
	OCT																		
1935	MAY													1	1				
	JUN	18	4					2	2					58	11				
	JUL	14	4	1	1			4	2	1	1			91	14	1	1		
	AUG	25	7					11	2	2	2			90	20				
	SEPT	1	1											23	6				
	OCT													1	1				
1936	MAY																		
	JUN													3	2				
	JUL	54	11					16	6	1	1			160	22				
	AUG	31	5					10	2	2	1			125	21				
	SEPT																		
	OCT																		
1937	MAY																		
	JUN	2	1											19	6				
	JUL	99	16	2	1			43	8	5	3			174	24	1	1		
	AUG	14	4	1	1			2	1	1	1			117	23				
	SEPT	1	1											16	7				
	OCT													1	1				
1938	MAY																		
	JUN	18	6					6	3	1	1			68	8				
	JUL	44	11	1	1			11	4	1	1			191	29				
	AUG	36	12					3	2					150	25				
	SEPT	2	2											21	8				
	OCT																		
1939	MAY																		
	JUN	4	1					2	1					27	4				
	JUL	91	12	4	2			38	5	11	3			170	16	3	2		
	AUG	51	7					23	4	3	2			108	13				
	SEPT	3	2											20	6				
	OCT													8	2				
1940	MAY																		
	JUN	1	1											1	1				
	JUL	44	11					12	4	1	1			151	21				
	AUG	11	4											64	11				
	SEPT																		
	OCT																		

Table A2. *Cont.*

Year	Month	TDI [27,29]		TDI [29,32]		TDI ≥ 32		HMDX [38,40]		HMDX [40,45]		HMDX ≥ 45		HI [32,44]		HI [41,54]		HI ≥ 54	
		Total Hours	Days	Total Hours	Days	Total Hours	Days	Total Hours	Days	Total Hours	Days	Total Hours	Days	Total Hours	Days	Total Hours	Days	Total Hours	Days
1941	MAY													4	1				
	JUN													2	2				
	JUL	64	11					23	6	2	1			153	23				
	AUG	44	8	2	2			22	2	3	2			130	22				
	SEPT																		
	OCT																		
1942	MAY													4	1				
	JUN	24	4					4	1	6	1			66	9	2	1		
	JUL	19	5					6	1	2	2			83	16				
	AUG	8	4											70	14				
	SEPT													12	5				
	OCT																		
1943	MAY																		
	JUN	2	1											17	4				
	JUL	23	7	1	1			6	1	2	1			84	14				
	AUG	35	10					7	3					131	23				
	SEPT	7	4											41	8				
	OCT													12	3				
1944	MAY													4	1				
	JUN	1	1											22	6				
	JUL	23	6					6	4					80	16				
	AUG	6	3					1	1					83	18				
	SEPT	3	2											41	9				
	OCT																		
1945	MAY													18	5				
	JUN	11	2					5	1					28	7				
	JUL	24	4	6	1			9	3	6	1			121	18	1	1		
	AUG	130	13	14	6			59	10	23	6			201	19	2	1		
	SEPT													25	5				
	OCT																		
1946	MAY																		
	JUN													32	10				
	JUL	41	7					17	5	2	1			140	20				
	AUG	40	8	1	1			8	2	1	1			199	26	1	1		
	SEPT	5	2					2	1					66	10				
	OCT																		
1947	MAY																		
	JUN													13	3				
	JUL	29	7					4	2					110	16				
	AUG	42	10					6	4	1	1			127	22				
	SEPT																		
	OCT																		
1948	MAY																		
	JUN	2	1											16	4				
	JUL	5	1											48	9				
	AUG	21	8					4	3					118	20				
	SEPT													6	1				
	OCT																		
1949	MAY																		
	JUN													6	2				
	JUL	13	2					2	2					75	13				
	AUG	4	1											41	7				
	SEPT													2	1				
	OCT																		

Table A2. *Cont.*

Year	Month	TDI [27,29)		TDI [29,32)		TDI ≥ 32		HMDX [38,40)		HMDX [40,45)		HMDX ≥ 45		HI [32,44)		HI [41,54)		HI ≥ 54	
		Total Hours	Days	Total Hours	Days	Total Hours	Days	Total Hours	Days	Total Hours	Days	Total Hours	Days	Total Hours	Days	Total Hours	Days	Total Hours	Days
1950	MAY													14	2				
	JUN	12	3					5	1					52	9				
	JUL	9	4					2	1					149	22				
	AUG	13	4					2	1					100	17				
	SEPT													22	7				
	OCT																		
1951	MAY																		
	JUN	1	1											18	5				
	JUL	18	5					6	3					90	16				
	AUG	68	8	6	2			24	5	10	3			148	18	2	1		
	SEPT	2	1											17	5				
	OCT																		
1952	MAY													2	2				
	JUN	8	2											39	7				
	JUL	12	4											56	13				
	AUG	91	16	7	2			32	7	20	5			224	27	5	2		
	SEPT	27	8					3	2					100	18				
	OCT																		
1953	MAY																		
	JUN													9	5				
	JUL	9	2					1	1					98	20				
	AUG	3	2											53	10				
	SEPT													3	3				
	OCT																		
1954	MAY																		
	JUN	2	1											54	11				
	JUL	51	11					16	4	1	1			144	23				
	AUG	26	8					8	2					119	23				
	SEPT	2	1											15	6				
	OCT																		
1955	MAY													4	2				
	JUN	1	1											32	7				
	JUL	15	6					3	2					79	17				
	AUG													23	6				
	SEPT																		
	OCT																		
1956	MAY																		
	JUN	4	1					2	1					38	7				
	JUL	46	7	4	1			21	2	7	3			115	16				
	AUG	60	13	1	1			20	6	1	1			174	28				
	SEPT	18	3					10	2					41	8				
	OCT																		
1957	MAY																		
	JUN	17	5					2	1	1	1			55	9				
	JUL	12	4											89	21				
	AUG	62	9					19	5	1	1			134	17				
	SEPT													1	1				
	OCT																		
1958	MAY																		
	JUN	6	2							1	1			22	4				
	JUL	9	4											101	18				
	AUG	46	7	1	1			22	3	9	3			133	18				
	SEPT																		
	OCT																		

Table A2. *Cont.*

Year	Month	TDI [27,29)		TDI [29,32)		TDI ≥ 32		HMDX [38,40)		HMDX [40,45)		HMDX ≥ 45		HI [32,44)		HI [41,54)		HI ≥ 54	
		Total Hours	Days	Total Hours	Days	Total Hours	Days	Total Hours	Days	Total Hours	Days	Total Hours	Days	Total Hours	Days	Total Hours	Days	Total Hours	Days
1959	MAY													3	1				
	JUN																		
	JUL	5	1											86	17				
	AUG													49	11				
	SEPT																		
	OCT																		
1960	MAY													1	1				
	JUN													20	5				
	JUL	3	3					1	1					86	17				
	AUG	10	4					2	2					89	16				
	SEPT													15	6				
	OCT																		
1961	MAY													1	1				
	JUN	3	1											19	4				
	JUL	11	4					4	4					60	11				
	AUG	7	4											81	13				
	SEPT													17	3				
	OCT																		
1962	MAY																		
	JUN													32	9				
	JUL	19	7					1	1					112	16				
	AUG	21	6					4	3					179	23				
	SEPT													26	6				
	OCT																		
1963	MAY																		
	JUN	11	4											57	9				
	JUL	15	3					1	1					102	16				
	AUG	50	7	4	2			17	3	7	2			152	22	2	1		
	SEPT	4	2					1	1					28	5				
	OCT																		
1964	MAY																		
	JUN	1	1											7	3				
	JUL	2	1											39	10				
	AUG	3	1											25	6				
	SEPT													1	1				
	OCT																		
1965	MAY																		
	JUN													21	7				
	JUL	33	10					13	2					127	16				
	AUG	1	1											37	7				
	SEPT													9	4				
	OCT																		
1966	MAY																		
	JUN													10	2				
	JUL	11	6					2	2					105	19				
	AUG	31	11					8	6					151	22				
	SEPT																		
	OCT																		
1967	MAY																		
	JUN													4	1				
	JUL	2	1											33	7				
	AUG	38	9					12	3	1	1			106	18				
	SEPT																		
	OCT																		

Table A2. *Cont.*

Year	Month	TDI [27,29)		TDI [29,32)		TDI ≥ 32		HMDX [38,40)		HMDX [40,45)		HMDX ≥ 45		HI [32,44)		HI [41,54)		HI ≥ 54	
		Total Hours	Days	Total Hours	Days	Total Hours	Days	Total Hours	Days	Total Hours	Days	Total Hours	Days	Total Hours	Days	Total Hours	Days	Total Hours	Days
1968	MAY													3	2				
	JUN	2	1					1	1					30	4				
	JUL	21	5					6	3					104	12				
	AUG	1	1											28	5				
	SEPT	1	1											3	2				
	OCT																		
1969	MAY	3	1					1	1					11	2				
	JUN	2	1											26	4				
	JUL													7	3				
	AUG	11	3											48	11				
	SEPT	1	1					1	1					9	4				
	OCT																		
1970	MAY																		
	JUN													10	5				
	JUL	1	1											60	15				
	AUG	9	3					3	1					81	17				
	SEPT													3	2				
	OCT																		
1971	MAY																		
	JUN	3	1					1	1					31	7				
	JUL	1	1											37	8				
	AUG	10	4											87	18				
	SEPT																		
	OCT																		
1972	MAY																		
	JUN													24	3				
	JUL	6	4											52	11				
	AUG	10	4											83	13				
	SEPT	2	1											3	1				
	OCT																		
1973	MAY													2	1				
	JUN													9	3				
	JUL	65	9	8	2			21	3	18	4			136	17				
	AUG													4	1				
	SEPT													7	2				
	OCT																		
1974	MAY																		
	JUN	2	2					1	1					20	6				
	JUL	13	3											93	14				
	AUG	3	2											50	13				
	SEPT													1	1				
	OCT																		
1975	MAY																		
	JUN													19	7				
	JUL	15	5					3	2					91	16				
	AUG	1	1											19	7				
	SEPT	1	1											16	6				
	OCT																		
1976	MAY																		
	JUN																		
	JUL	8	4											28	10				
	AUG	11	4							1	1			30	8				
	SEPT													1	1				
	OCT																		

Table A2. *Cont.*

Year	Month	TDI [27,29]		TDI [29,32)		TDI ≥ 32		HMDX [38,40)		HMDX [40,45)		HMDX ≥ 45		HI [32,44)		HI [41,54)		HI ≥ 54	
		Total Hours	Days	Total Hours	Days	Total Hours	Days	Total Hours	Days	Total Hours	Days	Total Hours	Days	Total Hours	Days	Total Hours	Days	Total Hours	Days
1977	MAY																		
	JUN													28	8				
	JUL	38	9					14	3	1	1			130	20				
	AUG	15	6					1	1					107	19				
	SEPT													1	1				
	OCT																		
1978	MAY																		
	JUN	11	3					1	1					57	9				
	JUL	46	11					9	5					124	17				
	AUG	5	1					1	1					34	6				
	SEPT																		
	OCT																		
1979	MAY																		
	JUN	6	3											40	6				
	JUL	32	7					7	3	1	1			80	13				
	AUG	16	3	1	1			1	1	1	1			63	12				
	SEPT	1	1											12	3				
	OCT																		
1980	MAY																		
	JUN													31	7				
	JUL	31	8	8	1			5	2	9	1			134	20	5	1		
	AUG	35	4	1	1			16	3	2	1			74	16				
	SEPT													1	1				
	OCT																		
1981	MAY																		
	JUN	24	3					10	2					57	7				
	JUL	7	2					1	1					47	10				
	AUG	2	1											47	13				
	SEPT													2	2				
	OCT																		
1982	MAY													3	1				
	JUN	38	5	5	1			14	2	8	1	1	1	71	8			1	1
	JUL	7	3					1	1					62	14				
	AUG	18	9					3	2					88	17				
	SEPT	10	3					3	1					21	3				
	OCT																		
1983	MAY																		
	JUN	5	1											7	2				
	JUL	25	6					4	3					80	17				
	AUG	8	3											40	9				
	SEPT													4	2				
	OCT																		
1984	MAY																		
	JUN													9	2				
	JUL	30	7					5	1					90	15				
	AUG	1	1											28	7				
	SEPT													5	2				
	OCT	1	1											6	4				
1985	MAY																		
	JUN	3	1											31	7				
	JUL	16	3					5	3					85	14				
	AUG	17	5	4	1			9	1	5	1			96	19				
	SEPT													12	3				
	OCT																		

Table A2. *Cont.*

Year	Month	TDI [27,29)		TDI [29,32)		TDI ≥ 32		HMDX [38,40)		HMDX [40,45)		HMDX ≥ 45		HI [32,44)		HI [41,54)		HI ≥ 54	
		Total Hours	Days	Total Hours	Days	Total Hours	Days	Total Hours	Days	Total Hours	Days	Total Hours	Days	Total Hours	Days	Total Hours	Days	Total Hours	Days
1986	MAY																		
	JUN	1	1											11	4				
	JUL	35	9					2	1					92	19				
	AUG	36	8	2	2			10	2	2	2			146	29				
	SEPT	1	1											27	7				
	OCT																		
1987	MAY																		
	JUN	27	6					4	2					75	12				
	JUL	86	6	53	8			33	4	65	5	3	3	186	15	33	7		
	AUG	68	12	5	4			22	5	7	4			131	19	1	1		
	SEPT	12	5											56	11				
	OCT																		
1988	MAY																		
	JUN	1	1											26	9				
	JUL	92	14	20	3			36	7	24	4			204	24	9	2		
	AUG	36	10					6	3					148	27				
	SEPT	15	4					2	2					32	6				
	OCT																		
1989	MAY																		
	JUN													3	2				
	JUL	8	4					3	2					69	15				
	AUG	2	1											108	23				
	SEPT													7	3				
	OCT																		
1990	MAY													8	2				
	JUN	13	3					4	1	1	1			41	8				
	JUL	76	11	9	4			26	3	14	7			178	23	3	2		
	AUG	6	3											79	16				
	SEPT	2	1					1	1					27	7				
	OCT																		
1991	MAY																		
	JUN	1	1											55	11				
	JUL	18	4											88	15				
	AUG	32	8					11	4					78	15				
	SEPT													2	1				
	OCT	2												12	2				
1992	MAY																		
	JUN	1	1											15	5				
	JUL	15	4					1	1					39	7				
	AUG	31	10	3	1			10	2	3	1			162	27	1	1		
	SEPT	2	2											16	4				
	OCT																		
1993	MAY													4	1				
	JUN	8	3					1	1					48	8				
	JUL	71	14					19	5					168	23				
	AUG	40	5	3	2			18	3	7	3			145	24	1	1		
	SEPT	3	1											22	5				
	OCT													1	1				
1994	MAY	10	4					1	1					40	6				
	JUN	9	3					2	2					50	8				
	JUL	41	10	1	1			8	3	1	1			117	17				
	AUG	98	13	8	4			45	4	19	7			212	26	2	2		
	SEPT	4	2											95	19				
	OCT													3	1				

Table A2. *Cont.*

Year	Month	TDI [27,29)		TDI [29,32)		TDI ≥ 32		HMDX [38,40)		HMDX [40,45)		HMDX ≥ 45		HI [32,44)		HI [41,54)		HI ≥ 54	
		Total Hours	Days	Total Hours	Days	Total Hours	Days	Total Hours	Days	Total Hours	Days	Total Hours	Days	Total Hours	Days	Total Hours	Days	Total Hours	Days
1995	MAY																		
	JUN	22	7	3	1	5	2	6	3	6	1	5	2	71	15	3	1	5	2
	JUL	33	10					8	4					142	23				
	AUG	17	6					2	1					79	18				
	SEPT	2	1											7	3				
	OCT																		
1996	MAY																		
	JUN	5	4											52	10				
	JUL	12	5					1	1					58	10				
	AUG	35	10					3	2					120	22				
	SEPT													3	1				
	OCT																		
1997	MAY													2	2				
	JUN	29	9					7	2					90	13				
	JUL	29	10					9	3	1	1			128	24				
	AUG	24	8											67	13				
	SEPT													3	2				
	OCT																		
1998	MAY																		
	JUN	18	3					9	2					57	12				
	JUL	62	11	20	3	1	1	17	3	20	4	5	1	195	23	14	3		
	AUG	86	20					27	10					208	28				
	SEPT																		
	OCT																		
1999	MAY																		
	JUN	4	4											85	17				
	JUL	98	20					43	10	2	2			205	28				
	AUG	118	11	24	6			56	4	44	9			215	18	11	5		
	SEPT													6	4				
	OCT													1	1				
2000	MAY													1	1				
	JUN	20	5					6	3					103	17				
	JUL	93	10	33	6			36	4	38	6	1	1	207	18	21	6		
	AUG	18	6					6		1	1			148	23				
	SEPT	9	4					2	1					43	8				
	OCT																		
2001	MAY													3	3				
	JUN	7	1					3	1					50	13				
	JUL	141	14	34	11			61	7	50	13			266	25	7	5		
	AUG	64	7	12	4			31	1	22	7			190	26	1	1		
	SEPT	2	2											29	8				
	OCT																		
2002	MAY																		
	JUN	15	3					4	2					60	11				
	JUL	70	16					14	7					202	27				
	AUG	20	5					7	2					80	17				
	SEPT													1	1				
	OCT																		
2003	MAY													7	2				
	JUN	10	6											122	20				
	JUL	58	12					20	4	2	1			189	30				
	AUG	72	15	1	1			14	4	2	1			212	28				
	SEPT	15	3					7	1					26	3				
	OCT																		

Table A2. *Cont.*

Year	Month	TDI [27,29)		TDI [29,32)		TDI ≥ 32		HMDX [38,40)		HMDX [40,45)		HMDX ≥ 45		HI [32,44)		HI [41,54)		HI ≥ 54	
		Total Hours	Days	Total Hours	Days	Total Hours	Days	Total Hours	Days	Total Hours	Days	Total Hours	Days	Total Hours	Days	Total Hours	Days	Total Hours	Days
2004	MAY																		
	JUN	48	13	7	1			20	5	9	3			111	18	1	1		
	JUL	130	16	46	7	8	3	59	9	52	9	15	3	216	20	33	6	3	1
	AUG	148	19	18	9			87	12	29	12	1	1	215	26	7	4		
	SEPT	20	4	1	1			11	3	1	1			32	7				
	OCT																		
2005	MAY																		
	JUN	2	2											32	9				
	JUL	174	13	37	12			85	4	58	17	1	1	254	19	17	8		
	AUG	113	17	33	7			45	7	42	8	4	2	212	26	22	5		
	SEPT	11	4					4	3					36	9				
	OCT																		
2006	MAY	27	4	2	1			13	2	4	2			54	8				
	JUN	104	6	36	10			40	2	52	9	1	1	157	11	18	6		
	JUL	107	10	24	7			46	6	38	9			157	15	14	5		
	AUG	165	8	116	22	1	1	73	3	129	20	14	5	271	17	78	14		
	SEPT	19	4					5	1	5	1			34	6				
	OCT													1	1				
2007	MAY													7	2				
	JUN	87	9	16	3			36	6	25	4			149	12	13	3		
	JUL	112	11	25	7			46	6	36	7			253	24	13	3		
	AUG	65	16					25	5					224	28				
	SEPT	12	2					6	1					23	6				
	OCT																		
2008	MAY	5	1					3	1					10	2				
	JUN	23	5					2	2					115	14				
	JUL	69	14					22	7	2	1			197	26				
	AUG	101	17	2	2			32	7	6	4			218	28				
	SEPT	18	7					1	1					64	13				
	OCT																		
2009	MAY													4	2				
	JUN	9	3					2	1					64	10				
	JUL	79	12	2	1			34	6	10	4			209	29				
	AUG	28	8					4	2					127	25				
	SEPT	10	3											25	5				
	OCT																		
2010	MAY													6	3				
	JUN	29	6					10	4					69	9				
	JUL	91	18					36	7	2	2			207	27				
	AUG	172	20	11	6			80	9	28	10			308	29				
	SEPT													11	5				
	OCT																		
2011	MAY																		
	JUN	2	1					1	1					29	6				
	JUL	60	10					14	4	1	1			205	26				
	AUG	32	7	1	1			11	2	1	1			159	26				
	SEPT	5	3											95	18				
	OCT																		
2012	MAY																		
	JUN	20	6					5	2					145	21				
	JUL	180	17	14	8			86	10	34	11			303	26	4	3		
	AUG	115	15	4	3			49	8	16	6			236	27				
	SEPT													48	13				
	OCT													5	1				

Table A2. *Cont.*

Year	Month	TDI [27,29)		TDI [29,32)		TDI ≥ 32		HMDX [38,40)		HMDX [40,45)		HMDX ≥ 45		HI [32,44)		HI [41,54)		HI ≥ 54	
		Total Hours	Days	Total Hours	Days	Total Hours	Days	Total Hours	Days	Total Hours	Days	Total Hours	Days	Total Hours	Days	Total Hours	Days	Total Hours	Days
2013	MAY																		
	JUN	5	3					1	1					71	12				
	JUL	22	5					9	2					135	21				
	AUG	9	4					1	1					140	27				
	SEPT	2	1											11	3				
	OCT																		
2014	MAY																		
	JUN	19	2	1	1			6	1	3	1			46	11				
	JUL	44	13					6	5					133	21				
	AUG	98	19					17	7	1	1			193	27				
	SEPT													6	2				
	OCT																		
2015	MAY																		
	JUN													22	5				
	JUL	70	10	1	1			33	5	5	2			233	27				
	AUG	67	17					22	7					191	26				
	SEPT	31	6					8	4					75	14				
	OCT																		
2016	MAY													2	1				
	JUN	53	9	1	1			21	4	3	2	1	1	125	16	1	1		
	JUL	43	13					4	3					214	29				
	AUG	56	15					10	6					174	23				
	SEPT													6	3				
	OCT																		
2017	MAY													4	2				
	JUN	49	6	5	1			25	4	7	2			94	11	2	1		
	JUL	85	9	14	4			36	4	23	5			203	22	7	2		
	AUG	62	11	1	1			14	3	3	1			182	24				
	SEPT	17	8					1	1					76	17				
	OCT																		
2018	MAY																		
	JUN													22	7				
	JUL	55	16					7	4					184	28				
	AUG	14	5					4	2					166	29				
	SEPT	4	2											42	7				
	OCT																		
2019	MAY																		
	JUN	14	5											106	16				
	JUL	72	13					24	7	1	1			190	23				
	AUG	111	16	7	4			40	10	14	4			213	25	1	1		
	SEPT	3		1	1			2		1	1			26	8				
	OCT																		
2020	MAY													22	5				
	JUN	4	1					1	1					28	6				
	JUL	54	9					18	3	2	2			184	27				
	AUG	27	8	1	1			6	1	2	1			171	26				
	SEPT	7	2											53	13				
	OCT													2	2				
2021	MAY													5	2				
	JUN	66	6	5	2			30	4	10	4			97	8	2	1		
	JUL	118	16	10	3			47	8	15	3			270	29	4	2		
	AUG	142	19	26	6			44	8	36	5	2	1	285	27	15	4		
	SEPT	6	2					2	2					32	7				
	OCT																		

Table A2. *Cont.*

Year	Month	TDI [27,29)		TDI [29,32)		TDI ≥ 32		HMDX [38,40)		HMDX [40,45)		HMDX ≥ 45		HI [32,44)		HI [41,54)		HI ≥ 54	
		Total Hours	Days	Total Hours	Days	Total Hours	Days	Total Hours	Days	Total Hours	Days	Total Hours	Days	Total Hours	Days	Total Hours	Days	Total Hours	Days
2022	MAY													4	3				
	JUN	25	8					1	1					106	19				
	JUL	87	16					21	6	3	3			202	23				
	AUG	52	10					13	5					155	26				
	SEPT	6	4											36	8				
	OCT																		
2023	MAY																		
	JUN	3	2											55	9				
	JUL	187	15	44	9			68	8	65	12	2	1	340	23	21	7		
	AUG	118	20					41	10	2	2			231	28				
	SEPT	2	1											30	7				
	OCT																		
2024	MAY																		
	JUN	81	19					16	5	3	1			226	27				
	JUL	168	24	7	3			67	10	17	7			352	30				
	AUG	140	23					48	11	5	3			274	30				
	SEPT	12	4					3	2					41	10				
	OCT																		

Appendix D

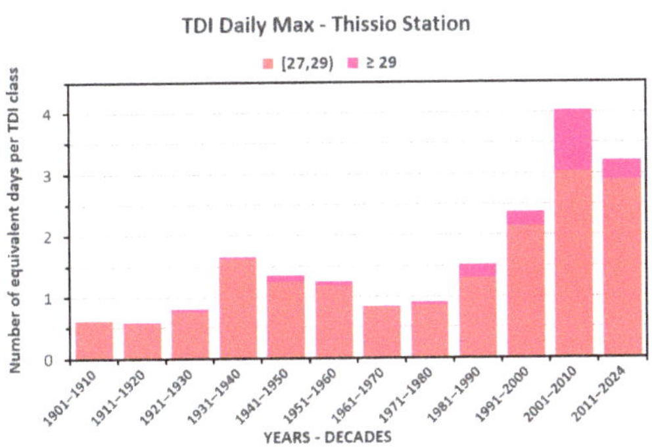

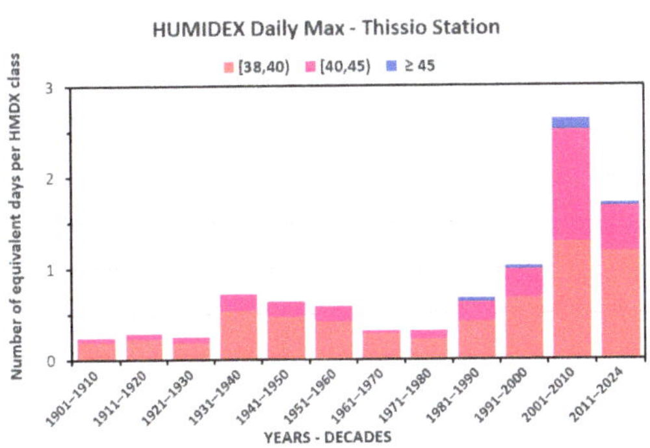

Figure A1. *Cont.*

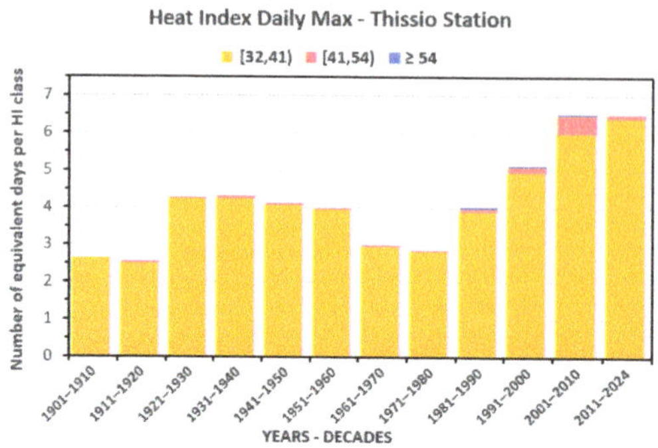

Figure A1. Decadal distribution of thermal discomfort conditions in Athens during the 1901–2024 period, based on the daily maximum values of the TDI (**up**), HMDX (**middle**), and HI (**down**). Each bar represents the total number of days per decade that fall within specific discomfort classes, expressed as equivalent day available in each decade.

References

1. Andersen, H.E.; Larsen, S.E.; Hoffman, C.C.; Jensen, T.S.; Rasmussen, E.K. Climate-Change Impacts on Hydrology and Nutrients in a Danish Lowland River Basin. *Sci. Total Environ.* **2006**, *365*, 223–237. [CrossRef]
2. Da, Y.; Xu, Y.; McCarl, B. Effects of Surface Ozone and Climate on Historical (1980–2015) Crop Yields in the United States: Implication for Mid-21st Century Projection. *Environ. Resour. Econ.* **2022**, *81*, 355–378. [CrossRef]
3. Jing, X.; Jiang, S.; Liu, H.; Li, Y.; He, J.S. Complex Relationships and Feedback Mechanisms between Climate Change and Biodiversity. *Biodivers. Sci.* **2022**, *30*, 22462. [CrossRef]
4. Shen, F.; Yang, L.; Zhang, L.; Guo, M.; Huang, H.; Zhou, C. Quantifying the Direct Effects of Long-Term Dynamic Land Use Intensity on Vegetation Change and Its Interacted Effects with Economic Development and Climate Change in Jiangsu, China. *J. Environ. Manag.* **2023**, *325*, 116562. [CrossRef] [PubMed]
5. Alimonti, G.; Mariani, L.; Prodi, F.; Ricci, R.A. A Critical Assessment of Extreme Events Trends in Times of Global Warming. *Eur. Phys. J. Plus* **2022**, *137*, 112. [CrossRef]
6. Clarke, B.; Otto, F.; Stuart-Smith, R.; Harrington, L. Extreme Weather Impacts of Climate Change: An Attribution Perspective. *Environ. Res. Clim.* **2022**, *1*, 012001. [CrossRef]
7. Amirkhani, M.; Ghaemimood, S.; von Schreeb, J.; El-Khatib, Z.; Yaya, S. Extreme Weather Events and Death Based on Temperature and CO_2 Emission—A Global Retrospective Study in 77 Low-, Middle- and High-Income Countries from 1999 to 2018. *Prev. Med. Rep.* **2022**, *28*, 101846. [CrossRef]
8. Laino, E.; Iglesias, G. Extreme Climate Change Hazards and Impacts on European Coastal Cities: A Review. *Renew. Sustain. Energy Rev.* **2023**, *184*, 113587. [CrossRef]
9. Laino, E.; Iglesias, G. Scientometric Review of Climate-Change Extreme Impacts on Coastal Cities. *Ocean Coast. Manag.* **2023**, *242*, 106709. [CrossRef]
10. Pachauri, R.K.; Reisinger, A. *Climate Change 2007: Synthesis Report. Contribution of Working Groups I, II and III to the Fourth Assessment Report of the Intergovernmental Panel on Climate Change*; IPCC: Geneva, Switzerland, 2007.
11. IPCC. *Climate Change 2014: Synthesis Report. Contribution of Working Groups I, II and III to the Fifth Assessment Report of the Intergovernmental Panel on Climate Change*; IPCC: Geneva, Switzerland, 2014.
12. Skendžić, S.; Zovko, M.; Živković, I.P.; Lešić, V.; Lemić, D. The Impact of Climate Change on Agricultural Insect Pests. *Insects* **2021**, *12*, 440. [CrossRef]
13. Nastos, P.T.; Matzarakis, A. Weather Impacts on Respiratory Infections in Athens, Greece. *Int. J. Biometeorol.* **2006**, *50*, 358–369. [CrossRef]
14. He, B.J.; Wang, J.; Liu, H.; Ulpiani, G. Localized Synergies between Heat Waves and Urban Heat Islands: Implications on Human Thermal Comfort and Urban Heat Management. *Environ. Res.* **2021**, *193*, 110584. [CrossRef]
15. Ren, Z.; Fu, Y.; Dong, Y.; Zhang, P.; He, X. Rapid Urbanization and Climate Change Significantly Contribute to Worsening Urban Human Thermal Comfort: A National 183-City, 26-Year Study in China. *Urban Clim.* **2022**, *43*, 101154. [CrossRef]
16. Isinkaralar, K.; Isinkaralar, O.; Bayraktar, E.P. Ecological and Health Risk Assessment in Road Dust Samples from Various Land Use of Düzce City Center: Towards the Sustainable Urban Development. *Water Air Soil Pollut.* **2024**, *235*, 84. [CrossRef]
17. Lin, Z.; Deng, S. A Study on the Thermal Comfort in Sleeping Environments in the Subtropics—Developing a Thermal Comfort Model for Sleeping Environments. *Build. Environ.* **2008**, *43*, 70–81. [CrossRef]

18. Kurmi, R.S.; Gupta, J.K. *A Textbook of Refrigeration and Air Conditioning*, 1st ed.; Eurasia Publishing House: New Delhi, India, 2006.
19. Isinkaralar, O.; Isinkaralar, K. Projection of Bioclimatic Patterns via CMIP6 in the Southeast Region of Türkiye: A Guidance for Adaptation Strategies for Climate Policy. *Environ. Monit. Assess.* **2023**, *195*, 1448. [CrossRef]
20. de Freitas, C.R.; Grigorieva, E.A. A Comparison and Appraisal of a Comprehensive Range of Human Thermal Climate Indices. *Int. J. Biometeorol.* **2015**, *61*, 487–512. [CrossRef]
21. McLaughlin, S.; Cooper, J.A.G. A Multi-Scale Coastal Vulnerability Index: A Tool for Coastal Managers? *Environ. Hazards* **2010**, *9*, 233–248. [CrossRef]
22. Rabby, Y.W.; Hossain, M.B.; Hasan, M.U. Social Vulnerability in the Coastal Region of Bangladesh: An Investigation of Social Vulnerability Index and Scalar Change Effects. *Int. J. Disaster Risk Reduct.* **2019**, *41*, 101329. [CrossRef]
23. Tasnuva, A.; Hossain, M.R.; Salam, R.; Islam, A.R.M.T.; Patwary, M.M.; Ibrahim, S.M. Employing Social Vulnerability Index to Assess Household Social Vulnerability of Natural Hazards: An Evidence from Southwest Coastal Bangladesh. *Environ. Dev. Sustain.* **2021**, *23*, 10223–10245. [CrossRef]
24. Pantusa, D.; D'Alessandro, F.; Frega, F.; Francone, A.; Tomasicchio, G.R. Improvement of a Coastal Vulnerability Index and Its Application along the Calabria Coastline, Italy. *Sci. Rep.* **2022**, *12*, 21959. [CrossRef]
25. Kovaleva, O.; Sergeev, A.; Ryabchuk, D. Coastal Vulnerability Index as a Tool for Current State Assessment and Anthropogenic Activity Planning for the Eastern Gulf of Finland Coastal Zone (the Baltic Sea). *Appl. Geogr.* **2022**, *143*, 102710. [CrossRef]
26. Oloyede, M.O.; Williams, A.B.; Ode, G.O.; Benson, N.U. Coastal Vulnerability Assessment: A Case Study of the Nigerian Coastline. *Sustainability* **2022**, *14*, 2097. [CrossRef]
27. Šimac, Z.; Lončar, N.; Faivre, S. Overview of Coastal Vulnerability Indices with Reference to Physical Characteristics of the Croatian Coast of Istria. *Hydrology* **2023**, *10*, 14. [CrossRef]
28. Esmaeili, R.; Gandomkar, A.; Nokhndan, M.H. Assessment of Comfortable Climate in Several Main Iranian Tourism Cities Using Physiologic Equivalence Temperature Index. *Phys. Geogr. Res. Q.* **2011**, *75*, 8–10. [CrossRef]
29. Thom, E.C. The Discomfort Index. *Weatherwise* **1959**, *12*, 57–61. [CrossRef]
30. Masterton, J.M.; Richardson, F.A. *Humidex: A Method of Quantifying Human Discomfort Due to Excessive Heat and Humidity*; Technical Report No. 45; Atmospheric Environment Service, Environment Canada: Downsview, ON, Canada, 1979.
31. Steadman, R.G. The Assessment of Sultriness. Part I: A Temperature–Humidity Index Based on Human Physiology and Clothing Science. *J. Appl. Meteorol.* **1979**, *18*, 861–873. [CrossRef]
32. Steadman, R.G. The Assessment of Sultriness. Part II: Effects of Wind, Extra Radiation and Barometric Pressure on Apparent Temperature. *J. Appl. Meteorol.* **1979**, *18*, 874–885. [CrossRef]
33. Rothfusz, L.P. *The Heat Index "Equation"*; Technical Attachment SR/SSD 90-23; National Weather Service, Southern Region Headquarters: Fort Worth, TX, USA, 1990.
34. Yousif, T.A.; Tahir, H.M. The Relationship between Relative Humidity and the Dew Point Temperature in Khartoum State, Sudan. *J. Appl. Ind. Sci.* **2013**, *1*, 20–23.
35. Epstein, Y.; Moran, D.S. Thermal Comfort and the Heat Stress Indices. *Ind. Health* **2006**, *44*, 388–398. [CrossRef]
36. Stathopoulou, M.I.; Cartalis, C.; Keramitsoglou, I.; Santamouris, M. Thermal Remote Sensing of Thom's Discomfort Index (DI): Comparison with In-Situ Measurements. *Remote Sens.* **2005**, *5983*, 131–139. [CrossRef]
37. Adegoke, O.O.; Dombo, T.P. Geospatial Modeling of Human Thermal Comfort in Akure Metropolis Using Thom's Discomfort Index. *Int. J. Environ. Bioenergy* **2019**, *14*, 40–55. [CrossRef]
38. Talukdar, M.S.J.; Hossen, M.S.; Baten, M.A. Trends of Outdoor Thermal Discomfort in Mymensingh: An Application of Thom's Discomfort Index. *J. Environ. Sci. Nat. Resour.* **2017**, *10*, 151–156. [CrossRef]
39. Ehsan, S.; Abbas, F.; Ibrahim, M.; Ahmad, B.; Farooque, A.A. Thermal Discomfort Levels, Building Design Concepts, and Some Heat Mitigation Strategies in Low-Income Communities of a South Asian City. *Int. J. Environ. Res. Public Health* **2021**, *18*, 2535. [CrossRef]
40. Monforte, P.; Ragusa, M.A. Evaluation of Bioclimatic Discomfort Trend in a Central Area of the Mediterranean Sea. *Climate* **2022**, *10*, 146. [CrossRef]
41. Maftei, C.; Buta, C. Application of Thermal Discomfort Indices for the Coastal Zone of Black Sea, in Dobrogea Region. *Ovidius Univ. Ann. Ser. Civ. Eng.* **2017**, *19*, 87–100. [CrossRef]
42. Azevedo, P.V.; Bezerra, P.T.C.; Leitão, M.M.V.B.R.; Santos, C.A.C. Thermal Comfort Level Assessment in Urban Area of Petrolina-PE County, Brazil. *Rev. Bras. Meteorol.* **2017**, *32*, 555–563. [CrossRef]
43. Diaconescu, E.; Sankare, H.; Chow, K.; Murdock, T.Q.; Cannon, A.J. A Short Note on the Use of Daily Climate Data to Calculate Humidex Heat-Stress Indices. *Int. J. Climatol.* **2023**, *43*, 837–849. [CrossRef]
44. Orimoloye, I.R.; Mazinyo, S.P.; Nel, W.; Iortyom, E.T. Climate Variability and Heat Stress Index Have Increasing Potential Ill-Health and Environmental Impacts in East London, South Africa. *Int. J. Appl. Eng. Res.* **2017**, *12*, 6910–6918.
45. Golbabaei, F.; Heidari, H.; Shamsipour, A. The Assessment of Thermal Conditions Using Humidex in Different Weather Conditions: A Case in Different Climates of Iran. *J. Health Sci. Surveill. Syst.* **2021**, *9*, 127–134. [CrossRef]

46. FallahGhalhari, G.; FarhangDehghan, S.; Asghari, M. Trend Analysis of Humidex as a Heat Discomfort Index Using Mann–Kendall and Sen's Slope Estimator Statistical Tests. *Environ. Health Eng. Manag. J.* **2022**, *9*, 165–176. [CrossRef]
47. Scoccimarro, E.; Fogli, P.G.; Gualdi, S. The Role of Humidity in Determining Scenarios of Perceived Temperature Extremes in Europe. *Environ. Res. Lett.* **2017**, *12*, 114029. [CrossRef]
48. Středová, H.; Středa, T.; Litschmann, T. Smart Tools of Urban Climate Evaluation for Smart Spatial Planning. *Morav. Geogr. Rep.* **2015**, *23*, 47–57. [CrossRef]
49. Infusino, E.; Caloiero, T.; Fusto, F.; Calderaro, G.; Brutto, A.; Tagarelli, G. Characterization of the 2017 Summer Heat Waves and Their Effects on the Population of an Area of Southern Italy. *Int. J. Environ. Res. Public Health* **2021**, *18*, 970. [CrossRef]
50. Dahl, K.; Licker, R.; Abatzoglou, J.T.; Declet-Barreto, J. Increased Frequency of and Population Exposure to Extreme Heat Index Days in the United States during the 21st Century. *Environ. Res. Commun.* **2019**, *1*, 075002. [CrossRef]
51. Michelozzi, P.; Accetta, G.; De Sario, M.; D'Ippoliti, D.; Marino, C.; Baccini, M.; Biggeri, A.; Anderson, H.R.; Katsouyanni, K.; Ballester, F.; et al. High Temperature and Hospitalizations for Cardiovascular and Respiratory Causes in 12 European Cities. *Am. J. Respir. Crit. Care Med.* **2009**, *179*, 383–389. [CrossRef] [PubMed]
52. Khalaj, B.; Lloyd, G.; Sheppeard, V.; Dear, K. The Health Impacts of Heat Waves in Five Regions of New South Wales, Australia: A Case-Only Analysis. *Int. Arch. Occup. Environ. Health* **2010**, *83*, 833–842. [CrossRef]
53. Kyselý, J.; Kim, J. Mortality during Heat Waves in South Korea, 1991 to 2005: How Exceptional Was the 1994 Heat Wave? *Clim. Res.* **2009**, *38*, 105–116. [CrossRef]
54. Bell, M.L.; O'Neill, M.S.; Ranjit, N.; Borja-Aburto, V.H.; Cifuentes, L.A.; Gouveia, N.C. Vulnerability to Heat-Related Mortality in Latin America: A Case-Crossover Study in São Paulo, Brazil, Santiago, Chile and Mexico City, Mexico. *Int. J. Epidemiol.* **2008**, *37*, 796–804. [CrossRef]
55. Angouridakis, V.E.; Makrogiannis, T.J. The Discomfort-Index in Thessaloniki, Greece. *Int. J. Biometeorol.* **1982**, *26*, 53–59. [CrossRef]
56. Tsitoura, M.; Tsoutsos, T.; Daras, T. Evaluation of Comfort Conditions in Urban Open Spaces. Application in the Island of Crete. *Energy Convers. Manag.* **2014**, *86*, 250–258. [CrossRef]
57. Matzarakis, A.; Mayer, H. Heat Stress in Greece. *Int. J. Biometeorol.* **1997**, *41*, 34–39. [CrossRef]
58. Kambezidis, H.D.; Psiloglou, B.E.; Varotsos, K.V.; Giannakopoulos, C. Climate Change and Thermal Comfort in Greece. *Climate* **2021**, *9*, 10. [CrossRef]
59. Kambezidis, H.D.; Psiloglou, B.E.; Kaskaoutis, D.G.; Karagiannis, D.; Petrinoli, K.; Gavriil, A.; Kavadias, K. Generation of typical meteorological years for 33 locations in Greece: Adaptation to the needs of various applications. *Theor. Appl. Clim.* **2020**, *141*, 1313–1330. [CrossRef]
60. Giles, B.D.; Balafoutis, C.; Maheras, P. Too Hot for Comfort: The Heatwaves in Greece in 1987 and 1988. *Int. J. Biometeorol.* **1990**, *34*, 98–104. [CrossRef]
61. Paliatsos, A.G.; Nastos, P.T. Relation between air pollution episodes and discomfort index in the greater Athens area, Greece. *Glob. Nest J.* **1999**, *1*, 91–97. [CrossRef]
62. Tselepidaki, I.; Santamouris, M.; Moustris, C.; Poulopoulou, G. Analysis of the Summer Discomfort Index in Athens, Greece, for Cooling Purposes. *Energy Build.* **1992**, *18*, 51–56. [CrossRef]
63. Bartzokas, A.; Lolis, C.J.; Kassomenos, P.A.; McGregor, G.R. Climatic Characteristics of Summer Human Thermal Discomfort in Athens and Its Connection to Atmospheric Circulation. *Nat. Hazards Earth Syst. Sci.* **2013**, *13*, 3271–3279. [CrossRef]
64. Matzarakis, A.; Nastos, P.T. Human-biometeorological assessment of heat waves in Athens. *Theor. Appl. Climatol.* **2011**, *105*, 99–106. [CrossRef]
65. Nastos, P.T.; Matzarakis, A. The Effect of Air Temperature and Human Thermal Indices on Mortality in Athens, Greece. *Theor. Appl. Climatol.* **2012**, *108*, 591–599. [CrossRef]
66. Pantavou, K.; Santamouris, M.; Asimakopoulos, D.; Theoharatos, G. Empirical Calibration of Thermal Indices in an Urban Outdoor Mediterranean Environment. *Build. Environ.* **2014**, *80*, 283–292. [CrossRef]
67. Pantavou, K.; Theoharatos, G.; Santamouris, M.; Asimakopoulos, D. Outdoor Thermal Sensation of Pedestrians in a Mediterranean Climate and a Comparison with UTCI. *Build. Environ.* **2013**, *66*, 82–95. [CrossRef]
68. Katavoutas, G.; Founda, D. Intensification of Thermal Risk in Mediterranean Climates: Evidence from the Comparison of Rational and Simple Indices. *Int. J. Biometeorol.* **2019**, *63*, 1251–1264. [CrossRef] [PubMed]
69. International Meteorological Organisation (IMO). *Provisional Guide to International Meteorological Instruments and Observing Practice*; IMO No. 78 Technical Report, Chapters 1 to 10; International Meteorological Organisation: Lausanne, Switzerland, 1950.
70. World Meteorological Organization (WMO). *Guide to Meteorological Instruments and Methods of Observation*, 7th ed.; WMO-No. 8 Technical Report; World Meteorological Organization (WMO): Geneva, Switzerland, 2008.
71. World Meteorological Organization (WMO). *Guide to Meteorological Instruments and Methods of Observation: Volume I—Measurements of Meteorological Variables*; WMO-No. 8; World Meteorological Organization (WMO): Geneva, Switzerland, 2024.
72. Eginitis, D. *The Climate of Greece, Part A*; P. Sakkelariou Editions: Athens, Greece, 1907; pp. 227–228. (In Greek)

73. Founda, D.; Kambezidis, H.D.; Petrakis, M.; Zanis, P.; Zerefos, C. A correction of the recent air-temperature record at the historical meteorological station of the National Observatory of Athens (NOA) due to instrument change. *Theor. Appl. Climatol.* **2009**, *97*, 385–389. [CrossRef]
74. McGregor, G.R.; Vanos, J.K. Heat: A primer for public health researchers. *Public Health* **2018**, *161*, 138–146. [CrossRef]
75. Psiloglou, B.E.; Larissi, I.K.; Petrakis, M.; Paliatsos, A.G.; Antoniou, A.; Viras, L.G. Case Studies on Summertime Measurements of O_3, NO_2, and SO_2 with a DOAS System in an Urban Semi-Industrial Region in Athens, Greece. *Environ. Monit. Assess.* **2013**, *185*, 6565–6576. [CrossRef]
76. Kallos, G.; Kassomenos, P.; Pielke, R. Synoptic and mesoscale weather conditions during air pollution episodes in Athens, Greece. *Bound. Layer Meteorol.* **1993**, *62*, 163–184. [CrossRef]
77. Kassomenos, P.; Kotroni, V.; Kallos, G. Analysis of the climatological and air quality observations from Greater Athens Area. *Atmos. Environ.* **1995**, *29*, 3671–3688. [CrossRef]
78. Tung, A. The City of the Gods Besieged. In *Preserving the World's Great Cities: The Destruction and Renewal of the Historic Metropolis*; Three Rivers Press: New York, NY, USA, 2001; pp. 260, 263, 265. ISBN 0-609-80815-X.
79. Hellenic Statistical Authority (ELSTAT). *2021 Population Census (Greek Language PDF)*; Hellenic Statistical Authority: Athens, Greece, 2021. Available online: https://www.statistics.gr/en/2021-census-pop-hous (accessed on 10 July 2025).
80. Hellenic Statistical Authority (ELSTAT). *Census 2022 Greece: Population and Housing Census (Greek Language PDF)*; Hellenic Statistical Authority: Athens, Greece, 2022. Available online: https://elstat-outsourcers.statistics.gr/Census2022_GR.pdf (accessed on 10 July 2025).
81. National Centre for Social Research. Panorama of Greek Census Data. Available online: https://panorama.statistics.gr/en/ (accessed on 10 July 2025).
82. Athens Social Atlas. Available online: https://www.athenssocialatlas.gr (accessed on 10 July 2025).
83. Metaxas, D.A.; Bartzokas, A.; Vitsas, A. Temperature Fluctuations in the Mediterranean Area during the Last 120 Years. *Int. J. Climatol.* **1991**, *11*, 897–908. [CrossRef]
84. Founda, D. Evolution of the Air Temperature in Athens and Evidence of Climatic Change: A Review. *Adv. Build. Energy Res.* **2011**, *5*, 7–41. [CrossRef]
85. American Society of Heating, Refrigeration and Air-Conditioning Engineers (ASHRAE). *Heating, Ventilating, Air-Conditioning, Guide No. 33*; ASHRAE: New York, NY, USA, 1955.
86. Thom, E.C. A New Concept for Cooling Degree-Days. *Air Cond. Heat. Vent.* **1957**, *54*, 73–80.
87. Giannakopoulos, C.; Kostopoulou, E.; Varotsos, K.V.; Tziotziou, K.; Plitharas, A. An Integrated Assessment of Climate Change Impacts for Greece in the Near Future. *Reg. Environ. Change* **2011**, *11*, 829–843. [CrossRef]
88. Rana, R.; Kusy, B.; Jurdak, R.; Wall, J.; Hu, W. Feasibility Analysis of Using Humidex as an Indoor Thermal Comfort Predictor. *Energy Build.* **2013**, *64*, 17–25. [CrossRef]
89. Steadman, R.G. A Universal Scale of Apparent Temperature. *J. Clim. Appl. Meteorol.* **1984**, *23*, 1674–1687. [CrossRef]
90. NOAA (National Oceanic and Atmospheric Administration). National Weather Service Glossary. 2009. Available online: http://www.weather.gov/glossary/ (accessed on 12 October 2012).
91. Steadman, R.G. Norms of Apparent Temperature in Australia. *Aust. Meteorol. Mag.* **1994**, *43*, 1–16.
92. NWS (National Weather Service). Meteorological Conversions and Calculations: Heat Index Calculator. 2011. Available online: https://www.wpc.ncep.noaa.gov/html/heatindex.shtml (accessed on 12 October 2011).
93. NOAA (National Oceanic and Atmospheric Administration). Heat: A Major Killer. 2012. Available online: http://www.nws.noaa.gov/om/heat/index.shtml (accessed on 12 October 2012).
94. United States National Weather Service, Pueblo, Colorado. Heat Index. Available online: https://web.archive.org/web/20110629041320/http://www.crh.noaa.gov/pub/heat.php (accessed on 10 July 2025).
95. The Intergovernmental Panel on Climate Change (IPCC). *Managing the Risks of Extreme Events and Disasters to Advance Climate Change Adaptation (SREX)*; Field, C.B., Barros, V., Stocker, T.F., Qin, D., Dokken, D.J., Ebi, K.L., Mastrandrea, M.D., Mach, K.J., Plattner, G.K., Allen, S.K., et al., Eds.; Cambridge University Press: New York, NY, USA, 2012. Available online: https://www.ipcc.ch/report/managing-the-risks-of-extreme-events-and-disasters-to-advance-climate-change-adaptation/ (accessed on 28 August 2025).
96. World Meteorological Organization & World Health Organization (WMO/WHO). *Heatwaves and Health: Guidance on Warning-System Development*; WMO No. 114 Technical Report, Chapter 7; McGregor, G.R., Bessemoulin, P., Ebi, K., Menne, B., Eds.; World Meteorological Organization: Geneva, Switzerland, 2015. Available online: https://www.who.int/publications/m/item/heatwaves-and-health{-}{-}guidance-on-warning-system-development (accessed on 28 August 2025).
97. D'Ippoliti, D.; Michelozzi, P.; Marino, C.; de'Donato, F.; Menne, B.; Katsouyanni, K.; Kirchmayer, U.; Analitis, A.; Medina-Ramón, M.; Paldy, A.; et al. The impact of heat waves on mortality in 9 European cities: Results from the EuroHEAT project. *Environ. Health* **2010**, *9*, 37. [CrossRef] [PubMed]
98. Habeeb, D.; Vargo, J.; Stone, B. Rising heat wave trends in large US cities. *Nat. Hazards* **2015**, *76*, 1651–1665. [CrossRef]

99. Spector, J.T.; Masuda, Y.J.; Wolff, N.H.; Calkins, M.; Seixas, N. Heat exposure and occupational injuries: Review of the literature and implications. *Curr. Environ. Health Rep.* **2019**, *6*, 286–296. [CrossRef] [PubMed]
100. Vicedo-Cabrera, A.M.; Scovronick, N.; Sera, F.; Royé, D.; Schneider, R.; Tobias, A.; Astrom, C.; Guo, Y.; Honda, Y.; Hondula, D.M.; et al. The burden of heat-related mortality attributable to recent human-induced climate change. *Nat. Med.* **2020**, *27*, 1027–1034. [CrossRef]
101. Perkins-Kirkpatrick, S.E.; Lewis, S.C. Increasing trends in regional heatwaves. *Nat. Commun.* **2020**, *11*, 3357. [CrossRef]
102. Schwingshackl, C.; Sillmann, J.; Vicedo-Cabrera, A.M.; Sandstad, M.; Aunan, K. Heat stress indicators in CMIP6: Estimating future trends and exceedances of impact-relevant thresholds. *Earth's Future* **2021**, *9*, e2020EF001885. [CrossRef]
103. Russo, S.; Dosio, A.; Graversen, R.G.; Sillmann, J.; Carrao, H.; Dunbar, M.B.; Vogt, J.V. Magnitude of extreme heat waves in present climate and their projection in a warming world. *J. Geophys. Res. Atmos.* **2014**, *119*, 12500–12512. [CrossRef]
104. Katavoutas, G.; Founda, D.; Varotsos, K.V.; Giannakopoulos, C. Diurnal Temperature Range and Its Response to Heat Waves in 16 European Cities—Current and Future Trends. *Sustainability* **2023**, *15*, 12715. [CrossRef]

climate

MDPI

Review

The Perceptions of Rural Australians Concerning the Health Impacts of Extreme Weather Events: A Scoping Review

Emily Vohralik [1], Jonathan Mond [1,2,3], I. Nyoman Sutarsa [1], Sally Hall Dykgraaf [1,*], Breanna Humber [1] and Sari Dewi [1,4]

1 Rural Clinical School, School of Medicine and Psychology, College of Science and Medicine, Australian National University, Acton 2601, Australia; emily.vohralik@anu.edu.au (E.V.); jon.mond@utas.edu.au (J.M.); sutarsa.nyoman@anu.edu.au (I.N.S.); breanna.humber@anu.edu.au (B.H.); sari.dewi@anu.edu.au (S.D.)
2 Centre for Rural Health, University of Tasmania, Hobart 7001, Australia
3 School of Medicine, Western Sydney University, Penrith 2751, Australia
4 Department of Public Health, Faculty of Medicine, Universitas Padjadjaran, Jatinangor 45363, Indonesia
* Correspondence: sally.hall@anu.edu.au

Abstract: Understanding rural communities' perceptions of the health impacts of extreme weather is vital for strengthening community resilience and adaptation strategies. This paper aimed to collate existing evidence on the perceptions of rural Australians regarding the health impacts of extreme weather events. A scoping review following PRISMA-ScR guidelines was conducted. Peer-reviewed empirical articles published up to 7 May 2025 were identified from Scopus, PubMed, and Web of Science. One author undertook two-step screening and data extraction, which was checked by another author, and data were analysed using a thematic approach. Of 242 non-duplicate articles screened, 34 were included, which discussed drought (n = 14), bushfire (n = 8), flood (n = 6), extreme heat (n = 4) or a combination of events (n = 2). Two main themes arose: (1) perceived severity, frequency and duration of extreme weather events; and (2) perceptions of health impacts. The second theme comprised six subthemes: mental health risks, social disconnectedness, disrupted connection to land, distress due to uncertainties, community resilience, and disproportionate effects on vulnerable groups. Evidence gaps included a lack of perspectives separated by gender and age and a shortage of voices of socio-economically disadvantaged groups. Future research should investigate how to understand rural communities' resilience to develop targeted adaptation and mitigation strategies.

Keywords: natural disasters; extreme weather events; perceptions; rural health; human health; health impact; Australia; climate change

1. Introduction

Climate change is increasing the frequency, intensity and/or duration of extreme weather events including extreme heat, drought, desertification, dust storms, floods and extreme precipitation [1]. Additionally, evidence that human influence has increased the occurrence of compound extreme events such as bushfires has strengthened [2]. Researchers have found that anthropogenic climate change, particularly its impact on temperature extremes, has increased the probability of an extreme fire season like the 2019–2020 Australian bushfires by at least 30% [2].

The health impacts of extreme weather events are inequitably distributed towards those experiencing existing disadvantage [1,3], particularly rural and remote communities. They include directly increased morbidity and mortality which are compounded by indirect socioeconomic impacts such as disruption to infrastructure, health systems and community social supports [1].

People living in rural areas are often adversely impacted by extreme weather events to a greater extent than those in urban areas because of numerous factors which increase their vulnerability [4,5]. Around 43% of the global population lives in rural areas [6], and in Australia, 28% people live outside major cities, while inequity in health outcomes and social determinants of health for people in outer regional, rural and remote areas is increasing [7–9]. Rural populations typically have a lower socioeconomic status than people living in cities, a higher burden of disease and poorer access to healthcare services [8–11]. Furthermore, economic dependence on agricultural and food production industries and the associated water sources make these populations prone to disruption by extreme weather events [4,12]. In addition, the physical proximity to bush and farmland increases risk of harm from bushfires and drought [13]. The geographical isolation of many rural towns further impairs access to essential services and the ability to cope with the impacts of extreme weather events [7,11,12,14].

Rural communities often rely on primary healthcare services for healthcare delivery particularly during natural disasters and to mitigate the health effects of climate change [15,16]. Climate-related risks and extreme weather events are negatively impacting primary care utilisation and health outcomes in rural and remote areas [11]. The increasing frequency, intensity and/or duration of climate change-related extreme weather events is likely to exacerbate healthcare inequities between rural and urban communities [7].

While Australia shares some commonalities with global trends in extreme weather events, particularly those linked to climate change, its unique geographical and climatic characteristics lead to specific challenges. Australia is experiencing more frequent and intense heatwaves, heavy rainfall events, and longer and more severe fire seasons [17]. Given that rural Australians are more dependent on natural resources likely to be impacted by extreme weather events than those in urban areas, it is important to understand how their health outcomes may be protected and improved. With increased extreme weather events in Australia, lessons learnt can be potentially applied in other similar rural areas outside Australia.

There is some existing research into the health impacts of drought [18–21], particularly on farmers [22], as well as extreme heat [23–25], floods [5,26,27] and bushfires [28–30] on rural Australian communities. Over a decade ago, researchers reviewed the effects of drought, heatwaves and bushfires on the health of older rural Australians, emphasising the need for collaborative community programmes to support the ageing, particularly targeted mental health interventions [19]. Recently, others have updated the review of the health impacts of extreme heat on older rural Australians, with only ten included articles demonstrated limited research in this area [23]. A meta-analysis has assessed the prevalence of psychological distress after bushfires [31]; however, it did not focus on rural Australians or community perceptions related to mental health or disaster recovery.

Research investigating the perceptions of rural populations can highlight the perceived severity of and susceptibility to climate change and will also identify their perceived benefits of change and barriers to action [32]. Perceptions, knowledge and beliefs as components of health literacy have been known to be associated with health outcomes, including help-seeking behaviour [33]. Furthermore, while rural populations depend more directly on their local environment, their perceived threat of climate change is lower than that of their urban counterpart [34]. Additionally, a focus on community capacity-building to promote resilience has been shown to be effective at increasing coping within the context of specific communities [29,35–37]. Considering rural communities' perceptions offers critical insight into how adaptation and mitigation strategies for future health impacts can be strengthened, as individuals' beliefs and experiences shape their risk perceptions of natural hazards and, in turn, their engagement with adaptive and mitigating behaviours [38, 39].

There is limited existing research related to rural Australians' perceptions at the intersection of health and climate change. Researchers in Australia have begun exploring perceptions about factors comprising rural community resilience [35], including after flooding [36], emphasising the importance of social capital as well as environmental and economic factors. Others have investigated the role of the media in influencing perceptions about climate change health risks amongst Australians, finding that exposure to and interest in climate news was important [40]. The impacts of climate change on rural health services have been assessed from the perspective of rural general practitioners in Australia [16,41] and more broadly in a globally focused scoping review [11]. Another global review synthesised perceptions of both the public and health professionals about the health implications of climate change, without a rural focus, finding low awareness of specific health harms and mixed desire to learn more [42]. Within Australia, youth mental health and climate change concerns have been explored but either not focused on rural settings [43] or not related to extreme weather events [44].

Overall, the perceptions of rural Australians concerning the health impacts of extreme weather events are less well-studied than the impacts per se, with the evidence remaining scattered and not well synthesised or mapped. A comprehensive understanding about these rural perceptions will facilitate strengthening of community resilience and development of targeted adaptive strategies by clarifying motivators for behavioural change and local health system needs [32,38]. Community-based health promotion programmes that take into account these perceptions are more likely to reach vulnerable groups and they may have more information and motivation to take part in the process of adaptation to extreme weather events [37].

This scoping review aimed to compile published literature on the perceptions of rural Australians regarding the health impacts of extreme weather events and to identify gaps in the current literature. The review highlights possible avenues for targeted adaptation and mitigation strategies that engage local communities and services to address the health impacts of future events.

2. Methods

This scoping review was conducted in accordance with the Preferred Reporting Items for Systematic Reviews and Meta-Analyses Extension for Scoping Reviews (PRISMA-ScR) guidelines [45]. The key concepts in this review were rural settings, extreme weather events, perceptions and health impacts. The Australian Bureau of Statistics (ABS) and the Australian Institute of Health and Welfare (AIHW) classify rurality in Australia into categories like Inner Regional, Outer Regional, Remote, and Very Remote [46]. Given the various definitions of rurality in different studies [47], an inclusive definition of 'not urban' was used for this review to capture Australians living in regional, rural or remote areas.

Extreme weather events, listed in Table 1, were those related to climate change with evidence for increasing frequency, intensity and duration due to global warming [1,2]. Perceptions were defined as subjective opinions expressed by individuals and communities in relation to awareness, knowledge, attitudes, and beliefs of how extreme weather events impacted their health, environment, and livelihood [48]. Health impacts were either general (including social), physical, mental or indirect socioeconomic impacts.

Table 1. Search terms identified using PICo and grouped into five themes.

Rural	Extreme Weather Event	Health Impact	Perceptions	Location
Rural population (MeSH) Rural Regional Remote Provincial Peri-rural Peri-urban	Extreme weather (MeSH) Natural disasters (MeSH) Extreme event Adverse climate event Extreme / severe heat / temperature Extreme / severe storm Heatwave Bushfire Extreme rain / rainfall / precipitation Flood, flooding Drought	Health	Health knowledge Health attitudes Health perceptions Risk perception Patient perceptions Community perceptions Public attitudes Perceptions Perspectives Concerns Beliefs Attitudes Awareness	Australia New South Wales, NSW Victoria, Vic Queensland, Qld South Australia, SA Tasmania, Tas Northern Territory, NT Australian Capital Territory, ACT Western Australia, WA

2.1. Search Strategy

The population, phenomena of interest, context (PICo) [49] approach was used to develop a search strategy. Search terms were identified using PICo by preliminary searches in PubMed and Scopus databases and noting the keywords of seminal articles. Relevant Medical Subject Headings (MeSH) terms were also included. Search terms were grouped into five themes: rural, extreme weather event, health impact, perceptions, location (Table 1). Search strings were formed by separating keywords with Boolean operators (see Supplementary Material S1). The initial search strategy was developed in Scopus and then adapted for PubMed and Web of Science databases. Additionally, a hand-search of cited articles within eligible studies was conducted to identify further relevant studies.

2.2. Selection Process

Search results were exported into EndNote 21 [50] where duplicates were removed and then imported into Covidence [51] for screening. Titles and abstracts were screened by EV and BH against inclusion and exclusion criteria and any queries were checked by SD or JM and resolved by discussion. Full text screening of eligible articles was performed by EV, and eight queries were resolved by discussion against inclusion and exclusion criteria between EV, BH and SD.

The review included English-language articles published in peer-reviewed journals up to 7 May 2025, with no date limit. Inclusion criteria were primary studies including quantitative, qualitative and mixed methods designs, non-urban location in Australia, extreme weather event and perceptions towards health impacts. Exclusion criteria were non-primary articles including reviews, study protocols, commentaries, letters and editorials, grey literature, rurality not specified, if the study could not differentiate urban and rural data, if the study was about climate change without mention of an extreme weather event, or the study did not mention perceptions about health.

2.3. Data Extraction and Synthesis

Data was extracted from included studies using a modified extraction form that aligned with review objectives and piloted on five studies to ensure all relevant results were extracted (see Supplementary Material S2). Data was extracted by EV and BH, and re-checked by SD, with discussion and revisiting the full text to clarify any differences. The following information was extracted: authors' names, publication year, study design and aims, extreme weather event, location, participant characteristics (such as age, gender, occupation, rurality) and key findings (perceptions of health impacts: general (including social) health, mental health, physical health, and indirect (socioeconomic) impacts).

Data were classified according to type of extreme weather event, perceptions of health impacts, gender, age and occupation. Given the inclusion of heterogenous study designs with both quantitative and qualitative findings, a thematic analysis of the articles was conducted [52]. This process provides a flexible yet systematic approach to integrating diverse data sources, allowing both numerical trends and narrative accounts to be examined together and distilled into comprehensive findings.

A hybrid deductive and inductive coding approach was employed. Extracted data were deductively coded using the predefined categories in the extraction form for key findings: 'perceptions of health impacts', described in the previous paragraph. Data were then further stratified by gender, age, occupation and vulnerability by tabulation in Microsoft Excel. An inductive approach was then used to identify more detailed codes that emerged from the data by grouping similar topics or keywords together. Draft descriptive themes were proposed by EV by identifying common or pertinent codes and grouping them into categories with similar meaning. Regular discussions between EV, SD and INS were conducted to check whether the themes were coherent, distinct, and well-supported by the data. Overall, final themes and subthemes were refined and presented narratively.

3. Results

A total of 373 articles were identified in the initial search. After duplicate removal and screening, a total of 34 articles were included in the analysis (Figure 1).

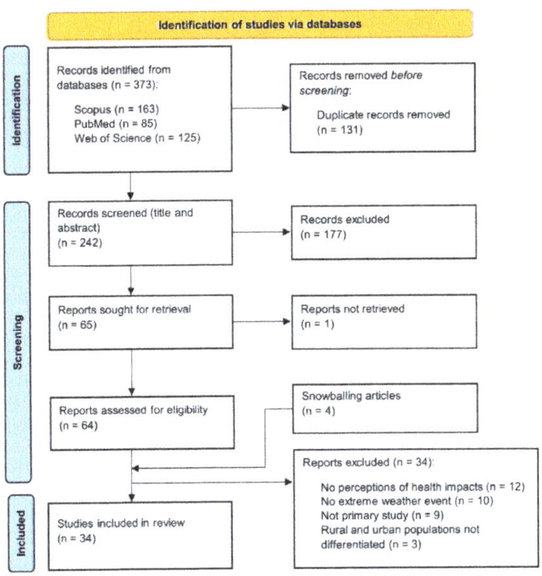

Figure 1. PRISMA flow diagram of scoping review.

3.1. Overview of Included Studies

Included articles were published between 2007 and 2025 and comprised a mix of qualitative (n = 19), quantitative (n = 10) and mixed methods (n = 5) studies. Almost half the studies originated from New South Wales (NSW) (n = 15), and others were from South Australia (SA) (n = 7), Victoria (Vic) (n = 6), Queensland (Qld) (n = 4), or included several states (n = 2). Specific extreme weather events reported in the included studies were drought (n = 14), bushfire (n = 8), flood (n = 6), extreme heat (n = 4), or a combination of events (n = 2) (see Table 2). The demographics of the study populations varied (Table 2). The number of participants ranged from two to 2004, with the study of two participants consisting of a series of interviews with two rural women's health workers to explore lived experiences of drought [53].

Table 2. Summary of characteristics of included studies.

Study Characteristic	Publication (n = 34)	References
Study design		
Qualitative	19	[30,36,53–69]
Semi-structured interviews	12	[30,53,55–58,60,63–66,68]
Interviews and focus groups	3	[36,62,69]
Forums	3	[54,59,61]
Interviews and informal discussions	1	[67]
Quantitative	10	[27,70–78]
Cross-sectional survey	10	[27,70–78]
Mixed methods	5	[79–83]
Survey and focus groups	4	[79–82]
Focus groups, interviews and meteorological data survey	1	[83]
Location		
New South Wales (NSW)	15	[27,54,58,60–62,65,70–72,76,77,79,81,82]
South Australia (SA)	7	[55,63,64,68,69,73,75]
Victoria (Vic)	6	[30,53,57,67,80,83]
Queensland (Qld)	4	[36,56,59,74]
SA and Vic	1	[78]
SA and NSW	1	[66]
Extreme weather event		
Drought	14	[53–56,59,61–63,72,73,75,77,81,82]
Bushfire	8	[30,58,60,64–67,80]
Flood	6	[27,36,71,74,76,79]
Extreme heat	4	[68,69,78,83]
Flood, bushfire and drought	1	[57]
Flood and drought	1	[70]
Age		
Children and adolescents	4	[54,64,81,82]
Young adults	3	[30,76,79]
Adults	26	[27,36,53,55–63,65–75,77,78,80]
Older adults	1	[83]
Occupation		
Farmers	6	[55,56,63,66,73,75]
Health workers or service providers	5	[53,57,58,65,68]
Both farmers and other	5	[36,59,62,67,69]
School students	4	[54,64,81,82]
Medical students	1	[79]
Other	3	[61,72,80]
Not specified	10	[27,30,60,70,71,74,76–78,83]
Gender		
Female	4	[36,53,57,59]
Male	0	
Mixed	30	[27,30,54–56,58,60–83]

Findings were presented under two main themes: (1) perceived severity, frequency and duration of extreme weather events; and (2) perceptions of health impacts (Figure 2). The second theme comprised six subthemes: (2.1) direct health risks and impacts (both physical and mental), (2.2) social disconnectedness and the effects on health, (2.3) reduced connection to land and the impacts on personal wellbeing, (2.4) increased health risks due to uncertainties, (2.5) increased community resilience before, during and after extreme weather events, and (2.6) disproportionate health impacts affecting vulnerable groups.

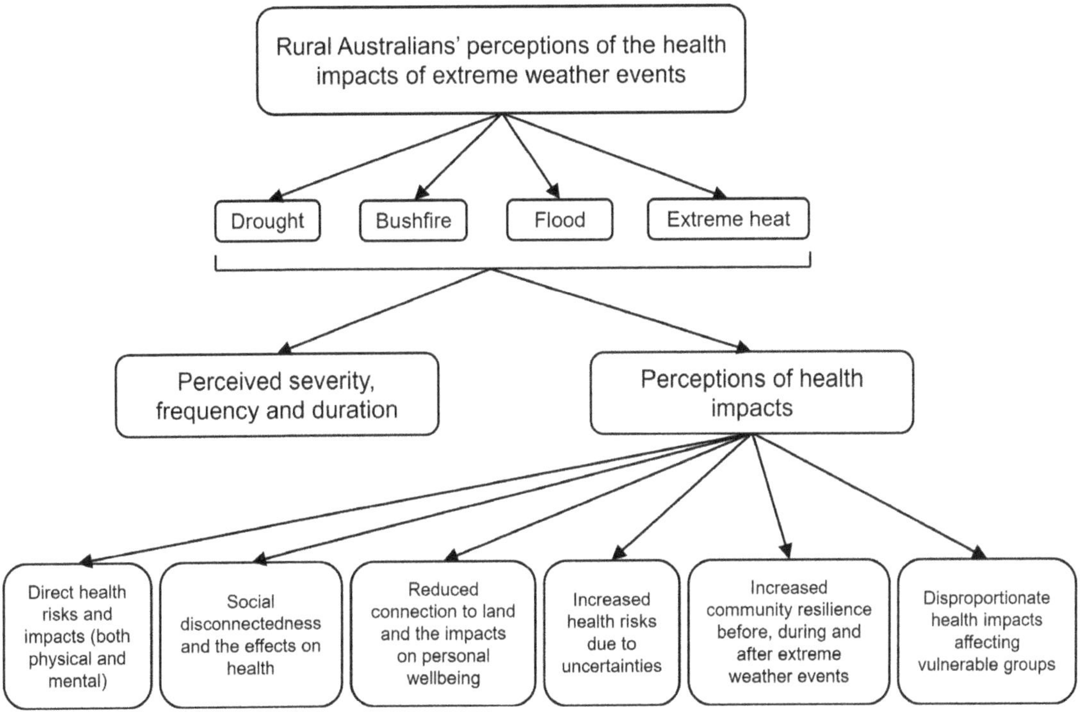

Figure 2. Concept map of scoping review themes and subthemes.

3.2. Perceived Severity, Frequency and Duration of Extreme Weather Events

Overall, as summed up in a 2020 survey of rural Australians [70], majority of participants were concerned that climate change impacts such as drought and flood will worsen in the future. Young people [76] and medical students [79] expressed surprise at the severity of flooding in the Northern Rivers' region of NSW, with greater concern for future flooding and climate change. They believed there were inadequate warnings for the severity of the situation which increased stress and reduced effective responses. Similarly, in a region of SA where drought was uncommon, farmers were surprised by the severity of drought [55].

An increase in heat exposure compared to the past was reported by almost half of the residents surveyed in both rural SA and Vic, with concern that heatwaves and health consequences would continue to increase into the future [78]. Other South Australians highlighted that increased intensity or duration of extreme heat made the experience more challenging and elevated the bushfire risk, but did not comment on heatwave trends over time [69]. In contrast, older people in Victoria who were sceptical about climate change did not recognise heat as a health threat and were confident in their ability to cope with future heatwaves as they had in the past [83].

Similarly, in a small NSW Northern Tablelands community belief about the frequency of future bushfires were linked to beliefs about climate change [60]. Some participants noted that bushfires are occurring more frequently than they used to, but those who believed in natural 'climate cycles' thought that bushfires will occur infrequently into the future and thus preparation efforts were reduced. Bushfire was perceived as a major threat to communities in Victoria [80] and was described as having catastrophic effects by NSW mental health workers which could have been reduced by listening to Aboriginal cultural land management practices [65].

Many articles described drought as prolonged, progressing and severe, with increased duration associated with greater levels of community-wide distress [53,54,56,62,70,82]. During the prolonged millennium drought adolescents in rural NSW began to recognise that climate change might be having an effect [81,82]. In a survey of 2004 NSW residents, respondents living in remote or very remote areas were 3.22 times more likely to think that drought was very or extremely likely to continue than people living in more geographically accessible areas [77].

3.3. Perceptions of Health Impacts

3.3.1. Direct Health Risks and Impacts (Both Physical and Mental)

Awareness of the health risks of extreme weather events varied among population groups. Older adults in rural NSW identified health-related suffering from climate change, particularly due to drought and flood, as a primary concern [70]. Conversely, peri-urban Victorians, including Landcare conservation volunteers who participate in community-led activities to protect and restore local environments, and new parents, perceived bushfire as a major community threat rather than an individual one and did not prioritise health issues [80]. Amongst Victorian residents over 55 years, 36% had heat-health concerns yet almost half did not consider themselves or others vulnerable to heat due to a sense of self-reliance [83]. Similarly, low awareness of heat risks was found in older age groups in regional SA and Victoria, especially those over 75 years, whereas younger adults and females showed higher awareness of heat impacts [78].

Both elderly and adult rural residents commonly perceived extreme heat as an inconvenience causing discomfort [69,83] and physical health issues like lethargy and heat stress [69,78]. Drought and extreme heat impacted outdoor enjoyment, particularly in rural farming communities [62,69]. A study of Central West NSW adults reported a negative impact of drought on their lives [72]. Interestingly, for children and adolescents in rural NSW, drought did not negatively impact their rural lifestyle, which was viewed as safer and more fun than city living [81,82].

Regarding mental health impacts, drought as a stressor was emphasised in most articles [54–56,61–63,70,72,73,75,81,82]. Farmers and rural businesspeople described drought stress in relation to financial pressure [55,62,63] and farmers reported concerns about the adverse effects of drought on their workload and mental health [55,56,62,63,73,75]. The emotional distress associated with witnessing the death of or killing their own livestock contributed to negative mental health effects for children [81,82] and farmers [56,66,67,73]. Rural Aboriginal people, farmers and health workers feared that prolonged drought may induce community-wide mood and behaviour issues [61,62].

In response to acute events like bushfire and flood, rural residents noted that mental health effects either arose immediately or were delayed, and many were long-term [27,60,65–67,71,76,79,80]. Immediate impacts included stress, anxiety and overwhelm, often exacerbated by communication difficulties and dealing with authorities during the disaster response [60,66,71,80]. Young people and rural medical students expressed trauma associated with severe flooding in the NSW Northern Rivers region [76,79]. Delayed psychological effects noted by rural healthcare workers [65,67] and Country Fire Association volunteers [80] surfaced months after bushfires when recovery programmes had often ended. Farmers [66], mental health workers [65] and young people [76] emphasised concerns about long-term mental health impacts such as residual anxiety and being easily triggered by weather changes. Pre-existing mental health issues were perceived to be exacerbated by drought [53,55] and flood [71,76], particularly notable in people with a disability [71].

Only six included articles discussed gender differences in relation to mental health effects of extreme weather events [56,57,67,75,81,82]. Local health workers found it difficult to reach men for counselling as they often considered themselves too busy or denied needing help [67]. Female farmers noted that their male partners were unavailable for emotional support, being preoccupied by disaster responses [57] or overwhelmed by their own negative mindset [56]. Interestingly, another study found no difference in psychological distress between male and female farmers during drought, but highlighted that females seek emotional social support more than males, who preferred venting of emotions [75]. Amongst rural NSW school students, males and females showed no difference in drought-related emotional distress [81], but a follow-up study in the same region reported higher distress in females [82].

In six studies, beliefs in self-reliance and stoicism were found to prevent discussions about mental health or seeking support during extreme weather events [53,56,60,62,64,67]. This led to a downplaying of mental health concerns despite residual stress, anxiety and post-traumatic stress disorder (PTSD) years after bushfires, with farmers struggling to talk about their mental health [60]. Within families, this culture of stoicism prevented children and parents from discussing their mental state after bushfires, impairing the seeking of social support and grieving losses [64]. Parents also worried about the negative emotional effects of drought [54] or bushfire [64] on their children, just as the children worried about their parents' mental health [54,64].

3.3.2. Social Disconnectedness and the Effects on Health

Participants of included studies perceived disruptions to social connectedness in personal relationships, families and communities due to extreme weather events. Drought strained family relationships and increased parental distress, which children and adolescents noticed [54,62,81,82]. Bushfires altered parent–child communication, with both parties becoming hypervigilant and withholding mental health concerns to avoid burdening each other [64]. Drought led women to seek off-farm work due to decreased farm income, potentially spending less time with family [53,55,62]. Some adolescents in rural NSW worked more on family farms during drought, reducing social interactions with friends and time at school [54].

Several studies reported shrinking rural communities due to drought [53,54,61,62,81,82] and reduced community activities [54,68,69,71,81], resulting in increased social isolation. Young people lost friends who moved away during drought, reducing their social networks [54,82]. Aboriginal communities in rural NSW worried about population skewing as working-age people left smaller towns, leaving behind the ill and elderly [61]. These socioeconomic effects contribute to rural demise, threatening the viability of smaller communities. Similar concerns were noted in SA regarding extreme heat [69].

Social connections within communities were often disrupted during or after extreme weather events. Perinatal women [57] and people with disabilities and their carers [71] experienced significant disruptions to social support networks post-flood. Some individuals in a remote Victorian community felt excluded post-bushfire due to perceived social divisiveness [67]. Mental health workers in rural NSW observed that bushfire disaster responses amplified both positive and negative human traits, increasing social tensions and vulnerabilities [65]. Farmers in SA found community work stressful, compounding farm-related stress during drought [73]. Aboriginal and financially disadvantaged participants reported lower community connectedness post-flood, compared to other residents in the

NSW Northern Rivers region [27]. In other rural NSW Aboriginal communities, drought was perceived to promote antisocial behaviour, negatively affecting social and emotional wellbeing and increasing community mistrust [61].

3.3.3. Reduced Connection to Land and the Impacts on Personal Wellbeing

A salient subtheme was how environmental damage and degradation harmed peoples' identity, sense of place and connection to the land. This was significant for Aboriginal communities in rural and remote NSW in the context of drought which harmed traditional family structure, culture and place [61]. For example, the drought-related loss of habitat and wildlife was reported to threaten traditional men's roles [61]. More broadly, Aboriginal participants described feelings of grief, loss and guilt because continuing severe drought was causing cultural and geographic dislocation, loss of country and identity [61]. Mental healthcare workers in rural NSW observed particularly devastating effects of bushfires on local Aboriginal communities who lost access to culturally significant lands [65].

Distress due to environmental changes and loss of sense of place, termed solastalgia, was a primary concern in some studies [62,70,81]. In rural farming communities, women in two studies reported distress [62] and a spiritual impact [55] associated with losing their gardens during drought. Extreme heat raised concerns about gardens and local parks in the context of water restrictions, highlighting the importance of the natural environment [69]. Others in farming communities described how the parched, dry landscape had a negative effect on mood, particularly for farmers, and made people reluctant to spend time outdoors [62]. Livestock farmers in Qld experienced feelings of personal failing due to progressing drought, destruction of the land, livestock death and declining business [56]. These farmers described a shift in the farming role from guardian of the land to a sole focus on survival [56]. During drought, farm labour increased and more time was spent managing livestock, water and fencing [55,59,73]. Bushfire was reported to change some farmers' identity and connection to the land, with the environmental changes a constant, unwanted reminder of the fire [66]. For young adults, bushfire altered their identities by making them more alienated from others [30].

3.3.4. Increased Health Risks Due to Uncertainties

Rural Australians raised concerns about uncertainties regarding potential climate change impacts, the future, a perceived lack of control, and finances. Worry about climate change was increased in medical students and young people who experienced flooding because it highlighted communities' vulnerabilities to extreme weather [76,79]. Rural healthcare workers observed that farmers experiencing drought worry about the effects of climate change on their livelihood and families, yet this may manifest as scepticism or denial [53]. Worry about the future was a common theme across drought [55,62,73,82] and flood [76]. For NSW farming communities, the most significant effects of drought on lifestyle and business were uncertainty and the inability to plan ahead [62]. Many rural Australians described a perceived lack of control during drought linked to unpredictable weather, finances and government regulations [62], particularly children and adolescents [54,81] and citrus farmers [63].

For many farmers and farm families, financial strain and uncertainty was a key concern arising from drought [54,55,59,62,73] and bushfire [67]. Financial pressures meant farm owners often had to reduce employee numbers [73] and Aboriginal farm employees described a differentially severe effect particularly due to lack of government compensation schemes [61]. Children and adolescents in rural farming communities worried about their family's financial situation at home [54,82]. Both adolescents and their parents worried about loss of future education and employment opportunities because of drought-related financial strain and increased need to work [54,62].

Young people reported financial strain from flooding caused by disruptions to housing, work, travel, food supplies and services which often negatively impacted mental health [76,79]. These would likely be amplified in people experiencing financial hardship; however, the article looking at this population only focused on the effects of flood on social capital [27]. In a large survey of rural NSW residents, other concerns about financial pressures included rising insurance, increased cost of living, difficulty growing crops, and lack of water security [70]. Some rural residents perceived that government policies to respond to extreme weather events, especially drought, exacerbated rural suffering through a lack of control and financial strain [54,62,70,73].

3.3.5. Increased Community Resilience Before, During and After Extreme Weather Events

Despite the frequently described harms to mental and physical health and disconnection from land and community, multiple studies emphasised community resilience in the context of extreme weather events. The belief that rural communities are resilient and able to cope well with adversity was emphasised across drought [56,81], extreme heat [68,69,78,83] and bushfire [30,58]. This belief was associated with a lack of concern about increasing extreme heat for some people [68], including farmers [69] and older Australians [83].

Young people described optimism and community resilience in the face of drought [81,82] or flood [76] due to increased community participation and volunteering. For example, adolescent school students perceived drought as a factor that brings people together and provides more resources for drought-affected schools [81,82]. Many students believed they were learning to cope with drought, with personal resilience enhanced by humour in the face of adversity and strong connections to community and school [81]. In another study of adolescents and young adults post-flood, most participants perceived strong community cohesion and connectedness due to involvement in the community-led clean-up and a new volunteer organisation established due to the flood [76].

Bushfires were reported to catalyse positive community change, promoting new community involvement for young adults such as bush regeneration [30]. These young adults perceived informal networks of friends and family as the most meaningful social supports post-bushfire and emphasised the importance of acknowledgement of their unique experience [30]. Parents noted positive relational changes in their children post-bushfire as they engaged in more prosocial behaviours [64]. Females in three studies particularly valued social connectedness to strengthen community resilience and helped promote it during and after disasters, such as organising social events to bring people together and temporarily alleviate stress [36,56,64]. Connecting with individuals with similar disaster experiences was another common facilitator of community optimism, allowing time to debrief about stressors and laugh together [30,58,76].

For healthcare workers, a positive effect of bushfires was the strengthening of supportive workplace relationships in hospitals supported by a formal 'Wellness Warriors' recovery programme in which 'pausing to listen' promoted meaningful interactions and a more positive atmosphere at work [58]. For these hospital staff [58] and other mental health staff [65], bushfires emphasised the importance of connecting with colleagues and renewed their sense of purpose and commitment to promoting health within their community.

Rural mental health staff noted more broadly that the bushfire disaster response amplified positive human traits, such as camaraderie and mutual support, whilst bringing out tensions and vulnerabilities as discussed in the previous section [65]. Some adults in rural and remote NSW mentioned increased community connectedness during drought, such as through church or farm family gatherings, but majority perceived that negative community impacts were more prevalent [54]. Other rural Victorians observed an increase in social cohesion post-bushfire with a sense of shared experience and co-operative clean-up, but pre-existing social divisions returned as people recovered [67].

Evidently, rural community resilience and positivity before, during and after extreme weather events is a nuanced topic which varies based on age, type of extreme event and an individual's perceived community connections. It is especially important given that community social connection was the strongest factor enabling coping in almost half of the included studies [30,36,55,56,62–64,66–69,71,76,79,81,83].

3.3.6. Disproportionate Health Impacts Affecting Vulnerable Groups

In 15 studies, participants expressed concerns about individuals or groups who were more vulnerable to increased hazard exposure or health impacts during extreme weather events [27,53,56,57,61,62,67–71,74,78,79,83]. People were concerned about the elderly, those with a chronic medical condition or an outdoor occupation being more susceptible to the health risks of extreme heat [68,69,78,83]. Female healthcare workers noted that rates of domestic violence increased during disasters which largely increased the harm to women's health [53,57]. Other groups of mothers identified as needing additional support during disasters included women with disabilities, pre-existing mental health challenges, First Nations families and young mothers [57]. During flooding, there were concerns for the safety of those who drive through floodwater [74] and volunteer for clean-up efforts [79].

Rural residents expressed concerns about the exacerbation of pre-existing socioeconomic disadvantage by extreme weather events, particularly drought and flood [61,70,71]. For rural Aboriginal communities this included reduced culturally appropriate housing, worsened overcrowding, and fewer local Aboriginal health services [61]. Loss of access to culturally significant lands was a major impact of drought [61] and bushfire [65] for Aboriginal communities, but local mental health workers perceived that its significance was not recognised in evacuation plans and service responses [65]. Six months after flooding in the NSW Northern Rivers, Aboriginal and financial hardship respondents experienced lower social capital and higher rates of distress and probable PTSD compared to other survey respondents [27].

People worried that those with the least resources, including people with a disability and carers, were more likely to live in flood-prone areas because they are generally more affordable [71]. People with a disability described how floods compounded physical and mental health symptoms, which further impeded access to essential services and the ability to navigate recovery systems [71]. People with a disability and carers experienced high levels of financial stress arising from lack of affordable accommodation post-flood, which caused some to live in unsafe housing, relocate and lose support networks, or become homeless [71]. Despite widespread concern about rural populations who are more vulnerable to the health impacts of extreme weather events, only a small number of included studies directly involved these groups.

4. Discussion

4.1. Overview of Findings

This is the first scoping review to summarise the perceptions of rural Australians towards the health impacts of extreme weather events. Included articles captured perceptions associated with drought, bushfire, flood, extreme heat and mixed events. Extreme weather events impact health directly and indirectly via socioeconomic factors and have been reviewed by others [1,11,23,84–86], with many focusing on mental health [3,18,21,26,31,87].

This review identified two central themes: first, perceptions on extreme weather in terms of severity, frequency, and duration; and second, perceptions on health impacts. The latter theme revealed six interrelated subthemes, drawing attention to the psychological toll of extreme events, including mental health risks and uncertainty-related distress; social disconnectedness and isolation; rural community resilience; disruptions to cultural and emotional connections to land; and the disproportionate vulnerability of specific groups, underscoring existing social and health inequities.

Rural Australians' beliefs about climate change are likely to influence their perceptions of severity and susceptibility to the impacts, as suggested in models of health behaviour such as the health belief model [32]. The nuance of peoples' beliefs should be considered in planning health-promotion programmes for adaptation and mitigation because beliefs about the human contribution to climate change influence pro-environmental behaviour relevant for adapting to increasing extreme weather events [88]. Our findings show that many rural Australians described the drought, flood and bushfire events they experienced as severe, with concern for worsening future events influenced by climate change [55,70,76,78,79]. However, in two included studies the majority of participants were sceptical about climate change, associated with lack of trust in government [70] and reliance of the elderly on personal experience to cope with heatwaves [83].

Similarly to this finding, a study on perceptions in rural United Kingdom shows that lower threats of climate change are perceived in the most rural locations [34]. Lower awareness of the impact of climate change and higher resilience within rural population is believed to explain this phenomenon. Grounded in social learning theory, adaptation and mitigation of extreme weather impacts can be strengthened through collaboration, knowledge sharing, and co-creation of practices among stakeholders [89]. By exchanging experiences and feedback, communities and systems can enhance collective adaptive capacity. In rural settings, for instance, integrating local strategies for managing drought or bushfire risks with scientific guidance can foster systemic resilience that extends beyond individual actions.

4.2. Wide-Ranging Mental Health Concerns

Notable across many subthemes was the perceived saliency of mental health impacts across varied rural Australian population groups and in response to different extreme weather events [27,53–57,60–67,70–73,75,76,78–82]. For example, drought was viewed as a stressor associated with uncertainty and as the origin of other stressors such as financial strain, occupational issues for farmers and social stress [54–56,61–63,70,72,73,75,81,82]. Researchers have noted the need to recognise the complexity of the socioeconomic impacts of drought when designing adaptive strategies, including economic impacts, social stress, water markets, a changing farming industry and rural demographic shifts [90].

Another subtheme highlighted the capacity for environmental destruction to negatively impact on connection to land, sense of identity, and mental wellbeing, an increasingly relevant harm of climate change termed solastalgia [91]. Other recent research has recognised the significance of the mental health impacts of drought on rural communities globally [92] and in Australia [20,21,90], particularly farmers [18,22], yet there are limited reviews on this topic as it can be challenging to study such a long-term exposure. Indigenous communities proposed that capturing the spirit of Aboriginal knowledge, traditions and arts to renew respect for the land during drought could be a valuable consideration for future policy [61].

For the more acute events of flood and bushfire, mental health impacts were seen to be divided into immediate and long-term or delayed effects. The exacerbation of pre-existing mental health conditions was common. Mental health symptoms could impair the ability to access support services and reciprocally, disrupted access to essential services and financial strain worsened mental wellbeing, especially for medical students [79] and people with disability [71]. A recent Australian review found that long-term psychological distress post-bushfire was experienced by 14% of the general population but was at least twice the rate in firefighters [31]. Similarly, a scoping review [85] and recent primary survey [5,26] have investigated factors associated with short- and long-term mental health impacts of flooding, such as rurality and socioeconomic disadvantage.

The emphasis on mental health highlights the need for more targeted, local supports for rural Australians before, during and after extreme weather events [93]. It is well known that rural communities are particularly vulnerable in terms of adverse mental health outcomes including suicide [94]. Participatory action research which involves participants with researchers throughout design, conduct and evaluation of research can aid the implementation of effective community-based prevention strategies, for example, suicide prevention, by empowering communities to promote action [95]. It is important that mental health support options offered are appropriate to the needs of the population groups within the specific community and relevant to the extreme weather event(s) being experienced [93,96–98]. For example, farming men's mental health was best supported through informal social settings in a rural Victorian town post-bushfire [67].

In the NSW Snowy Valleys bushfire response, mental health clinicians noted that mental health support for Aboriginal communities occurred predominantly outside of mainstream services [65]. Other rural Aboriginal communities in NSW valued community-based Aboriginal health promotion programmes to support general social and emotional health during drought, such as annual 'footie knockouts' aimed at youth [61]. Education was suggested by adolescents and farming communities in relation to mental illness during drought [54,55,62]. A recent qualitative study found that farmers prefer online mental health interventions that seem relevant and authentic with a casual tone whilst reflecting the diversity of the farming population [96].

4.3. Social Disconnection and Resilience

Social disconnectedness as a subtheme encapsulates the perception that extreme weather events have negative effects on community social health, reducing participation, increasing social isolation and harming mental health. Yet despite this, it was common for rural Australians to view their community as resilient before, during and after extreme weather events. Other research has found that social networks influenced mental health outcomes after a bushfire [28,29] and that social participation and networks are central to effective bushfire preparedness [99]. Therefore, adaptation strategies for extreme weather events should prioritise community-led planning to enable rural communities to strengthen their connectedness as this may also benefit mental health [35,36,93,95].

Harnessing the notion of resilience and optimism may more effectively engage communities for future adaptation and mitigation in response to increasing extreme weather events in rural areas [37]. Factors that promoted resilience in one rural Australian community included social networks and support, positive outlook, learning, infrastructure and support services, innovative economy and leadership [35]. More recent Australian research shows that rural communities value community-led, collective and democratic action in approaches to mitigate the mental health effects of extreme weather events [93]. These approaches promoted belonging, informal social connectedness, hope and optimism. Examples included local climate change mitigation groups with a focus on caring for the land, community water usage monitoring during drought, and getting together to support firefighters in bushfire season [93]. Others have noted that more research is needed on existing adaptive capacity and knowledge of rural communities [7,90] as well as understanding how rural people define their 'sense of community' [99].

4.4. Evidence Gaps

This review comprised perceptions of rural populations from only four states of Australia (NSW, Vic, SA and Qld), and did not capture perceptions of people from Tasmania, Western Australia, the Northern Territory or the Australian Capital Territory. Valuable insights relating to specific population groups in these states and territories have therefore been missed. These may include how hazard risk perceptions vary with socioeconomic status in these areas of Australia, and issues unique to the numerous remote farming and remote Indigenous communities across Western Australia and the Northern Territory. Most of the included studies also did not differentiate their results based on rurality of their participants. Therefore, we could not map and analyse whether perceptions vary between outer regional, remote, and very remote areas. This information would support targeted specific policy recommendations for each setting. Another possible factor that might have skewed the findings of this review was the overrepresentation of drought compared to other extreme weather events. This likely reflects a bias in the literature due to the chronicity of drought in Australia, including the decade-long drought experienced from 2000 to 2010 [18], allowing for more time to produce studies on drought.

Although a small number of studies looked at perceptions of children, adolescents, older Australians and females, most did not separate the perceptions between gender and age, limiting this cross-analysis of health impacts of extreme weather events. This is an important gap to address because women, young children and the elderly are known to be more vulnerable to the impacts of climate change [1,20,23]. There is, however, an Australian review being conducted on climate change adaptations to protect maternal and child health [100]. Additionally, males can have different preferences for mental health support or be difficult to reach [56,67]. Therefore, separating study findings into gendered perspectives has implications for designing more targeted strategies to increase engagement and mitigate health impacts of future extreme weather events. This gap on age- and gender-specific perspectives may persist in the literature if studies do not collect or present individual data specifically linked to gender and age as this may not be feasible in large-scale quantitative or qualitative data analysis.

The review found limited evidence exploring perceptions of vulnerable rural groups such as the elderly [83], people with a disability [71] or chronic health condition [68,69,78,83], females at risk of domestic violence or young mothers [53,57], those in financial hardship [27], and Aboriginal people [27,61], highlighting under-researched areas. Participants in 15 included studies [27,53,56,57,61,62,67–71,74,78,79,83] expressed concerns that vulnerable groups are disproportionately affected by the health impacts of extreme weather events. This reveals a research gap and a need for more targeted local financial and social supports that are accessible to vulnerable populations. The limited representation of socioeconomically disadvantaged groups possibly stems from factors relating to study design, participant recruitment and lack of time for people to participate due to socioeconomic stresses. Future research should aim to involve rural socioeconomically disadvantaged populations, for example, through community-academic partnerships with researchers with lived experience, community involvement in study design and purposive participant recruitment [7,26,95].

4.5. Strengths and Limitations

This scoping review was the first to assess the wide-ranging, yet often similar, perceptions of rural Australians concerning the health impacts of extreme weather events, filling a research gap. A key strength was the use of a comprehensive strategy including different study designs, allowing multiple types of data to be evaluated. Instead of characterising health impacts, we focused on perceptions to understand attitudes and beliefs of rural populations which can warrant adaptation and mitigation strategies suitable for rural populations.

Despite this review being novel and comprehensive, there are a number of limitations. In this review, only one reviewer performed screening and data extraction; however, any queries were checked by a second reviewer and resolved by discussion. It is possible that more bias was introduced during the screening and data extraction process than if two authors had performed all steps. The heterogeneous data from included studies made it difficult to combine; however, synthesising the data using thematic analysis provided robust qualitative findings. The broad scope of the review may have sacrificed depth of cross-analysis of extracted data and missed identifying, comparing and contrasting more specific trends in the findings.

5. Conclusions

This scoping review summarises rural Australians' perceptions of health impacts from extreme weather events. It highlights many concerns about the diverse impacts of drought, bushfire, flood, and extreme heat on mental health, connection to land and community, and on vulnerable populations. The review underscores the importance of tailoring future adaptation and mitigation strategies to the unique needs of each rural community including their knowledge and beliefs, considering the substantial diversity in demographics and occupations. Mental health impacts, particularly from prolonged drought, are a significant concern, with many studies noting stressors associated with social impacts and financial strain. The review emphasises the need for effective mental health support and community social connectedness to enhance coping.

Future research should aim to understand rural communities' existing adaptive capacity, definitions of 'sense of community' and preferred adaptive strategies to guide community-led adaptation. Longitudinal studies of perception changes over time and how they influence health outcomes will help to support longer-term behavioural changes necessary for adaptation to more frequent extreme weather events. Participatory research methods should be utilised to engage rural socioeconomically disadvantaged populations, such as community-academic partnerships [7,26,95]. Future research should also aim to stratify findings by age and gender and consider rural populations across all of Australia to enable targeted adaptive and mitigation strategies that consider the diversity between population groups.

Supplementary Materials: The following supporting information can be downloaded at: https://www.mdpi.com/article/10.3390/cli13090180/s1, Supplementary Material S1: Final database searches performed; Supplementary Material S2: Data extraction form.

Author Contributions: Conceptualization, E.V., J.M. and S.D.; methodology, E.V., J.M., I.N.S. and S.D.; formal analysis, E.V., J.M., B.H., I.N.S. and S.D.; investigation, E.V. and S.D.; writing—original draft preparation, E.V.; writing—review and editing, E.V., J.M., B.H., I.N.S., S.H.D. and S.D.; supervision, J.M., S.H.D. and S.D. All authors have read and agreed to the published version of the manuscript.

Funding: This research received no external funding.

Conflicts of Interest: The authors declare no conflicts of interest.

Abbreviations

The following abbreviations are used in this manuscript:

PRISMA-ScR	Preferred Reporting Items for Systematic Reviews and Meta-Analyses Extension for Scoping Reviews
PICo	Population, phenomena of interest, context
MeSH	Medical Subject Headings
NSW	New South Wales
Vic	Victoria
SA	South Australia
Qld	Queensland
PTSD	Post-traumatic stress disorder

References

1. Ebi, K.L.; Vanos, J.; Baldwin, J.W.; Bell, J.E.; Hondula, D.M.; Errett, N.A.; Hayes, K.; Reid, C.E.; Saha, S.; Spector, J.; et al. Extreme Weather and Climate Change: Population Health and Health System Implications. *Annu. Rev. Public Health* **2021**, *42*, 293–315. [CrossRef]
2. van Oldenborgh, G.J.; Krikken, F.; Lewis, S.; Leach, N.J.; Lehner, F.; Saunders, K.R.; Van Weele, M.; Haustein, K.; Li, S.; Wallom, D.; et al. Attribution of the Australian bushfire risk to anthropogenic climate change. *Nat. Hazards Earth Syst. Sci.* **2021**, *21*, 941–960. [CrossRef]
3. Hayes, K.; Blashki, G.; Wiseman, J.; Burke, S.; Reifels, L. Climate change and mental health: Risks, impacts and priority actions. *Int. J. Ment. Health Syst.* **2018**, *12*, 28. [CrossRef] [PubMed]
4. Nelson, R.; Kokic, P.; Crimp, S.; Martin, P.; Meinke, H.; Howden, S.M.; de Voil, P.; Nidumolu, U. The vulnerability of Australian rural communities to climate variability and change: Part II—Integrating impacts with adaptive capacity. *Environ. Sci. Policy* **2010**, *13*, 18–27. [CrossRef]
5. Matthews, V.; Longman, J.; Berry, H.L.; Passey, M.; Bennett-Levy, J.; Morgan, G.G.; Pit, S.; Rolfe, M.; Bailie, R.S. Differential Mental Health Impact Six Months After Extensive River Flooding in Rural Australia: A Cross-Sectional Analysis Through an Equity Lens. *Front. Public Health* **2019**, *7*, 367. [CrossRef] [PubMed]
6. World Bank Group. Rural Population (% of Total Population). 2025. Available online: https://data.worldbank.org/indicator/SP.RUR.TOTL.ZS?end=2023&start=1960&view=chart (accessed on 20 June 2025).
7. Bailie, R. Climate-Related Natural Disasters: Reflections on an Agenda for Rural Health Research. *Int. J. Environ. Res. Public Health* **2023**, *20*, 5553. [CrossRef]
8. Flavel, J.; Kedzior, S.G.; Isaac, V.; Cameron, D.; Baum, F. Regional health inequalities in Australia and social determinants of health: Analysis of trends and distribution by remoteness. *Rural Remote Health* **2024**, *24*, 1–11. [CrossRef]
9. Australian Institute of Health and Welfare. Rural and Remote Health. 2024. Available online: https://www.aihw.gov.au/reports/rural-remote-australians/rural-and-remote-health (accessed on 26 May 2025).
10. Beard, J.R.; Tomaska, N.; Earnest, A.; Summerhayes, R.; Morgan, G. Influence of socioeconomic and cultural factors on rural health. *Aust. J. Rural Health* **2009**, *17*, 10–15. [CrossRef]
11. Dewi, S.P.; Kasim, R.; Sutarsa, I.N.; Hunter, A.; Dykgraaf, S.H. Effects of climate-related risks and extreme events on health outcomes and health utilization of primary care in rural and remote areas: A scoping review. *Fam. Pract.* **2023**, *40*, 486–497. [CrossRef]
12. Bi, P.; Parton, K.A. Effect of climate change on Australian rural and remote regions: What do we know and what do we need to know? *Aust. J. Rural Health* **2008**, *16*, 2–4. [CrossRef]
13. Liu, J.C.; Pereira, G.; Uhl, S.A.; Bravo, M.A.; Bell, M.L. A systematic review of the physical health impacts from non-occupational exposure to wildfire smoke. *Environ. Res.* **2015**, *136*, 120–132. [CrossRef]
14. Romanello, M.; Di Napoli, C.; Drummond, P.; Green, C.; Kennard, H.; Lampard, P.; Scamman, D.; Arnell, N.; Ayeb-Karlsson, S.; Ford, L.B.; et al. The 2022 report of the Lancet Countdown on health and climate change: Health at the mercy of fossil fuels. *Lancet* **2022**, *400*, 1619–1654. [CrossRef]
15. Burns, P.L.; Douglas, K.A.; Hu, W. Primary care in disasters: Opportunity to address a hidden burden of health care. *Med. J. Aust.* **2019**, *210*, 297–299.e1. [CrossRef]
16. Purcell, R.; McGirr, J. Preparing rural general practitioners and health services for climate change and extreme weather. *Aust. J. Rural Health* **2014**, *22*, 8–14. [CrossRef]
17. Xu, R.; Yu, P.; Liu, Y.; Chen, G.; Yang, Z.; Zhang, Y.; Wu, Y.; Beggs, P.J.; Zhang, Y.; Boocock, J.; et al. Climate change, environmental extremes, and human health in Australia: Challenges, adaptation strategies, and policy gaps. *Lancet Reg. Health West. Pac.* **2023**, *40*, 100936. [CrossRef]
18. Edwards, B.; Gray, M.; Hunter, B. The Impact of Drought on Mental Health in Rural and Regional Australia. *Soc. Indic. Res.* **2015**, *121*, 177–194. [CrossRef]

19. Horton, G.; Hanna, L.; Kelly, B. Drought, drying and climate change: Emerging health issues for ageing Australians in rural areas. *Australas. J. Ageing* **2010**, *29*, 2–7. [CrossRef] [PubMed]
20. Hanigan, I.C.; Schirmer, J.; Niyonsenga, T. Drought and Distress in Southeastern Australia. *EcoHealth* **2018**, *15*, 642–655. [CrossRef]
21. Obrien, L.V.; Berry, H.L.; Coleman, C.; Hanigan, I.C. Drought as a mental health exposure. *Environ. Res.* **2014**, *131*, 181–187. [CrossRef]
22. Austin, E.K.; Handley, T.; Kiem, A.S.; Rich, J.L.; Lewin, T.J.; Askland, H.H.; Askarimarnani, S.S.; Perkins, D.A.; Kelly, B.J. Drought-related stress among farmers: Findings from the Australian Rural Mental Health Study. *Med. J. Aust.* **2018**, *209*, 159–165. [CrossRef]
23. Harvey, G.; Bain-Donohue, S.; Dewi, S.P. The impact of extreme heat on older regional and rural Australians: A systematic review. *Aust. J. Rural Health* **2024**, *32*, 216–226. [CrossRef]
24. O'Donnell, E.; Honan, B.; Quilty, S.; Schultz, R. The Effect of Heat Events on Prehospital and Retrieval Service Utilization in Rural and Remote Areas: A Scoping Review. *Prehosp. Disaster Med.* **2021**, *36*, 782–787. [CrossRef]
25. Jegasothy, E.; McGuire, R.; Nairn, J.; Fawcett, R.; Scalley, B. Extreme climatic conditions and health service utilisation across rural and metropolitan New South Wales. *Int. J. Biometeorol.* **2017**, *61*, 1359–1370. [CrossRef]
26. Longman, J.M.; Bennett-Levy, J.; Matthews, V.; Berry, H.L.; Passey, M.E.; Rolfe, M.; Morgan, G.G.; Braddon, M.; Bailie, R. Rationale and methods for a cross-sectional study of mental health and wellbeing following river flooding in rural Australia, using a community-academic partnership approach. *BMC Public Health* **2019**, *19*, 1255. [CrossRef]
27. Matthews, V.; Longman, J.; Bennett-Levy, J.; Braddon, M.; Passey, M.; Bailie, R.S.; Berry, H.L. Belonging and Inclusivity Make a Resilient Future for All: A Cross-Sectional Analysis of Post-Flood Social Capital in a Diverse Australian Rural Community. *Int. J. Environ. Res. Public Health* **2020**, *17*, 7676. [CrossRef]
28. Gibbs, L.; Waters, E.; Bryant, R.A.; Pattison, P.; Lusher, D.; Harms, L.; Richardson, J.; MacDougall, C.; Block, K.; Snowdon, E.; et al. Beyond Bushfires: Community, Resilience and Recovery—A longitudinal mixed method study of the medium to long term impacts of bushfires on mental health and social connectedness. *BMC Public Health* **2013**, *13*, 1036. [CrossRef]
29. Bryant, R.A.; Gallagher, H.C.; Gibbs, L.; Pattison, P.; MacDougall, C.; Harms, L.; Block, K.; Baker, E.; Sinnott, V.; Ireton, G.; et al. Mental Health and Social Networks After Disaster. *Am. J. Psychiat.* **2016**, *174*, 277–285. [CrossRef] [PubMed]
30. Forbes, R.J.; Jones, R.; Reupert, A. In the wake of the 2009 Gippsland fires: Young adults' perceptions of post-disaster social supports. *Aust. J. Rural Health* **2012**, *20*, 119–125. [CrossRef] [PubMed]
31. Zhang, Y.; Workman, A.; Russell, M.A.; Williamson, M.; Pan, H.; Reifels, L. The long-term impact of bushfires on the mental health of Australians: A systematic review and meta-analysis. *Eur. J. Psychotraumatol.* **2022**, *13*, 2087980. [CrossRef] [PubMed]
32. Alyafei, A.; Easton-Carr, R. The Health Belief Model of Behavior Change. In *StatPearls [Internet]*; StatPearls Publishing: Treasure Island, FL, USA, 2024. Available online: https://www.ncbi.nlm.nih.gov/books/NBK606120/ (accessed on 22 August 2025).
33. Berkman, N.D.; Sheridan, S.L.; Donahue, K.E.; Halpern, D.J.; Crotty, K. Low health literacy and health outcomes: An updated systematic review. *Ann. Intern. Med.* **2011**, *155*, 97–107. [CrossRef]
34. Tenbrink, T.; Willcock, S. Place attachment and perception of climate change as a threat in rural and urban areas. *PLoS ONE* **2023**, *18*, e0290354. [CrossRef]
35. Buikstra, E.; Ross, H.; King, C.A.; Baker, P.G.; Hegney, D.; McLachlan, K.; Rogers-Clark, C. The components of resilience—Perceptions of an Australian rural community. *J. Community Psychol.* **2010**, *38*, 975–991. [CrossRef]
36. Madsen, W.; O'Mullan, C. Perceptions of community resilience after natural disaster in a rural Australian town. *J. Community Psychol.* **2016**, *44*, 277–292. [CrossRef]
37. Ebi, K.L.; Semenza, J.C. Community-Based Adaptation to the Health Impacts of Climate Change. *Am. J. Prev. Med.* **2008**, *35*, 501–507. [CrossRef] [PubMed]
38. Ferrer, R.; Klein, W.M. Risk perceptions and health behavior. *Curr. Opin. Psychol.* **2015**, *5*, 85–89. [CrossRef]
39. Parsons, M.; Lykins, A.D. Cultural worldviews and the perception of natural hazard risk in Australia. *Environ. Hazards* **2023**, *22*, 29–50. [CrossRef]
40. Thaker, J.; Richardson, L.M.; Holmes, D.C. Australians' perceptions about health risks associated with climate change: Exploring the role of media in a comprehensive climate change risk perception model. *J. Environ. Psychol.* **2023**, *89*, 102064. [CrossRef]
41. Purcell, R.; McGirr, J. Rural health service managers' perspectives on preparing rural health services for climate change. *Aust. J. Rural Health* **2018**, *26*, 20–25. [CrossRef]
42. Hathaway, J.; Maibach, E.W. Health Implications of Climate Change: A Review of the Literature About the Perception of the Public and Health Professionals. *Curr. Environ. Health Rep.* **2018**, *5*, 197–204. [CrossRef]
43. Lykins, A.D.; Parsons, M.; Craig, B.M.; Cosh, S.M.; Hine, D.W.; Murray, C. Australian Youth Mental Health and Climate Change Concern After the Black Summer Bushfires. *EcoHealth* **2023**, *20*, 3–8. [CrossRef]
44. Boyd, C.P.; Jamieson, J.; Gibson, K.; Duffy, M.; Williamson, M.; Parr, H. Eco-anxiety among regional Australian youth with mental health problems: A qualitative study. *Early Interv. Psychiatry* **2024**, *18*, 1022–1031. [CrossRef] [PubMed]

45. Page, M.J.; McKenzie, J.E.; Bossuyt, P.M.; Boutron, I.; Hoffmann, T.C.; Mulrow, C.D.; Shamseer, L.; Tetzlaff, J.M.; Akl, E.A.; Brennan, S.E.; et al. The PRISMA 2020 statement: An updated guideline for reporting systematic reviews. *BMJ* **2021**, *372*, n71. [CrossRef] [PubMed]
46. Australian Bureau of Statistics. Remoteness Areas—Australian Statistical Geography Standard (ASGS) Edition 3. Canberra 2023. Available online: https://www.abs.gov.au/statistics/standards/australian-statistical-geography-standard-asgs-edition-3/jul2021-jun2026/remoteness-structure/remoteness-areas (accessed on 20 June 2025).
47. Nelson, K.S.; Nguyen, T.D.; Brownstein, N.A.; Garcia, D.; Walker, H.C.; Watson, J.T.; Xin, A. Definitions, measures, and uses of rurality: A systematic review of the empirical and quantitative literature. *J. Rural Stud.* **2021**, *82*, 351–365. [CrossRef]
48. National Institute of Mental Health. Perception and Understanding of Self. 2025. Available online: https://www.nimh.nih.gov/research/research-funded-by-nimh/rdoc/constructs/perception-and-understanding-of-self (accessed on 20 June 2025).
49. Munn, Z.; Stern, C.; Aromataris, E.; Lockwood, C.; Jordan, Z. What kind of systematic review should I conduct? A proposed typology and guidance for systematic reviewers in the medical and health sciences. *BMC Med. Res. Methodol.* **2018**, *18*, 5. [CrossRef]
50. The EndNote Team. EndNote. In *EndNote*, 20th ed.; Clarivate: Philadelphia, PA, USA, 2013.
51. *Covidence Systematic Review Software*; Veritas Health Innovation: Melbourne, Australia, 2024; Available online: www.covidence.org (accessed on 25 March 2024).
52. Thomas, J.; Harden, A. Methods for the thematic synthesis of qualitative research in systematic reviews. *BMC Med. Res. Methodol.* **2008**, *8*, 45. [CrossRef]
53. Anderson, D. Enduring drought then coping with climate change: Lived experience and local resolve in rural mental health. *Rural Soc.* **2009**, *19*, 340–352. [CrossRef]
54. Carnie, T.L.; Berry, H.L.; Blinkhorn, S.A.; Hart, C.R. In their own words: Young people's mental health in drought-affected rural and remote NSW. *Aust. J. Rural Health* **2011**, *19*, 244–248. [CrossRef]
55. Greenhill, J.; King, D.; Lane, A.; MacDougall, C. Understanding resilience in South Australian farm families. *Rural Soc.* **2009**, *19*, 318–325. [CrossRef]
56. Healing, K.; Lowrie, D. Exploring the occupational experiences of livestock farmers during drought: A narrative inquiry. *Aust. J. Rural Health* **2023**, *31*, 855–865. [CrossRef]
57. Hine, R.H.; Mitchell, E.; Whitehead-Annett, L.; Duncan, Z.; McArdle, A. Natural disasters and perinatal mental health: What are the impacts on perinatal women and the service system? *J. Public Health* **2024**, *32*, 805–812. [CrossRef]
58. Knezevic, A.; Olcoń, K.; Smith, L.; Allan, J.; Pai, P. Wellness Warriors: A qualitative exploration of healthcare staff learning to support their colleagues in the aftermath of the Australian bushfires. *Int. J. Qual. Stud. Health Well-Being* **2023**, *18*, 2167298. [CrossRef]
59. Madsen, W.; Judd, J.; Williams, S.L.; McKenzie, F.; Deagon, J.; Ames, K. Time as a Social and Environmental Determinant of Health for Rural Women. *Health Educ. Behav.* **2020**, *47*, 468–473. [CrossRef] [PubMed]
60. Pike, C.E.; Lykins, A.D.; Bartik, W.; Tully, P.J.; Cosh, S.M. Climate Change in Rural Australia: Natural Hazard Preparedness and Recovery Needs of a Rural Community. *Climate* **2024**, *12*, 57. [CrossRef]
61. Rigby, C.W.; Rosen, A.; Berry, H.L.; Hart, C.R. If the land's sick, we're sick: The impact of prolonged drought on the social and emotional well-being of Aboriginal communities in rural New South Wales. *Aust. J. Rural Health* **2011**, *19*, 249–254. [CrossRef] [PubMed]
62. Sartore, G.M.; Kelly, B.; Stain, H.; Albrecht, G.; Higginbotham, N. Control, uncertainty, and expectations for the future: A qualitative study of the impact of drought on a rural Australian community. *Rural Remote Health* **2008**, *8*, 950. [CrossRef]
63. Staniford, A.K.; Dollard, M.F.; Guerin, B. Stress and help-seeking for drought-stricken citrus growers in the Riverland of South Australia. *Aust. J. Rural Health* **2009**, *17*, 147–154. [CrossRef]
64. Stribley, L.; Krishnamoorthy, G.; Dallinger, V.; Ma, J.; Nielsen, T.; Bryce, I.; Rees, B.; Morse, A.; Rogers, M.; Burton, L. Post-Disaster Social Connectedness in Parent–Child Dyads: A Qualitative Investigation of Changes in Coping and Social Capital of Rural Australian Families Following Bushfires. *Br. J. Soc. Work* **2024**, *55*, 5–24. [CrossRef]
65. Thomas, M.; Martin, E.; Isobel, S. 'Profound personal and professional impacts': A qualitative study of clinician experiences of a mental health disaster response to Australia's black summer bushfires. *Aust. J. Rural Health* **2024**, *32*, 959–968. [CrossRef]
66. van der Kruk, S.R.; Gunn, K.M. 'It sort of broke me': A thematic analysis of the psychological experiences and coping strategies employed by Australian fire-affected farmers. *Aust. J. Rural Health* **2024**, *32*, 299–310. [CrossRef]
67. Whittaker, J.; Handmer, J.; Mercer, D. Vulnerability to bushfires in rural Australia: A case study from East Gippsland, Victoria. *J. Rural Stud.* **2012**, *28*, 161–173. [CrossRef]
68. Williams, S.; Bi, P.; Newbury, J.; Robinson, G.; Pisaniello, D.; Saniotis, A.; Hansen, A. Extreme heat and health: Perspectives from health service providers in rural and remote communities in South Australia. *Int. J. Environ. Res. Public Health* **2013**, *10*, 5565–5583. [CrossRef]

69. Williams, S.; Hanson-Easey, S.; Robinson, G.; Pisaniello, D.; Newbury, J.; Saniotis, A.; Bi, P. Heat adaptation and place: Experiences in South Australian rural communities. *Reg. Environ. Change* **2017**, *17*, 273–283. [CrossRef]
70. Austin, E.K.; Rich, J.L.; Kiem, A.S.; Handley, T.; Perkins, D.; Kelly, B.J. Concerns about climate change among rural residents in Australia. *J. Rural Stud.* **2020**, *75*, 98–109. [CrossRef]
71. Bailie, J.; Matthews, V.; Bailie, R.; Villeneuve, M.; Longman, J. Exposure to risk and experiences of river flooding for people with disability and carers in rural Australia: A cross-sectional survey. *BMJ Open* **2022**, *12*, e056210. [CrossRef] [PubMed]
72. Chan, K.; Xue, H.; Carlson, J.; Gray, J.M.; Bailey, J.; Vines, R. Impact of COVID-19 on lifestyle and mental wellbeing in a drought-affected rural Australian population. *Rural Remote Health* **2022**, *22*, 18. [CrossRef] [PubMed]
73. Fennell, K.M.; Jarrett, C.E.; Kettler, L.J.; Dollman, J.; Turnbull, D.A. "Watching the bank balance build up then blow away and the rain clouds do the same": A thematic analysis of South Australian farmers' sources of stress during drought. *J. Rural Stud.* **2016**, *46*, 102–110. [CrossRef]
74. Franklin, R.C.; King, J.C.; Aitken, P.J.; Leggat, P.A. "Washed away"-assessing community perceptions of flooding and prevention strategies: A North Queensland example. *Nat. Hazards* **2014**, *73*, 1977–1998. [CrossRef]
75. Gunn, K.M.; Kettler, L.J.; Skaczkowski, G.L.A.; Turnbull, D.A. Farmers' stress and coping in a time of drought. *Rural Remote Health* **2012**, *12*, 16.
76. Luk, M.; Longman, J. Young people's experiences of the Northern Rivers 2017 flood and its effects on their mental health. *Aust. J. Rural Health* **2024**, *32*, 343–353. [CrossRef]
77. Raphael, B.; Taylor, M.; Stevens, G.; Barr, M.; Gorringe, M.; Agho, K. Factors associated with population risk perceptions of continuing drought in Australia. *Aust. J. Rural Health* **2009**, *17*, 330–337. [CrossRef]
78. Williams, S.; Hanson-Easey, S.; Nitschke, M.; Howell, S.; Nairn, J.; Beattie, C.; Wynwood, G.; Bi, P. Heat-health warnings in regional Australia: Examining public perceptions and responses. *Environ. Hazards* **2019**, *18*, 287–310. [CrossRef]
79. Bailie, J.; Reed, K.; Matthews, V.; Scott, K.M.; Ahern, C.; Bailie, R. Volunteering as prosocial behaviour by medical students following a flooding disaster and impacts on their mental health: A mixed-methods study. *Med. Educ.* **2023**, *58*, 430–442. [CrossRef] [PubMed]
80. Burns, R.; Robinson, P.; Smith, P. From hypothetical scenario to tragic reality: A salutary lesson in risk communication and the Victorian 2009 bushfires. *Aust. N. Z. J. Public Health* **2010**, *34*, 24–31. [CrossRef] [PubMed]
81. Dean, J.; Stain, H.J. The impact of drought on the emotional well-being of children and adolescents in rural and remote New South Wales. *J. Rural Health* **2007**, *23*, 356–364. [CrossRef]
82. Dean, J.G.; Stain, H.J. Mental health impact for adolescents living with prolonged drought. *Aust. J. Rural Health* **2010**, *18*, 32–37. [CrossRef]
83. Loughnan, M.E.; Carroll, M.; Tapper, N. Learning from our older people: Pilot study findings on responding to heat. *Australas. J. Ageing* **2014**, *33*, 271–277. [CrossRef]
84. Alderman, K.; Turner, L.R.; Tong, S. Floods and human health: A systematic review. *Environ. Int.* **2012**, *47*, 37–47. [CrossRef]
85. Fernandez, A.; Black, J.; Jones, M.; Wilson, L.; Salvador-Carulla, L.; Astell-Burt, T.; Black, D.; Ebi, K.L. Flooding and Mental Health: A Systematic Mapping Review. *PLoS ONE* **2015**, *10*, e0119929. [CrossRef]
86. Stanke, C.; Kerac, M.; Prudhomme, C.; Medlock, J.; Murray, V. Health effects of drought: A systematic review of the evidence. *PLoS Curr.* **2013**, *5*, ecurrents.dis.7a2cee9e980f91ad7697b570bcc4b004. [CrossRef]
87. Cruz, J.; White, P.C.L.; Bell, A.; Coventry, P.A. Effect of Extreme Weather Events on Mental Health: A Narrative Synthesis and Meta-Analysis for the UK. *Int. J. Environ. Res. Public Health* **2020**, *17*, 8581. [CrossRef]
88. Leviston, Z.; Walker, I. Beliefs and Denials About Climate Change: An Australian Perspective. *Ecopsychology* **2012**, *4*, 277–285. [CrossRef]
89. Ensor, J.; Harvey, B. Social learning and climate change adaptation: Evidence for international development practice. *WIREs Clim. Change* **2015**, *6*, 509–522. [CrossRef]
90. Kiem, A.S.; Austin, E.K. Drought and the future of rural communities: Opportunities and challenges for climate change adaptation in regional Victoria, Australia. *Glob. Environ. Change* **2013**, *23*, 1307–1316. [CrossRef]
91. Galway, L.P.; Beery, T.; Jones-Casey, K.; Tasala, K. Mapping the solastalgia literature: A scoping review study. *Int. J. Environ. Res. Public Health* **2019**, *16*, 2662. [CrossRef] [PubMed]
92. Reyes García, V.; García-Del-Amo, D.; Porcuna-Ferrer, A.; Schlingmann, A.; Abazeri, M.; Attoh, E.M.N.A.N.; Ávila, J.V.d.C.; Ayanlade, A.; Babai, D.; Benyei, P.; et al. Local studies provide a global perspective of the impacts of climate change on Indigenous Peoples and local communities. *Sustain. Earth Rev.* **2024**, *7*, 1. [CrossRef]
93. Longman, J.; Braddon, M.; Verlie, B.; Schlosberg, D.; Hampshire, L.; Hawke, C.; Noonan, A.; Saurman, E. Building resilience to the mental health impacts of climate change in rural Australia. *J. Clim. Change Health* **2023**, *12*, 100240. [CrossRef]
94. Grattidge, L.; Hoang, H.; Mond, J.; Visentin, D.; Lees, D.; Auckland, S. The Community's Role in Rural Youth Suicide Prevention: Perspectives from the Field. *Aust. J. Rural Health* **2025**, *33*, e70024. [CrossRef]

95. Grattidge, L.; Purton, T.; Auckland, S.; Lees, D.; Mond, J. Participatory action research in suicide prevention program evaluation: Opportunities and challenges from the National Suicide Prevention Trial, Tasmania. *Aust. N. Z. J. Public Health* **2021**, *45*, 311–314. [CrossRef]
96. Gunn, K.M.; Barrett, A.; Hughes-Barton, D.; Turnbull, D.; Short, C.E.; Brumby, S.; Skaczkowski, G.; Dollman, J. What farmers want from mental health and wellbeing-focused websites and online interventions. *J. Rural Stud.* **2021**, *86*, 298–308. [CrossRef]
97. Handley, T.E.; Kay-Lambkin, F.J.; Inder, K.J.; Lewin, T.J.; Attia, J.R.; Fuller, J.; Perkins, D.; Coleman, C.; Weaver, N.; Kelly, B.J. Self-reported contacts for mental health problems by rural residents: Predicted service needs, facilitators and barriers. *BMC Psychiatry* **2014**, *14*, 249. [CrossRef]
98. Morrissey, S.A.; Reser, J.P. Natural disasters, climate change and mental health considerations for rural Australia. *Aust. J. Rural Health* **2007**, *15*, 120–125. [CrossRef]
99. Fairbrother, P.; Tyler, M.; Hart, A.; Mees, B.; Phillips, R.; Stratford, J.; Toh, K. Creating "Community"? Preparing for Bushfire in Rural Victoria. *Rural Sociol.* **2013**, *78*, 186–209. [CrossRef]
100. Lokmic-Tomkins, Z.; Bhandari, D.; Watterson, J.; Pollock, W.E.; Cochrane, L.; Robinson, E.; Su, T.T. Multilevel interventions as climate change adaptation response to protect maternal and child health: A scoping review protocol. *BMJ Open* **2023**, *13*, e073960. [CrossRef]

climate

MDPI

Article

The Interplay Between Climate Change Exposure, Awareness, Coping, and Anxiety Among Individuals with and Without a Chronic Illness

Shiri Shinan-Altman [1,*] and Yaira Hamama-Raz [2]

[1] Louis and Gabi Weisfeld School of Social Work, Bar Ilan University, Ramat-Gan 52900, Israel
[2] School of Social Work, Ariel University, Ariel 40700, Israel; yairahr@ariel.ac.il
* Correspondence: shiri.altman@biu.ac.il; Tel.: +972-37384546; Fax: +972-37384042

Abstract: Climate change poses a significant threat to mental health, including the emergence of climate change anxiety (CCA). In this study, we examined whether exposure to climate-related events was associated with higher CCA through the mediating roles of climate change awareness and ecological coping strategies and whether these pathways differed by chronic illness status. In February 2025, 600 Israeli adults (50% female; mean age $\approx$ 50) completed an online self-report questionnaire assessing climate change exposure, awareness, coping, and anxiety. Data were analyzed using moderated mediation models, controlling for gender, age, and education. Greater climate change exposure was associated with an increased awareness and higher use of problem-focused coping, which, in turn, predicted elevated CCA. Meaning-focused coping was not associated with anxiety overall; however, among the participants without a chronic illness, it was linked to higher CCA. Climate change awareness alone was not associated with anxiety. A significant serial mediation was found via awareness and problem-focused coping, and a moderated mediation was found via meaning-focused coping among those without a chronic illness. Coping strategies play a key role in climate change anxiety. Although health status may influence this process, tailored interventions should prioritize coping styles in climate adaptation efforts.

Keywords: climate change anxiety; climate change awareness; ecological coping; chronic illness; moderated mediation model

1. Introduction

Climate change is widely recognized as one of the most significant global challenges of the 21st century, with profound implications not only for ecosystems and economies but also for human mental health and psychological well-being [1,2]. As the frequency and intensity of climate-related phenomena—such as extreme heatwaves, droughts, floods, and wildfires—continue to rise, so too does public concern, contributing to the emergence of climate change anxiety (CCA) [3]. This form of anxiety is characterized by persistent worry, fear, and distress related to the anticipated and ongoing impacts of climate change and has been associated with functional impairments and diminished well-being [4].

Growing awareness of climate change and its consequences can heighten people's emotional and cognitive responses, particularly when individuals perceive themselves or their communities as being at risk [5]. Studies have shown that exposure to climate-related events is associated with greater climate change awareness [6] and anxiety [7]. At the same time, psychological adaptation and resilience depend heavily on how individuals cope with the stress posed by such global environmental threats [8].

The current study was grounded in the transactional model of stress and coping developed by Lazarus and Folkman [9], which conceptualizes stress as a dynamic process arising from the interaction between individuals and their environment. According to this model, when individuals perceive a threat (primary appraisal), they evaluate their coping resources (secondary appraisal) and adopt strategies accordingly. In the context of climate change, exposure to environmental stressors serves as the initiating condition (primary appraisal), which has been hypothesized to increase awareness of climate-related threats. This awareness can influence individuals' emotional responses (e.g., anxiety) both directly and indirectly through their coping strategies, such as meaning-focused or problem-focused coping. However, theories such as the Extended Parallel Process Model (EPPM) [10] and the Risk Perception Attitude (RPA) framework [11] emphasize that threat perception alone is insufficient to determine emotional outcomes—that is, self-efficacy and the type of coping strategy employed moderate this relationship. Building on these models, we hypothesized that individuals who were more exposed to climate-related events would report greater awareness and engagement in coping strategies, which would, in turn, influence their experience of CCA. We incorporated chronic illness as a moderating variable, given that individuals with chronic health conditions may appraise climate threats differently due to heightened vulnerability, potentially altering both coping styles and emotional outcomes.

Ecological coping strategies, including meaning-focused coping (e.g., deriving hope from collective environmental efforts), de-emphasizing the seriousness of climate change (e.g., minimizing the perceived severity or personal relevance of climate threats), and problem-focused coping (e.g., engaging in environmentally responsible behaviors), have emerged as crucial psychological mechanisms in response to climate-related stressors [12,13]. However, the impact of coping strategies on CCA is not uniform. Although some strategies may alleviate distress, others may inadvertently intensify it [13,14]. For example, engaging in problem-focused coping without sufficient environmental efficacy or societal support can lead to increased anxiety and feelings of helplessness [15].

Moreover, recent studies have emphasized the importance of considering individual vulnerability factors, such as physical health status [16,17]. People living with chronic illnesses may be more psychologically sensitive to environmental disruptions due to their heightened health risks and higher perceived threat [17,18]. These individuals often rely on ongoing health-related coping efforts, which may amplify emotional responses to additional stressors such as climate change. This intersection between physical and mental vulnerability suggests that chronic illnesses may act as a moderating factor in the relationship between climate-related variables and psychological outcomes [17].

Despite increasing interest in climate change-related psychological responses [19–21], empirical studies exploring the relationships among climate change exposure, awareness, coping strategies, and anxiety remain limited. Moreover, few studies have examined how these relationships might differ among individuals with chronic health conditions—a population likely to be disproportionately affected by climate-related stressors [17,22]. Understanding these dynamics is critical for developing effective mental health interventions and communication strategies tailored to diverse subgroups within the population.

The aim of the current study was to address this gap by investigating the associations between climate change exposure, climate change awareness, and coping strategies (problem-focused coping, meaning-focused coping, and de-emphasizing the seriousness of climate change) in relation to CCA. Furthermore, we examined whether having a chronic illness moderated these relationships. By applying a moderated mediation model and drawing on Lazarus and Folkman's [9] theoretical framework, this study provides a perspective on the psychological processes underlying CCA and offers evidence-based insights for supporting vulnerable groups. This study's theoretical model is illustrated in Figure 1.

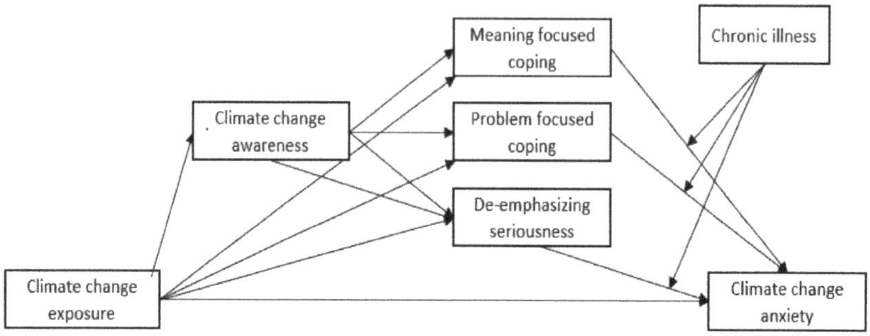

Figure 1. The study's theoretical model.

We hypothesized that:

1. Higher climate change exposure would be associated with higher climate change awareness, which, in turn, would be associated with higher CCA.
2. Higher climate change exposure would be associated with a greater use of meaning-focused and problem-focused coping, which, in turn, would be associated with higher CCA. However, higher climate change exposure would be associated with a lower use of de-emphasizing the seriousness of climate change, which, in turn, would be associated with higher CCA.
3. Higher climate change exposure would be associated with higher climate change awareness, which, in turn, would be associated with a greater use of meaning-focused and problem-focused coping and a lower use of de-emphasizing the seriousness of climate change, all of which would subsequently be associated with higher CCA.
4. Having a chronic illness would moderate the associations between climate change awareness and coping strategies on the one hand and CCA on the other hand, such that these associations would be stronger for people with a chronic illness than for people without one.

2. Materials and Methods

2.1. Participants and Procedure

Participants were recruited from an Israeli internet panel (*iPanel*) consisting of about 100,000 Israelis. This panel adheres to Israel's Central Bureau of Statistics (CBS) in key demographic factors, including age, gender, marital status, and education, which represent the general population [23]. A quota sampling approach was used with quotas meeting the Israeli national census data for sex, employment status, and self-reported health as specified by the CBS census data (see Supplementary Table S1 for the Israeli population demographic values [24]). Furthermore, to ensure a gender balance, stratified sampling by biological sex was employed. The research team requested an equal distribution of male and female participants (50% each), consistent with the national census data provided by the CBS. The inclusion criteria were being 18 years of age or older and being fluent in Hebrew. This study was approved by the second author's university's institutional review board (IRB; approval No. 20241226). Data collection took place during February 2025.

The sample size was calculated using Kenny's calculation [25] for a mediation analysis. Assuming a minimum $\beta = 0.15$ for the direct effects in the model, $\alpha = 0.05$, and power = 0.80, a sample of 451 participants was required for the indirect effect. According to

G*Power 3 [26], for multiple regression analysis with a moderate-low effect size of $f^2 = 0.10$, $\alpha = 0.05$, and power = 0.90, and up to 20 predictors (including interactions), the minimum required sample size was $N = 278$.

The study sample included 600 Israeli citizens, half men and half women, with an average age of about 50 years. About three-fourths were married or in a relationship, and about three-fourths had children. The participants had an average of about 15 years of education. About three-fourths of the participants were employed, with perceived income levels ranging from below the national average to above the national average. About half of the participants were secular (hiloni), and most of the others were traditional (masorti), reflecting common identity categories in Israel, which vary by degree of religious practice [27]. Most participants assessed their health status as good to excellent, and yet over half reported having a chronic illness (Table 1). Of those with chronic illnesses, about a third had more than one illness. Common chronic illnesses were allergies (20%), high blood pressure (19%), diabetes (10%), asthma (9%), and cancer (6%). A chronic illness was diagnosed on average at about 18 years of age (*SD* = 14.08), and most participants received medications for their illnesses.

Table 1. Demographic characteristics (*N* = 600).

Variable	Categories	Values
Gender, *n* (%)	Male	300 (50.0)
	Female	300 (50.0)
Age, *M* (*SD*), range		49.83 (16.71), 20–80
Marital status, *n* (%)	Single	96 (16.0)
	Married, in a relationship	438 (73.0)
	Divorced, widowed, other	66 (11.0)
Children, *n* (%)	Yes	
Number of children, *M* (*SD*), range		2.20 (1.72), 0–11
Years of education, *M* (*SD*), range		14.85 (2.80), 8–26
Employment, *n* (%)	Yes	450 (75.0)
Income, compared to the general Israeli population *n* (%) (*n* = 586)	Below average	204 (34.8)
	Average	159 (27.1)
	Above average	233 (38.1)
Religiosity, *n* (%)	Secular	311 (51.8)
	Traditional	187 (31.2)
	Religious	102 (17.0)
Health status, *n* (%)	Bad or not so good	73 (12.2)
	Good	402 (67.0)
	Excellent	125 (20.8)
Chronic illness, *n* (%)	Yes	323 (53.8)

2.2. Measures

The participants completed the following battery of self-report questionnaires:

Climate change exposure was assessed through seven questions composed by the researchers. These questions referred to climate change-related events in Israel to which the participants had been exposed in the past year. The listed events, specifically, were extreme heat waves, severe floods, prolonged droughts, sand and dust storms, heavy rainfall and thunderstorms, wildfires/vegetation fires due to hot and dry weather, and sharp temperature fluctuations (i.e., extreme differences between hot and cold days). Responses were coded as 0—no, 1—yes, and the total score was the sum of the events to which the participant had been exposed, ranging from 0–7.

Climate change awareness was assessed by the Climate Change Awareness scale [6]. This questionnaire includes 13 items (e.g., "I am aware that floods will occur if the temperatures of the world increase") scored on a 5-point Likert scale ranging from 1 (strongly disagree) to 5 (strongly agree). Higher scores reflect higher climate change awareness. Cronbach's alpha of the original questionnaire was 0.88 [6], and in the present study, it was 0.90.

Ecological coping was assessed with a questionnaire developed by Ojala [12]. It consists of 15 items divided into three sub-scales: problem-focused coping (three items, e.g., "I think about what I myself can do"); de-emphasizing the seriousness of climate change (six items, e.g., "I can't be bothered to care about climate change"); and meaning-focused coping (six items, e.g., "I have faith in people engaged in environmental organizations"). Answers are rated on a five-point Likert scale ranging from 1 (not true at all) to 5 (completely true). A mean item score was calculated for each coping strategy. Cronbach's alphas of the original questionnaire were 0.76 for problem-focused coping, 0.75 for de-emphasizing the seriousness of climate change, and 0.76 for meaning-focused coping [12]. In the present study, Cronbach's alphas were 0.80 for problem-focused coping, 0.72 for de-emphasizing the seriousness of climate change, and 0.67 for meaning-focused coping.

Climate change anxiety was assessed via the *CCA Scale* (CCAS) [3]. This scale includes 13 items (e.g., "Thinking about climate change makes it difficult for me to concentrate") scored on a 5-point Likert scale, from 1 (never) to 5 (almost always). A higher score reflects higher climate change anxiety. Cronbach's alpha of the original questionnaire was 0.87 [3], and in the present study, it was 0.94.

Sociodemographic characteristics were gender, age (years), marital status (single, married, in a relationship, divorced, widowed, other), number of children, years of education, employment status (yes/no), income (participants' perceived income levels), and religiosity (secular, traditional, religious/Orthodox). Self-rated health was assessed with one question: "In general, how would you rate your health?" Responses ranged from 1 (bad) to 4 (excellent) [28]. In addition, the participants were asked to mark from a list of chronic diseases whether they had been diagnosed with any of these diseases (yes/no) and, if so, in what year they had received the diagnosis.

3. Data Analysis

Data were analyzed with SPSS version 29. Descriptive statistics were used for participants' demographic and background characteristics and for the study variables. Cronbach's α was calculated for internal consistency. Climate change exposure was positively skewed (skewness = 0.71, SE = 0.10), as was CCA (skewness = 2.17, SE = 0.10). Both were logarithmically transformed. Pearson correlations were calculated between the study variables. Associations between the background variables and the study variables were analyzed with Pearson correlations and t-tests to identify the variables that should be controlled for when assessing the study model. The study model was examined with the PROCESS procedure custom model [29], controlling for gender, age, and years of education. The independent variable was climate change exposure; the mediators were climate change awareness and coping strategies; the moderator was chronic illness (yes/no); and the dependent variable was CCA. The variables were standardized, and mediation and moderation were examined within the model using 5000 bootstrap samples and a 95% confidence interval. Significant interactions were interpreted with simple slopes.

4. Results

4.1. Descriptive Results

Climate change exposure varied. One-fifth of the participants reported no exposure (n = 129, 21.5%), another one-fifth reported being exposed to one phenomenon (n = 133, 22.2%), close to one quarter reported being exposed to two phenomena (n = 143, 23.8%), one quarter was exposed to three or four phenomena (n = 153, 25.5%), and the others were exposed to at least five phenomena (n = 42, 7.0%).

The mean of climate change awareness was moderate, but the mean of CCA was low. The means for ecological coping strategies were moderate to moderate-low (Table 2). Having a chronic illness was associated only with higher CCA. Higher climate change exposure was associated with higher climate change awareness, higher use of problem-focused strategies, lower de-emphasizing of the seriousness of climate change, and higher CCA. Climate change awareness was positively associated with meaning-focused and problem-focused coping, negatively associated with de-emphasizing the seriousness of climate change, and positively associated with CCA. Finally, meaning-focused and problem-focused coping were positively associated with CCA, and de-emphasizing the seriousness of climate change was negatively associated with CCA.

Table 2. Means, standard deviations, and Pearson correlations for the study variables ($N = 600$).

Variables	*M (SD)*	1.	2.	3.	4.	5.	6.	7.
1. Chronic illness (yes)	0.54 (0.50)	1						
2. Climate change exposure	1.95 (1.60)	0.07	1					
3. Climate change awareness	3.48 (0.70)	0.05	0.33 ***	1				
4. Meaning-focused coping	2.87 (0.56)	−0.02	0.01	0.28 ***	1			
5. Problem-focused coping	2.39 (0.81)	0.04	0.24 ***	0.49 ***	0.39 ***	1		
6. De-emphasizing seriousness	2.31 (0.64)	−0.01	−0.31 ***	−0.56 ***	0.02	−0.35 ***	1	
7. Climate change anxiety	1.32 (0.49)	0.13 **	0.15 **	0.25 ***	0.24 ***	0.54 ***	−0.12 **	1

** $p < 0.01$, *** $p < 0.001$ Note. Ranges: Chronic illness 0/1; climate change exposure 0–7; climate change awareness 1–5; ecological coping 1–5; climate change anxiety 1–4. The Bonferroni criterion was applied- $p = 0.002$.

It should be noted that de-emphasizing the seriousness of climate change was rather highly associated with climate change awareness ($r = -0.56$, $p < 0.001$). Indeed, both variables had rather high variance inflation factor (VIF) values in the regression model (de-emphasizing $VIF = 1.65$, awareness $VIF = 1.89$), rather low tolerance values (0.60 and 0.53), and a high condition index (index = 31.87), indicating possible collinearity. Further, the negative association between de-emphasizing and CCA in the correlation matrix turned positive in the regression model ($\beta = 0.08$, $p = 0.082$), suggesting a suppressor variable. In order to verify the suppressor variable, another regression model was calculated, showing that the contribution of de-emphasizing the seriousness of climate change to anxiety beyond the demographic and background variables (i.e., gender, age, years of education, chronic illness, and climate change exposure) was indeed negative ($\beta = -0.08$, $p = 0.059$) and not significant. Thus, de-emphasizing the seriousness of climate change was excluded from further analyses.

4.1.1. Associations with the Demographic Variables

Gender was associated with problem-focused coping such that women reported using these strategies ($M = 2.46$, $SD = 0.81$) more than men ($M = 2.32$, $SD = 0.81$), $t(598) = 2.24$, $p = 0.026$, and $d = 0.18$. Age was negatively associated with exposure ($r = -0.17$, $p < 0.001$) and positively associated with awareness ($r = 0.10$, $p = 0.011$), meaning-focused coping ($r = 0.16$, $p < 0.001$), problem-focused coping ($r = 0.11$, $p = 0.008$), and CCA ($r = 0.13$, $p = 0.001$). Years of education were positively associated with exposure ($r = 0.12$, $p = 0.003$) and negatively associated with meaning-focused coping ($r = -0.11$, $p = 0.008$). To a small extent, years of education were negatively associated with CCA, as well ($r = -0.08$, $p = 0.047$). Other demographic variables had low variances or were not meaningfully associated with the study variables. Thus, the study model was analyzed while controlling for gender (1—male, 0—female), age, and years of education.

4.1.2. The Study Model

The results of the study's model are shown in Figure 2.

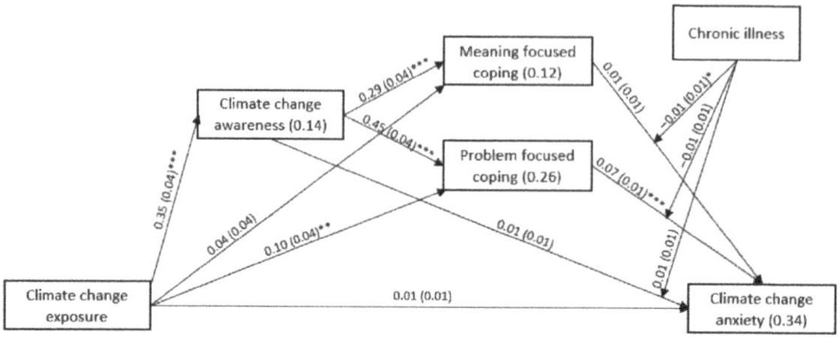

Figure 2. The mediating role of climate change awareness and coping strategies, and the moderating role of chronic illness in the association between climate change exposure and climate change anxiety. *Note:* Values on arrows: β, values within rectangles: R^2. * $p < 0.05$, ** $p < 0.01$, *** $p < 0.001$.

The results showed that higher climate change exposure was associated with higher climate change awareness ($p < 0.001$), but climate change awareness was not significantly associated with CCA ($p = 0.302$). Indeed, the indirect effect was not significant (for those with a chronic illness, effect = 0.002, $SE = 0.019$, 95% $CI = -0.037, 0.038$; and for those without a chronic illness, effect = −0.034, $SE = 0.022$, 95% $CI = -0.083, 0.006$). The first hypothesis was thus not supported.

Further, higher climate change exposure was associated with a higher use of problem-focused coping ($p = 0.006$), which, in turn, was associated with higher CCA ($p < 0.001$). In this case, the indirect effect was significant (for those with a chronic illness, effect = 0.062, $SE = 0.025$, 95% $CI = 0.014, 0.112$; and for those without a chronic illness, effect = 0.050, $SE = 0.020$, 95% $CI = 0.012, 0.091$). This part of the second hypothesis was thus supported. Climate change exposure was not associated with the use of meaning-focused coping ($p = 0.305$), and meaning-focused coping was not associated with CCA ($p = 0.485$), leaving this indirect effect non-significant (for those with a chronic illness, effect = 0.002, $SE = 0.004$, 95% $CI = -0.004, 0.013$; and for those without a chronic illness, effect = −0.005, $SE = 0.007$, 95% $CI = -0.020, 0.007$). This part of the second hypothesis was thus not supported.

As shown in Figure 2, higher climate change exposure was associated with higher climate change awareness ($p < 0.001$), which, in turn, was associated with a higher use of both meaning-focused and problem-focused coping ($p < 0.001$). Subsequently, a higher use of problem-focused coping was associated with higher CCA ($p < 0.001$), yet the association between meaning-focused coping and anxiety was not significant ($p = 0.485$). As a result, the total indirect effect for problem-focused coping was significant (for those with a chronic illness, effect = 0.094, $SE = 0.016$, 95% $CI = 0.064, 0.127$; and for those without a chronic illness, effect = 0.075, $SE = 0.016$, 95% $CI = 0.048, 0.111$), and the indirect effect for meaning-focused coping was dependent on the moderating role of chronic illness. The third hypothesis was thus partly supported.

The moderating role of chronic illness was found to be significant regarding the association between meaning-focused coping and CCA ($\beta = -0.091$, $SE = 0.037$, $p = 0.014$, 95% $CI = -0.164, -0.018$). Interpreting the significant interaction with simple slopes revealed that the association between meaning-focused coping and CCA was significant among participants without a chronic illness (effect = 0.124, $SE = 0.055$, $p = 0.024$, 95% $CI = 0.016, 0.233$) and was not significant among participants with a chronic illness (effect = −0.058, $SE = 0.051$, $p = 0.257$, 95% $CI = -0.158, 0.042$) (see Figure 3).

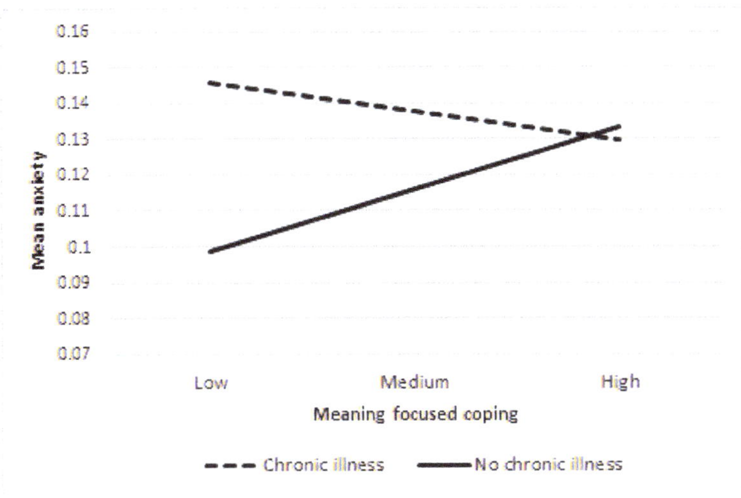

Figure 3. The moderating effect of chronic illness on meaning-focused coping and climate change anxiety. *Note.* Mean anxiety, logarithmically transformed and standardized.

The interactions for climate change awareness and problem-focused coping were not significant ($p = 0.185$, 95% $CI = -0.025$, 0.128, and $p = 0.152$, 95% $CI = -0.022$, 0.140, respectively), leaving the direct association between awareness and anxiety insignificant for the whole sample, and the association between problem-focused coping and anxiety positive and significant for the whole sample. Given that the fourth hypothesis was that the association between meaning-focused coping and CCA would be stronger for participants with a chronic illness, this hypothesis was not supported.

In sum, a significant serial mediation was found for climate change exposure, climate change awareness, problem-focused coping, and CCA regarding the whole sample (for those with a chronic illness, effect = 0.094, $SE = 0.016$, 95% $CI = 0.064$, 0.127; and for those without a chronic illness, effect = 0.075, $SE = 0.016$, 95% $CI = 0.048$, 0.111). A significant moderated serial mediation was found for climate change exposure, climate change awareness, meaning-focused coping, and CCA (effect = −0.019, $SE = 0.008$, 95% $CI = -0.037$, −0.004), where higher climate change exposure was associated with higher climate change awareness, thereby being associated with a higher use of meaning-focused coping, and subsequently being associated with higher CCA among participants without a chronic illness.

5. Discussion

In the current study, we aimed to explore the interplay between climate change exposure, climate change awareness, and climate change coping strategies in reference to CCA while also examining the differences between individuals with and without a chronic illness. The findings reveal a complex picture in which greater climate change exposure was found to be associated with higher climate change awareness. However, awareness alone did not predict CCA, suggesting that the use of coping strategies, particularly problem-focused coping, was a key mechanism linking exposure to CCA. Notably, the role of meaning-focused coping varied by health status, with its association with CCA reaching significance only among participants without a chronic illness.

The study's findings did not support our first hypothesis: Although greater climate change exposure was associated with higher climate change awareness, awareness was not significantly related to CCA. This finding contradicts previous findings showing that individuals who are more aware of the consequences of climate change tend to experience higher levels of climate concern [30] and anxiety [31]. A possible explanation might be related to Israel's regional context and weather, especially considering that this study was conducted in February 2025. During this period, participants were unlikely to be exposed to extreme weather events, such as heatwaves or wildfires, which are more common in the summer. Without immediate environmental stressors, climate change may be felt as less threatening, reducing emotional responses [22]. Additionally, ongoing sociopolitical and security concerns in Israel may have competed for emotional resources, diminishing the salience of climate-related worries [8]. This phenomenon, referred to as the "finite pool of worry," suggests that individuals have a limited emotional capacity to respond to multiple concurrent threats [32]. Consistent with this explanation, our data showed relatively low levels of climate change anxiety and limited variability in this variable, which may have constrained the detection of expected associations in our hypothesized model. Finally, climate change may be perceived by many Israelis as a global rather than local issue, further distancing this phenomenon from personal emotional impact [33]. In this context, climate change awareness might not uniformly translate into emotional responses such as CCA.

The study's findings partially supported our second hypothesis. Specifically, greater climate change exposure was associated with a higher use of problem-focused coping, which, in turn, was linked to higher levels of CCA. This pattern aligns with previous research [12,34], which has shown that attempts to take action in the face of a global and overwhelming threat (as identified with problem-focused coping) may amplify emotional distress rather than reduce it. A possible explanation for this finding is that problem-focused coping, although generally effective in controllable situations, may backfire in the context of a vast and persistent issue such as climate change. When individuals engage in concrete actions (e.g., recycling, reducing carbon use) but perceive little systemic change or global impact, feelings of helplessness and emotional burden can intensify [12,35]. Thus, greater engagement may paradoxically increase distress, particularly among those who are more informed or emotionally invested. This explanation is supported by Lazarus and Folkman's transactional theory of stress and coping [9], which states that problem-focused coping is adaptive in the short run but might be maladaptive in the long run. Continuous attempts to "solve" an unsolvable issue can lead to frustration and emotional fatigue while limiting opportunities for emotional processing or acceptance, which are essential for long-term psychological adjustment.

In contrast, the part of the second hypothesis regarding an indirect path through meaning-focused coping was not supported: Namely, climate change exposure did not predict meaning-focused coping, nor was this coping strategy significantly associated with CCA. This finding differs from prior studies suggesting that meaning-focused coping can promote resilience and psychological well-being in the context of climate-related concerns [35,36]. It could be that meaning-focused coping requires a deeper level of cognitive and emotional processing that is not necessarily triggered by exposure alone. Cultural context may also play a role; the ongoing sociopolitical stressors in Israel may have overshadowed abstract, global threats such as climate change, rendering meaning-focused coping less accessible or less commonly used in relation to climate change. From a theoretical standpoint, Lazarus and Folkman's [9] model suggests that meaning-focused coping is more likely to emerge when individuals appraise the stressor as uncontrollable and have moved beyond efforts to manage the stressor directly. If participants in the current study had still been oriented toward problem-solving, they may not have shifted toward meaning-focused coping.

Our third hypothesis was partially confirmed: Namely, greater climate change exposure was associated with greater climate change awareness, which, in turn, was associated with an increased use of meaning-focused coping strategies (e.g., reframing and meaning-making) and problem-focused coping strategies (e.g., direct actions to mitigate stressors). This finding aligns with recent findings suggesting that direct experiences with climate change phenomena (e.g., extreme weather) can enhance individuals' environmental awareness and perception of risk [37,38] and that heightened climate change awareness is associated with an increased use of meaning-focused coping strategies and problem-focused coping strategies [39]. Interestingly, our results revealed distinct outcomes for these coping strategies. Namely, problem-focused coping was associated with elevated CCA, reflecting that direct engagement with climate issues seems to expose individuals to the severity of environmental threats, thereby increasing their anxiety due to a heightened perception of urgency and potential consequences [12,35]. On the other hand, we found no significant relationship between meaning-focused coping and CCA. This finding might suggest that individuals employing meaning-focused coping strategies do not show heightened anxiety despite an increased awareness of climate threats. The indication may be that meaning-focused coping strategies contribute to emotional resilience that buffers against heightened emotional distress [40,41].

Finally, contrary to our expectations, the fourth hypothesis was not confirmed: A significant serial mediation emerged between climate change exposure, climate change awareness, problem-focused coping, and climate change anxiety for the entire sample (and not for a specific group). A possible explanation stems from Daeninck and colleagues' [35] finding among university students, which demonstrated that problem-focused coping was the most commonly endorsed strategy, although the prior literature on coping suggested that problem-focused coping might not be sustainable for individually intractable problems. Thus, it might be that both participants with and without chronic illnesses used problem-focused coping to address the problem directly, such as taking action (e.g., activism, lifestyle changes, pro-environmental behaviors) or seeking information. Interestingly, among participants without a chronic illness, our findings revealed that higher climate change exposure led to increased climate change awareness, subsequently fostering greater use of meaning-focused coping strategies, which unexpectedly was associated with higher CCA. This finding contrasts with the previously discussed protective or buffering role often attributed to meaning-focused coping strategies [12]. One possible explanation for this finding is that individuals without chronic illnesses might have less experience in regularly employing coping strategies to handle ongoing health-related stressors, making them potentially less skilled at effectively mitigating anxiety. Alternatively, it is possible that meaning-focused coping initially increases anxiety due to heightened cognitive engagement with existential dimensions of climate threats, which healthy individuals might find particularly disruptive. In contrast, for chronically ill individuals, meaning-focused coping may serve as a more adaptive strategy, perhaps because physical limitations reduce the feasibility of problem-focused coping, thereby increasing reliance on meaning-based approaches that they are more practiced in using.

6. Limitations

Several limitations should be noted. First, this study's cross-sectional design limits the ability to determine causality. Although we modeled coping as a predictor of anxiety, it is also plausible to suggest that anxiety triggers coping. Alternative models [42,43] should be explored in future longitudinal research to clarify the directionality of these relationships. Second, the study sample was drawn from an online panel, which, despite being structured to reflect the national population in key demographics, may not have fully represented the broader Israeli public. Participants in online panels may differ from the general population in terms of digital literacy, years of education, and motivation to participate in research. Notably, the sample had a slightly higher average age and education level than national averages, which may have introduced some sampling bias and limited the generalizability of the findings. Third, we do not have data on the extent of climate change denial in Israel. Although such denial is not believed to be widespread, varying levels of belief and engagement may have influenced participants' responses and should be considered a potential limitation. Moreover, participants who engage in online surveys could have different personal characteristics from those who do not participate in such panels. In line with this notion, a previous systematic review and meta-analysis showed that personality traits were related to climate change attitudes [44]. Thus, it might be that those people who are willing to participate in online panels have specific personality traits; therefore, caution, again, is required in generalizing the present findings. Fourth, all data were collected through self-reported measures, which can introduce biases such as social desirability or recall. Fifth, this study was conducted during a specific season (winter), in a specific regional weather context (Israel), and in a specific sociopolitical context (wartime), thus potentially limiting the generalizability of the findings to other regions or seasons. Different cultural, political, and environmental contexts may yield different results, suggesting the replication of the present study model. Finally, we did not assess participants' perceptions of self-efficacy, which has been shown in models such as the EPPM [10] and RPA framework [11] to moderate the relationship between risk perception and anxiety. This omission may have limited our understanding of the mechanisms underlying individual responses to climate threats.

7. Implications

The present findings have several implications. First, the consistent link found between greater climate change exposure and greater climate change awareness underscores the critical importance of enhancing public awareness through transparent communication about environmental risks.

Second, the observed positive association between problem-focused coping and CCA highlights the need for design interventions promoting coping strategies to manage climate anxiety. These interventions should carefully balance encouragement of direct action (e.g., pro-environmental behavior) with strategies to emotionally manage the anxiety that arises from confronting an overwhelming global issue. Moreover, the lack of a significant direct link between meaning-focused coping and CCA suggests a potentially protective or stabilizing function for this coping strategy. Meaning-focused coping strategies such as reframing, acceptance, or finding personal meaning may help individuals maintain emotional balance when confronting climate threats.

Third, the moderated serial mediation found specifically among participants without a chronic illness highlights the need for differentiated intervention strategies between populations. Public health messaging is recommended to be sensitive to individual differences, potentially offering tailored guidance and psychological support to healthy populations who might otherwise appear less vulnerable to climate-related distress. Policymakers and practitioners are advised to recognize that one-size-fits-all coping strategies are insufficient and potentially harmful. Future policies should integrate multi-dimensional approaches combining awareness enhancement with diversified coping skills training, explicitly addressing the emotional complexities and paradoxical outcomes identified in the current study.

8. Conclusions

The present study highlights the complex relationships among climate change exposure, awareness, coping strategies, and anxiety. Our results show that climate change awareness was not associated with CCA. Coping strategies play critical and sometimes paradoxical roles: problem-focused coping may heighten anxiety, whereas meaning-focused coping might offer emotional stability but could also amplify anxiety among specific populations, such as individuals without chronic illnesses. Thus, tailored and differentiated psychological interventions and educational programs are essential for effectively addressing the emotional impacts of climate change across diverse population groups. Further research is recommended to explore the long-term psychological trajectories of different coping strategies across varied population groups.

Supplementary Materials: The following supporting information can be downloaded at: https://www.mdpi.com/article/10.3390/cli13060124/s1: Table S1. Demographic Comparison Table.

Author Contributions: Conceptualization, S.S.-A. and Y.H.-R.; methodology, S.S.-A. and Y.H.-R.; writing—original draft preparation, S.S.-A. and Y.H.-R.; writing—review and editing, S.S.-A. and Y.H.-R.; supervision, S.S.-A. and Y.H.-R. All authors have read and agreed to the published version of the manuscript.

Funding: This research received no external funding.

Data Availability Statement: The data that support the findings of this study are available from the corresponding author upon request.

Conflicts of Interest: The authors declare no conflicts of interest.

References

1. Abbass, K.; Qasim, M.Z.; Song, H.; Murshed, M.; Mahmood, H.; Younis, I. A review of the global climate change impacts, adaptation, and sustainable mitigation measures. *Environ. Sci. Pollut. Res.* **2022**, *29*, 42539–42559. [CrossRef] [PubMed]
2. Paz, S.; Díaz, J.; Negev, M.; Linares, C. Climate change and global health. In *Handbook of Epidemiology*; Springer: New York, NY, USA, 2024; pp. 1–35.
3. Clayton, S.; Karazsia, B. Development and validation of a measure of climate change anxiety. *J. Environ. Psychol.* **2020**, *69*, 101434. [CrossRef]
4. Pihkala, P. Anxiety and the ecological crisis: An analysis of eco-anxiety and climate anxiety. *Sustainability* **2020**, *12*, 7836. [CrossRef]
5. Schneiderbauer, S.; Pisa, P.F.; Delves, J.L.; Pedoth, L.; Rufat, S.; Erschbamer, M.; Thaler, T.; Carnelli, F.; Granados-Chahin, S. Risk perception of climate change and natural hazards in global mountain regions: A critical review. *Sci. Total Environ.* **2021**, *784*, 146957. [CrossRef]
6. Gönen, Ç.; Deveci, E.Ü.; Aydede, M.N. Development and validation of climate change awareness scale for high school students. *Environ. Dev. Sustain.* **2023**, *25*, 4525–4537. [CrossRef]
7. Maran, D.A.; Begotti, T. Media exposure to climate change, anxiety, and efficacy beliefs in a sample of Italian university students. *Int. J. Environ. Res. Public Health* **2021**, *18*, 9358. [CrossRef]
8. Hamama-Raz, Y.; Shinan-Altman, S. Does climate change worry decrease during armed conflicts? *Climate* **2024**, *12*, 162. [CrossRef]
9. Lazarus, R.S.; Folkman, S. *Stress, Appraisal, and Coping*; Springer Publishing Company: New York, NY, USA, 1984.
10. Witte, K. Putting the fear back into fear appeals: The Extended Parallel Process Model. *Commun. Monogr.* **1992**, *59*, 329–349. [CrossRef]
11. Rimal, R.N.; Real, K. Perceived risk and efficacy beliefs as motivators of change: Use of the Risk Perception Attitude (RPA) framework to understand health behaviors. *Hum. Commun. Res.* **2003**, *29*, 370–399. [CrossRef]
12. Ojala, M. How do children cope with global climate change? Coping strategies, engagement, and well-being. *J. Environ. Psychol.* **2012**, *32*, 225–233. [CrossRef]
13. Van Valkengoed, A.M.; Steg, L. The climate anxiety compass: A framework to map the solution space for coping with climate anxiety. *Dialogues Clim. Change* **2024**, *1*, 39–48. [CrossRef]
14. Helm, S.V.; Li, X.; Curran, M.A.; Barnett, M.A. Coping profiles in the context of global environmental threats: A person-centered approach. *Anxiety Stress Coping* **2022**, *35*, 609–622. [CrossRef]

15. Chan, H.W.; Tam, K.P.; Clayton, S. Testing an integrated model of climate change anxiety. *J. Environ. Psychol.* **2024**, *97*, 102368. [CrossRef]
16. Rocque, R.J.; Beaudoin, C.; Ndjaboue, R.; Cameron, L.; Poirier-Bergeron, L.; Poulin-Rheault, R.A.; Fallon, C.; Tricco, A.C.; Witteman, H.O. Health effects of climate change: An overview of systematic reviews. *BMJ Open* **2021**, *11*, e046333. [CrossRef] [PubMed]
17. Shinan-Altman, S.; Hamama-Raz, Y. Factors associated with pro-environmental behaviors in Israel: A comparison between participants with and without a chronic disease. *Int. J. Environ. Res. Public Health* **2022**, *19*, 13123. [CrossRef]
18. Lindsay, S.; Hsu, S.; Ragunathan, S.; Lindsay, J. The impact of climate change-related extreme weather events on people with pre-existing disabilities and chronic conditions: A scoping review. *Disabil. Rehabil.* **2023**, *45*, 4338–4358. [CrossRef]
19. Ojala, M.; Cunsolo, A.; Ogunbode, C.A.; Middleton, J. Anxiety, worry, and grief in a time of environmental and climate crisis: A narrative review. *Annu. Rev. Environ. Resour.* **2021**, *46*, 35–58. [CrossRef]
20. Steg, L. Psychology of climate change. *Annu. Rev. Psychol.* **2023**, *74*, 391–421. [CrossRef]
21. Van Valkengoed, A.M.; Steg, L.; De Jonge, P. Climate anxiety: A research agenda inspired by emotion research. *Emotion Rev.* **2023**, *15*, 258–262. [CrossRef]
22. Clayton, S. Climate anxiety: Psychological responses to climate change. *J. Anxiety Disord.* **2020**, *74*, 102263. [CrossRef]
23. Bodas, M.; Siman-Tov, M.; Kreitler, S.; Peleg, K. Psychological correlates of civilian preparedness for conflicts. *Disaster Med. Public Health Prep.* **2017**, *11*, 451–459. [CrossRef] [PubMed]
24. Israel Central Bureau of Statistics. *Statistical Abstract of Israel 2023—No. 74*; CBS: Jerusalem, Israel, 2023. Available online: https://www.cbs.gov.il/he/publications/DocLib/isr_in_n/isr_in_n23e.pdf (accessed on 23 April 2025).
25. Kenny, D.A. MedPower: An Interactive Tool for the Estimation of Power in Tests of Mediation [Computer Software]. 2017. Available online: https://davidakenny.shinyapps.io/MedPower/ (accessed on 30 March 2025).
26. Faul, F.; Erdfelder, E.; Buchner, A.; Lang, A.G. Statistical power analyses using G*Power 3.1: Tests for correlation and regression analyses. *Behav. Res. Methods* **2009**, *41*, 1149–1160. [CrossRef] [PubMed]
27. Pew Research Center. Israel's Religiously Divided Society [Internet]. 8 March 2016. Available online: https://www.pewresearch.org/religion/2016/03/08/israels-religiously-divided-society/ (accessed on 30 May 2025).
28. Benyamini, Y.; Blumstein, T.; Lusky, A.; Modan, B. Gender differences in the self-rated health–mortality association: Is it poor self-rated health that predicts mortality or excellent self-rated health that predicts survival? *Gerontologist* **2003**, *43*, 396–405. [CrossRef]
29. Hayes, A.F. *Introduction to Mediation, Moderation, and Conditional Process Analysis*, 3rd ed.; The Guilford Press: New York, NY, USA, 2022.
30. Lee, K.; Gjersoe, N.; O'Neill, S.; Barnett, J. Youth perceptions of climate change: A narrative synthesis. *Wiley Interdiscip. Rev. Clim. Chang.* **2020**, *11*, e641. [CrossRef]
31. Ediz, Ç.; Yanik, D. The effects of climate change awareness on mental health: Comparison of climate anxiety and hopelessness levels in Turkish youth. *Int. J. Soc. Psychiatry* **2023**, *69*, 2157–2166. [CrossRef]
32. Weber, E.U. Experience-based and description-based perceptions of long-term risk: Why global warming does not scare us (yet). *Clim. Change* **2006**, *77*, 103–112. [CrossRef]
33. McDonald, R. Perceived temporal and geographic distance and public opinion about climate change. *Oxford Res. Encycl. Clim. Sci.* **2016**. Available online: https://oxfordre.com/climatescience/view/10.1093/acrefore/9780190228620.001.0001/acrefore-9780190228620-e-308 (accessed on 5 April 2025).
34. Ballew, M.T.; Uppalapati, S.S.; Myers, T.; Carman, J.; Campbell, E.; Rosenthal, S.A.; Kotcher, J.E.; Leiserowitz, A.; Maibach, E. Climate change psychological distress is associated with increased collective climate action in the U.S. *npj Clim. Action* **2024**, *3*, 88. [CrossRef]
35. Daeninck, C.; Kioupi, V.; Vercammen, A. Climate anxiety, coping strategies and planning for the future in environmental degree students in the UK. *Front. Psychol.* **2023**, *14*, 1126031. [CrossRef]
36. Ojala, M. Hope and climate change: The importance of hope for environmental engagement among young people. *Environ. Educ. Res.* **2012**, *18*, 625–642. [CrossRef]
37. Liu, Y.; Liu, J.; Su, Y. Low-carbon awareness and behaviors: Effects of exposure to climate change impact photographs. *SAGE Open* **2021**, *11*. [CrossRef]
38. Ricart, S.; Gandolfi, C.; Castelletti, A. Climate change awareness, perceived impacts, and adaptation from farmers' experience and behavior: A triple-loop review. *Reg. Environ. Change* **2023**, *23*, 82. [CrossRef]
39. Rikner Martinsson, A.; Ojala, M. Patterns of climate-change coping among late adolescents: Differences in emotions concerning the future, moral responsibility, and climate-change engagement. *Clim. Change* **2024**, *177*, 125. [CrossRef]
40. Folkman, S. The case for positive emotions in the stress process. *Anxiety Stress Coping* **2008**, *21*, 3–14. [CrossRef]
41. Ojala, M. Regulating worry, promoting hope: How do children, adolescents, and young adults cope with climate change? *Int. J. Environ. Sci. Educ.* **2012**, *7*, 537–561.

42. Prentice-Dunn, S.; Rogers, R.W. Protection motivation theory and preventive health: Beyond the health belief model. *Health Educ. Res.* **1986**, *1*, 153–161. [CrossRef]
43. Kothe, E.J.; Ling, M.; North, M.; Klas, A.; Mullan, B.A.; Novoradovskaya, L. Protection motivation theory and pro-environmental behaviour: A systematic mapping review. *Aust. J. Psychol.* **2019**, *71*, 411–432. [CrossRef]
44. Cipriani, E.; Frumento, S.; Gemignani, A.; Menicucci, D. Personality traits and climate change denial, concern, and proactivity: A systematic review and meta-analysis. *J. Environ. Psychol.* **2024**, *95*, 102277. [CrossRef]

climate

MDPI

Article

Clean Air Benefits and Climate Penalty: A Health Impact Analysis of Mortality Trends in the Mid-South Region, USA

Chunrong Jia [1,*], Hongmei Zhang [1], Namuun Batbaatar [1], Abu Mohd Naser [1], Ying Li [2] and Ilias Kavouras [1]

1 School of Public Health, University of Memphis, Memphis, TN 38152, USA; hzhang6@memphis.edu (H.Z.); nbtbatar@memphis.edu (N.B.); atitu@memphis.edu (A.M.N.); kavouras@memphis.edu (I.K.)

2 Department of Environmental and Occupational Health and Safety Sciences, College of Public Health, East Tennessee State University, Johnson City, TN 37614, USA; liy005@etsu.edu

* Correspondence: cjia@memphis.edu; Tel.: +1-901-678-2655

Abstract: The lowering air pollution in the US has brought significant health benefits; however, climate change may offset the benefits by increasing the temperature and worsening air quality. This study aimed to estimate the mortality changes due to air pollution reductions and evaluate the potential climate penalty in the Mid-South Region of the US. Daily concentrations of $PM_{2.5}$ and ozone measured at local monitoring stations in 1999–2019 were extracted from the US Environmental Protection Agency's Air Quality System. Meteorological data for the same period were obtained from the National Oceanic and Atmospheric Administration's Local Climatological Data. Annual average age-adjusted all-cause mortality rates (MRs) were downloaded from the US Centers for Disease Control and Prevention's WONDERS Databases. MRs attributable to exposure to $PM_{2.5}$, ozone, and high temperatures in warm months were estimated using their corresponding health impact functions. Using Year 1999 as the baseline, contributions of environmental changes to MR reductions were calculated. Results showed that annual average concentrations of $PM_{2.5}$ and ozone decreased by 46% and 23% in 2019, respectively, compared with the base year; meanwhile, the mean daily temperature in the warm season fluctuated and displayed an insignificant increasing trend (Kendall's tau = 0.16, $p = 0.30$). MRs displayed a significant decreasing trend and dropped by 215 deaths100,000 person-year in 2019. Lower $PM_{2.5}$ and ozone concentrations were estimated to reduce 59 and 30 deaths/100,000 person-year, respectively, contributing to 23% and 17% of MR reductions, respectively. The fluctuating temperatures had negligible impacts on mortality changes over the two-decade study period. This study suggests that improved air quality may have contributed to mortality reductions, while the climate penalty effects appeared to be insignificant in the Mid-South Region.

Keywords: air pollution; climate change; all-cause mortality; health impact; climate penalty; Mid-South

1. Introduction

Clean air is crucial for promoting better health, given the widely recognized harmful impacts of air pollution [1]. The levels of criteria pollutants have steadily declined by 22–90% between 1970 and 2022, due to the Clean Air Act, its amendments and derivative federal and state rules, and technological advances to reduce emissions from industrial and mobile sources [2]. Consequently, Americans are now exposed to lower air pollution, reducing the risk of death and other diseases. Studies have quantified the positive impacts, e.g., a 10 $\mu g/m^3$ decrease in average $PM_{2.5}$ concentrations increased the life expectancy of the US population by 0.61 years during 1980–2000 [3] and 0.35 years during 2000–2007 [4].

https://doi.org/10.3390/cli13030045

A follow-up study emphasized that the reduction in air pollution-related mortality was primarily due to decreases in ambient fine particulate matter ($PM_{2.5}$) [5]. Improved air quality is particularly beneficial for children's health, including improvements in lung function, asthma, and low birth weight [6–8]. The benefits are not only reported for the US but extend to countries worldwide [9,10].

Climate change, unfortunately, undermines the health benefits of air quality improvements. The direct environmental impacts of climate change are increased frequency, magnitude, and duration of extreme temperature events [11]. The interaction between climate change and air pollution is projected to impose a "climate penalty", defined as the amplification effect on air pollutants by climate change and the associated excess health risks [12]. Climate penalty mainly occurs to ozone (O_3), $PM_{2.5}$, black carbon, and nitrogen oxides, despite the ongoing efforts to reduce their emissions [13,14]. Ozone is most sensitive to temperature variation. Although the national average ozone concentration decreased from 88 ppb in 1999 to 65 ppb in 2019 [2], studies projected a potential increase of up to 7 ppb in some US regions by 2050 due to rising temperatures [15]. Similarly, temperature changes also affect $PM_{2.5}$ concentrations, with estimates suggesting a 2–3 $\mu g/m^3$ increase in the eastern US during the summertime by 2050 [16]. As a result, climate change has both direct and indirect impacts on human health. An increase of 1 °C in summer mean daily temperature is associated with a 2.2% mortality increase and indirectly associated with 1.0% and 3.2% increases in $PM_{2.5}$- and ozone-related mortality, respectively [17].

The cumulative impacts of air pollution and climate change warrant urgent attention in research and practice. The scientific community is interested in predicting future impacts with hypothetical climate change scenarios on the global and national scales [18]. Limaye et al. [19] estimated that warmer average apparent temperatures could cause over 10,000 additional annual deaths in the eastern US by 2050. Neumann et al. [20] projected that wildfire emissions would cause 1300–1600 $PM_{2.5}$-related premature deaths per year in the western US by 2050. Wilson et al. [21] estimated that mortality attributed to ozone elevation will increase by 14.2% in the US in 2035. A global study suggested that global greenhouse gas reductions in Representative Concentration Pathway 4.5 (RCP4.5) could avoid 16,000 $PM_{2.5}$-related all-cause deaths and 8000 O_3-related respiratory deaths per year in the US [22]. Limited studies have retrospectively evaluated the cumulative impacts for a specific region in the US. In practice, health impacts are assessed using tools such as the World Health Organization (WHO)'s AirQ+ and the US Environmental Protection Agency (EPA)'s BenMAP to evaluate pollution control policies and programs [23]; however, these assessments often focus only on $PM_{2.5}$, without simultaneous consideration of climate penalty effects.

This study aimed to evaluate the impacts of air pollution reduction and temperature increase on mortality changes over the past two decades in the Mid-South Region of the US. This region, like many other US metropolitan areas, has had declining mortality rates in the past decades before the COVID-19 pandemic [24]. With the recent 20 years of vital and environmental data, this study was aimed at understanding the contributions to mortality changes from air quality improvements as well as the potential climate penalty in this region. This motive responded to the need for local decision-making in affected communities [25], a gap in predominant global and national analyses on this topic [26].

2. Methods

2.1. Study Area

The Mid-South is an informally defined region of the US anchored by the City of Memphis. Memphis is the largest city in the State of Tennessee and the 20th largest city in the US. As the demographic and environmental data are collected at the county level, this study used data in three neighboring urban counties in the Mid-South: Shelby County, Tennessee (TN), DeSoto County, Mississippi (MS), and Crittenden County, Arkansas (AR), as displayed Figure 1. Table S1 summarizes the key demographic and socioeconomic statistics of Shelby, DeSoto, and Crittenden Counties in Year 2019.

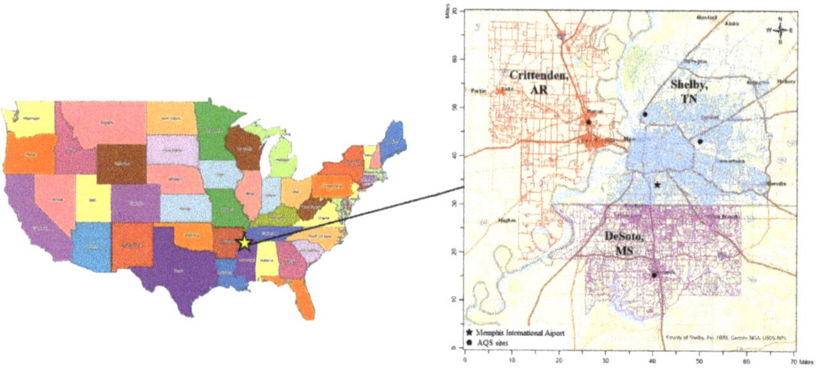

Figure 1. Map of the Mid-South Metropolitan region, USA.

2.2. *Vital and Environmental Data*

All the vital and environmental data for this study were extracted from publicly available online databases. The start and end years for data extraction were initially set as 1980 and 2019. The US EPA archived air quality data starting in 1980, but $PM_{2.5}$ data were only available beginning in 1999. The COVID-19 pandemic started in this region in March 2020 and was known to increase the mortality rate dramatically. Considering the data availability, this study covered the period between 1999 and 2019.

County-level demographic data, including mortality rates, were downloaded from the US Centers for Disease Control and Prevention's (CDC) WONDER online databases [27]. To match the original studies that determined relative risks (RRs), health measures used natural-cause mortality (ICD codes: A00-R99) for $PM_{2.5}$ [28], all-cause mortality (ICD codes: A00-U99) for ozone [29], and all-cause mortality excluding accidental deaths (V01-Y98) for temperature [30,31]. Although the ICD codes were different for these three mortality rates, the actual rates were the same for any county in any year. The crude mortality rates were converted to age-adjusted mortality rates, if not available in the original datasets, by applying age-specific rates in the study population to Year 2000 standardized US age distribution [32,33]. For convenience, the mortality rate (MR) in this study refers to the annual average age-adjusted mortality rate. MR was expressed as the number of deaths per 100,000 people-years, written as deaths/PY* following a convenient way by EPA [34]. For any year, the MR of the three counties was computed as the weighted average of the three counties' MRs by weighing their population sizes. As Shelby County had a far larger population (Table S1), the weighted average statistics were driven by and similar to Shelby County data.

For air pollution data, annual average concentrations of $PM_{2.5}$ and ozone were extracted from EPA's Air Quality System [35] for 1999–2019. It should be noted that the annual average concentrations of ozone were calculated from daily 8 h maximum concentrations. DeSoto County had only one monitoring station (Hernando Station, AQS Site ID: 28-033-0002), and Crittenden County had only one station (Marion Station, AQS Site ID: 05-035-0005), too. Shelby County had multiple air monitoring stations. The Memphis National CORE Station (AQS Site ID: 47-157-0075) was a comprehensive air monitoring station that started operation in 2011, and the data collected at this site were used for the period 2011–2019. Data measured at the Frayser Station (AQS Site ID: 47-157-0021) were used for the period 1999–2010. Our previous study showed small variability in criteria air pollutant concentrations in this region [36], and thus, it was reasonable to use air pollution data collected at one site in each county and average them to represent the overall air pollution level in the Mid-South Region.

Meteorological data for the same period were obtained from the National Oceanic and Atmospheric Administration's (NOAA) Local Climatological Data [37]. There were no weather stations that had validated meteorological data in DeSoto County or Crittenden County. Thus, the meteorological data collected at the Memphis International Airport (Network ID: WBAN:13893: Latitude/Longitude: 35.05639°, −89.9864°) were used for this region. Weather station data have been used to study the impacts of temperature on mortality in urban areas in the US [38,39]. The Memphis International Airport is situated at the center of the entire study area (Figure 1) and can broadly represent the climate of this region. The hourly temperature data (in °C) at this airport were downloaded from the NOAA database [37], and the daily average temperatures were computed for the following analyses.

2.3. Trend Analysis

The long-term trends of MRs, air pollution levels, and ambient summertime temperatures during the study period of 1999 through 2019 were examined by plotting these parameters against year. The significance of trends was examined using the Mann–Kendall test, which yielded a Kendall tau correlation between parameter and time. In this study, a positive correlation indicated an increasing trend, and vice versa. A p-value of <0.05 indicated a statistically significant trend.

2.4. Impacts of Environmental Factors on Mortality

The US EPA has been using the health impact function approach to evaluate the health impacts of air pollution on mortality [40]. In any specific year, the MR attributable to exposure to $PM_{2.5}$ ($MR_{PM2.5}$) or ozone (MR_{O3}) is calculated as

$$MR_{PM2.5\ (or\ O3)} = MR \times (e^{\beta \cdot \Delta x} - 1) \quad (1)$$

where β is the coefficient for the exposure–health association, and Δx is the difference between an exposure and the threshold level. In this study, Δx was calculated as the annual average concentration of $PM_{2.5}$ or ozone minus 0, assuming air pollution is always detrimental. MRs attributable to other criteria air pollutants were not calculated due to the lack of exposure–mortality relationships. The β value was derived from the relative risk (RR) of the effect determined in previous epidemiological studies conducted in representative US populations. Table 1 summarizes the RRs, β values, and their data sources. Model (1) assumes there was no safe threshold of air pollution exposure for adverse health effects, supported by the recent World Health Organization's Air Quality Guidelines [41].

High temperatures during the warm season, i.e., between 1 May and 30 September, elevate the daily mortality rate if the daily average temperature is above the optimum temperature of 15.6 °C (60 F°) [42]. The daily mortality rate attributable to the elevated temperature is calculated as [31]

$$\text{Daily } MR_{Temp} = \text{Daily MR} \times (e^{\beta \cdot \Delta x} - 1) \quad (2)$$

where β is the coefficient for the heat–mortality association (Table 1), and Δx is the temperature increase from 15.6 °C. Equation (2) was applied to days between 1 May and 30 September inclusively for any year. The daily mortality rate was unavailable and thus was calculated by dividing the annual MR by 365 (or 366 for a leap year), an approach applied in previous health impact studies [31,43]. Days with an average temperature below 15.6 °C were assumed to have no detrimental effects due to heat. In any year, the elevated daily mortality rates were added to yield the annual MR related to heat, i.e., MR_{Temp}.

Table 1. Exposure–response relationships determined in previous cohort studies.

Environmental Parameter	Mortality	Relative Risk (95% Confidence Interval)	β (95% CI)	Source
$PM_{2.5}$ [1]	Natural-cause mortality, excluding accidents, suicides, and homicides	1.08 (1.06, 1.09) per 10 μg/m^3	0.0077 (0.0058, 0.0086) per 1 μg/m^3	[28]
Ozone [2]	All-cause mortality	1.02 (1.01, 1.03) per 10 ppb	0.00198 (0.0010, 0.0030) per 1 ppb	[29]
Temperature in the warm season [3]	All-cause mortality, excluding accidental deaths	1.0009 (1.0000, 1.0018) per 1 °C [4]	0.0009 (0.0000, 0.0018) per 1 °C	[42]

[1] Measured as annual average $PM_{2.5}$ concentrations (μg/m^3). [2] Measured as annual average 8 h daily maximum ozone concentrations (ppb). [3] Daily average ambient temperature (°C). [4] The RR for the Southeast Region was used, as this study area belongs to the Southeast Region.

As seen in the results, MR displayed a downward trend in this area. Using Year 1999 as the baseline, the change of the MRs in a later year (ΔMRs) was calculated by subtracting the MR in 1999 from the MR in that particular year. Similarly, changes in environment-related MRs, i.e., $\Delta MR_{PM2.5}$, ΔMR_{O3}, and ΔMR_{Temp}, were calculated by subtracting $MR_{PM2.5\ (O3,\ Temp)}$ in 1999 from $MR_{PM2.5\ (O3,\ Temp)}$ in a later year. The contribution of an environmental ($PM_{2.5}$, ozone, or temperature) change to ΔMR in a year was expressed as a percentage:

$$Contrib_{PM2.5\ (or\ O3,\ Temp)} = \Delta MR_{PM2.5\ (O3,\ Temp)}/\Delta MR \times 100\% \tag{3}$$

For the Mid-South Region, the environment-related MRs and their contributions to ΔMR were computed by weighing their population sizes. As Shelby County had a predominantly large population (Table S1), the Mid-South results were similar to those using only Shelby County data.

3. Results

3.1. Trends in Air Pollution and Temperature

$PM_{2.5}$ and ozone levels had apparent declining trends in the Mid-South Region during the study period, as displayed in Figure 2. Annual average concentrations of $PM_{2.5}$ and ozone were 8.3 μg/m^3 and 41 ppb in 2019, 46% and 23% down from 1999, respectively. The Mann–Kendall tests showed both $PM_{2.5}$ and ozone trends were statistically significant (p-values < 0.01). These trends were in accordance with the national trends: the national annual average $PM_{2.5}$ concentration dropped from 13.5 μg/m^3 in 2000 (1999 data were not available) to 7.7 μg/m^3 in 2019, and the national average ozone concentration (The national annual average is the average of the annual 98th percentile of daily 8 h maximum across the sites, and our annual average ozone concentration is the average of daily 8 h max concentrations. Thus, the national annual averages and our annual averages differed in their values, but both showed distinctly decreasing trends) from 84 ppb in 1999 to 65 ppb in 2019 [2]. Warm-season temperatures fluctuated over the years and had an insignificant positive association with time (Kendall's tau = 0.16, p = 0.30). The three bordering counties had similar levels and trends of $PM_{2.5}$ concentrations (Figure S1). Ozone concentrations were similar in Shelby and DeSoto Counties but were lower in Crittenden County; they all displayed downward trends in these three counties.

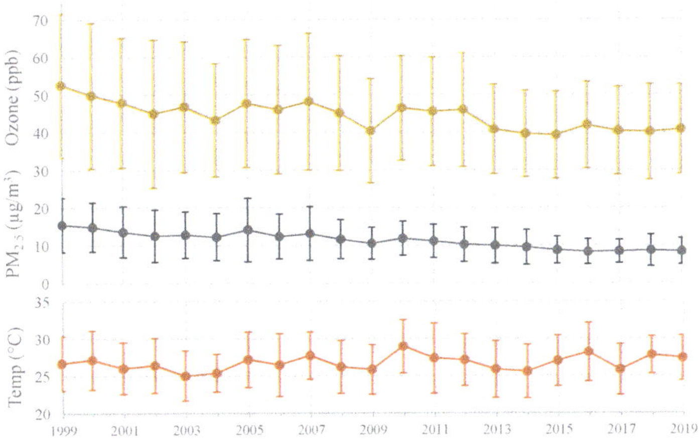

Figure 2. Annual average $PM_{2.5}$ concentrations (in μg/m^3), annual average 8 h max daily ozone concentrations (in ppb), and average ambient temperatures (in °C) during the warm season (May–September) in the Mid-South Region between 1999 and 2019. The error bars indicate standard deviations.

3.2. All-Cause and Environment-Related Mortality Rates

The all-cause mortality rate (MR) in the Mid-South Region showed a declining trend over the two decades between 1999 and 2019 (Figure 3A). MR and year had a significant negative association (Kendall's tau = −0.86, $p < 0.0001$). The MR was 1003 deaths/PY* in 1999 and decreased to 788 deaths/PY* in 2019, with some small fluctuations in between. By 2019, MR was 215 deaths/PY* less than that in 1999, a 21% decrease. In individual counties, MRs consistently showed an order of Crittenden > Shelby > DeSoto and exhibited generally decreasing trends (Figure S2). The Kendall's tau coefficients between MR and year were −0.84 ($p < 0.0001$), −0.50 ($p = 0.0014$), and −0.71 ($p < 0.0001$) in Shelby, DeSoto, and Crittenden Counties, respectively, suggesting statistically significant decreasing trends.

MRs attributable to $PM_{2.5}$ and ozone displayed downward trends, as a result of the decreasing air pollution levels. $MR_{PM2.5}$ dropped from 126 deaths/PY* in 1999 to 67 deaths/PY* in 2019 (Figure 3B), and the trend was significant (Kendall's tau = −0.88, $p < 0.0001$). MR_{O3} decreased from 117 deaths/PY* in 1999 to 87 deaths/PY* in 2019 (Figure 3C), and the trend was significant (Kendall's tau = −0.79, $p < 0.0001$), too. Due to the variation in the original exposure–effect relationships, $MR_{PM2.5}$ and MR_{O3} both had large confidence intervals. In individual counties, $MR_{PM2.5}$ showed a consistent order of Crittenden > Shelby > DeSoto (Figure S3), and MR_{O3} was similar in Shelby and DeSoto Counties but higher in Crittenden County (Figure S4). Comparing Figures 3B and 3C, $PM_{2.5}$-related mortality was greater than ozone-related mortality before 2010 but smaller afterward, suggesting more health benefits from $PM_{2.5}$ reduction over the past two decades.

The impact of high temperatures on mortality was much lower than those of air pollution (Figure 3D). MR_{Temp} ranged between 3.5 and 4.7 deaths/PY* during the study period. MR_{Temp} showed a positive but insignificant association with time (Kendall's tau = 0.21, $p = 0.17$). The MR_{Temp} averaged 4.0 deaths/PY* in the Mid-South Region and Shelby County. Considering a population of around 1 million in Shelby County, the total number of deaths due to high temperatures was about 40 deaths per year. Our previous study based on actual daily death numbers in Shelby County also reported 40 deaths/year on heatwave days between 2008 and 2017 [44]. This agreement between the modeled and observed death numbers, to some extent, suggested the validity of the impact assessment model Equation (1).

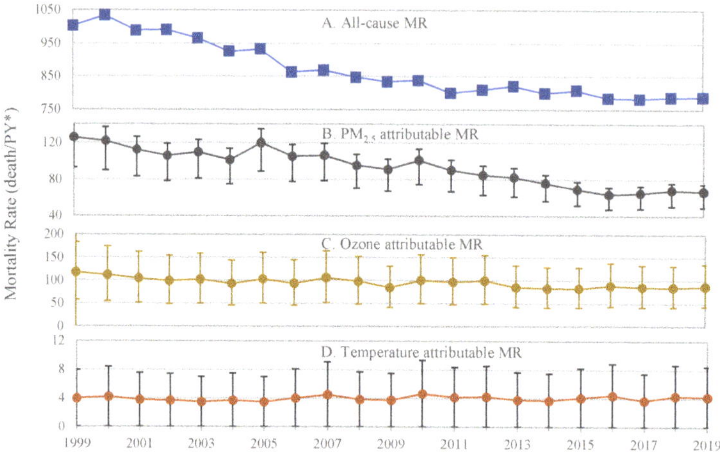

Figure 3. Total mortality rates (MRs) and MRs attributable to $PM_{2.5}$ ($MR_{PM2.5}$), ozone (MR_{O3}), and warm-season temperature (MR_{Temp}) in the Mid-South Metropolitan Region between 1999 and 2019. The errors bars of $MR_{PM2.5}$, MR_{O3}, and MR_{Temp} were upper and lower 95% confidence intervals (CIs) calculated from the 95% CIs of β values in Table 1.

3.3. *Contributions of Environmental Factors to MR Changes*

The improved air quality had sizable contributions to the reductions in all-cause MRs. The lowered $PM_{2.5}$ concentrations reduced MRs by 4–62 deaths/PY* in 2000 through 2019, compared to the baseline MR in 1999. The contributions of $\Delta MR_{PM2.5}$ to ΔMR drastically fluctuated in 2000 through 2002, which then showed a limited range of 8–44% afterwards (Figure 4A). The average contribution of $\Delta MR_{PM2.5}$ to ΔMR was 23% between 2003 and 2019. Contributions of $\Delta MR_{PM2.5}$ to ΔMR displayed similar patterns in individual counties (Figure S5).

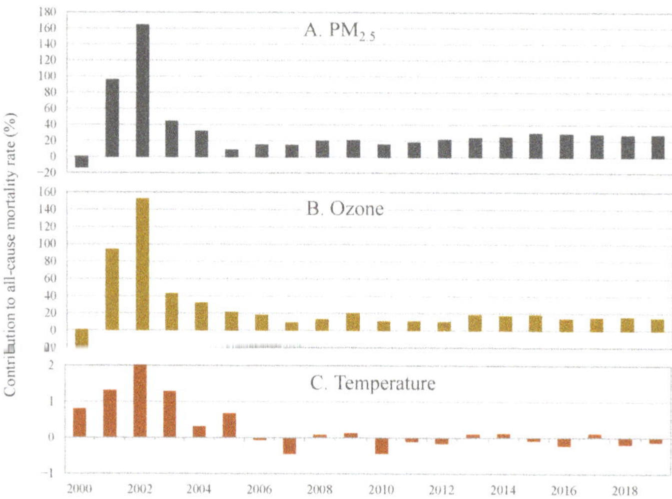

Figure 4. Contributions of $PM_{2.5}$, ozone, and temperature-related ΔMRs to observed ΔMRs in the Mid-South Metropolitan Region. Notes: The baseline year was 1999.

Reduced ozone concentrations were estimated to reduce MRs by 6–34 deaths/PY* in 2000 through 2019. Similar to $PM_{2.5}$, the contributions of ozone-related MR reductions (ΔMR_{O3}) to ΔMRs varied drastically between 2000 and 2002, but had a limited range of 9–43% in the later years (Figure 4B). Between 2003 and 2019, ΔMR_{O3} contributed 17% of ΔMR on average in the Mid-South, and 18%, 12%, and 9% to ΔMRs in Shelby, DeSoto, and Crittenden Counties, respectively (Figure S6).

The contributions of temperature-related MR changes (ΔMR_{Temp}) to ΔMR were overall small, ranging from −0.4% to 2.0% (Figure 4C). The average contribution of ΔMR_{Temp} to ΔMR was 0 on average for the period of 2003 through 2019, meaning a negligible contribution of high temperatures on mortality in warm seasons.

4. Discussion

This study documented the declining mortality, improved air quality, and fluctuating but generally unchanged warm-season temperatures in the Mid-South Region two decades before the COVID-19 pandemic. Changes in the death risk factors drive the change in population mortality [45]. Smoking, poor diet, physical inactivity, and alcohol consumption were the leading causes of death in the US in 2000 [46]. The declining mortality rates in the US were attributable to improvements in public health, medical advances, and behavioral changes [47], e.g., reductions in tobacco smoking accounted for 40% of fall in US male cancer mortality [48]. Numerous studies have confirmed air pollution as a risk factor for mortality [28,49]; however, there is no direct analysis showing that the decreases in environmental stressors lead to a decline in mortality.

This retrospective health impact assessment confirmed the health benefits of air pollution reduction reported in previous studies. Most health impact studies only estimate death numbers due to $PM_{2.5}$ exposure on the international [50], national [51], and local [52] scales, without quantifying contributions to the actual mortality. Only a few studies quantified the contributions of air pollution improvement to mortality reductions in the US, and the estimates varied greatly. An Arkansas-based study reported that 50–100% reduction of non-accidental mortality was attributable to $PM_{2.5}$ reductions in Crittenden County between 2000 and 2010 [53]. A modeling study showed that 5.0%, 5.2%, and 5.0% of deaths were attributable to 40% reduction of $PM_{2.5}$ levels in TN, MS, and AR, respectively [54]. A national analysis of $PM_{2.5}$ trends in the US reported that the average annual $PM_{2.5}$-related decrease in cardiovascular mortality rate (CMR) contributed 11% of the total CMR reduction between 1990 and 2010 [34]. Globally, the reduction in deaths due to PM improvement contributed 18% of the all-cause death reduction from 2007 to 2017 [55]. The estimated 23% contribution to mortality reduction by $PM_{2.5}$ reduction in the Mid-South was comparable to the range of the reported contributions.

The tropospheric ozone situation was more complex, as ozone is a secondary air pollutant strongly affected by temperature shifts. Increasing temperatures accelerate atmospheric reactions and, in turn, increase ground ozone levels; however, the climate penalty on ozone has high site-to-site heterogeneity in the US [56]. A few studies predicted ozone-related mortality burden attributable to climate changes by modeling, e.g., Bell et al. [57] predicted 0.11–0.27% increase in daily total mortality due to elevated ozone levels by 2050, and Wilson et al. [21] predicted that mortality would increase by 7.7% and 14.2% attributed to ozone exceeding 40 ppb and 75 ppb, respectively, by 2030. In the Mid-South, the improvement in air pollution outweighed the climate penalty on ozone, evidenced by the significantly decreasing ozone trends. Our results were consistent with a recent study that reported a significant reduction in climate penalty effects on $PM_{2.5}$ and ozone due to efficient emission control strategies [58].

The changes in heat-related mortality, using average daily temperatures in warm seasons, were negligible over the study period for two reasons. First, the health impact of heat was small in this region based on the exposure–response relationship. Our results showed that heat-related mortality was about 25 times lower than air pollution-related mortality (Figure 3). The Mid-South is not a traditionally vulnerable heat area, and the correlation between anomalously high temperatures and human health is more ambiguous [59]. A national analysis reported decreased deaths attributable to summertime heat in Memphis from 1987 to 2005, due mainly to the higher population resilience to heat over time [55]. The global disease burden analysis found mortality due to environmental heat and cold dropped from 0.9 deaths/PY* in 2005 to 0.7 deaths/PY* in 2015, suggesting a lack of association between mortality and climate change [60]. Second, climate change is a long-term change in the state of the climate that persists for decades or longer. Two decades may not be long enough to show significant temperature increases, evidenced by the generally flat temperatures in warm seasons between 1999 and 2019 (Figure 2).

The analyses in this study had multiple uncertainties in every component of the health impact function. A significant source of uncertainty derived from the coefficients for the exposure–health association (βs in Table 1) taken from national or regional cohort studies. For this study, the best available β values were adopted for $PM_{2.5}$, ozone, and temperature from the literature for all-cause mortality. The validity of Model (2) and its parameters could be evidenced by the good agreement between modeled and observed heat-related death numbers in Shelby County. The calculations did not factor in changes in the exposure–response relationship over time. Nationally, there was a remarkable decline in the US temperature–mortality relationship, thanks to wide uses of residential air conditioning [61]. Numerous factors are known to impact mortality rates, e.g., healthcare system improvements, medical treatment advancements, and public health interventions, which were not considered in this study. This health impact assessment also had several limitations. The joint effects from co-exposure to multiple environmental stressors are still inconsistent [62], and the exposure–health associations in original studies accounted for other environmental factors. Thus, each stressor's impact was presented separately. Evaluating health impacts for ethnic and socioeconomic subgroups would be an interesting study addressing health and environmental disparities. National analyses using county-level data indicated that $PM_{2.5}$ attributable cardiovascular mortality burden was 3.5 times higher for black people than for white people [63], and $PM_{2.5}$-related cardiovascular mortality reductions were highest in counties with moderate to high deprivation [34]. However, this task was unachievable for this study due to a lack of mortality data for subpopulations. Our study covered the urban counties of the Mid-South but not the rural counties. In the US, trends, causes, and risks of mortality in rural (nonmetropolitan) areas differed substantially from those in urban areas [64,65]. It would be interesting to address sociodemographic disparities and urban–rural gaps in future health impact studies.

This study presented several methodological and practical novelties. First, it put together the health impacts of changes in atmospheric $PM_{2.5}$, ozone, and temperature. Previous studies either estimated the positive impacts of air quality improvements or the negative impacts of climate change. Our methodology offered a novel perspective to assess the benefits of clean air and the penalties of climate change simultaneously. Second, this study presented the only and first detailed health impact assessment for the Mid-South Region. Most air pollution and climate impact studies have been conducted at national or global levels that include hundreds or thousands of cities. Those studies can offer a broad overview of health impacts but lack the detailed results necessary for local environmental decision-making. The Mid-South Region is particularly unique, characterized by a high percentage of disadvantaged populations, numerous sources of air pollution, and poor health outcomes. No specific environmental health impact assessment has been conducted for this area, so our findings help fill this knowledge gap. More regional and local impact assessments are warranted to complement large-scale national and global studies. Third, from the practical perspective, our study demonstrated a health impact framework using publicly available health and environmental data and multi-pollutant impact models. Local regulatory agencies and interested stakeholders can easily replicate this framework to evaluate the health impacts from changing air pollution and climate in their respective areas. Our results evidenced the effectiveness and benefits of air pollution control in the Mid-South Region. These findings endorsed the contributions to and achievements in environmental protection of local professionals, who otherwise seldom gain professional recognition or credibility.

5. Conclusions

To our knowledge, this is the first study that simultaneously examined clean air benefits and climate penalties in the Mid-South Region of the US. The long-term environmental data showed significant reductions in annual average concentrations of $PM_{2.5}$ (46%) and ozone (23%) and in mortality rate (21%) in 2019 compared with the baseline year 1999. Average temperatures in warm seasons did not display significant increasing or decreasing trends. Health impact assessment showed that $PM_{2.5}$ and ozone reductions contributed to 23% and 17% of the mortality reduction, respectively, but warm-season temperature changes had an average null contribution to mortality changes. These results suggested that air quality improvement might have brought health benefits in the Mid-South in the past two decades; meanwhile, the climate penalty seemed negligible during the same period. This study demonstrated a health assessment framework that local governments can adopt to evaluate the health benefits of pollution control and endorse environmental professionals' contributions.

Supplementary Materials: The following supporting information can be downloaded at: https://www.mdpi.com/article/10.3390/cli13030045/s1, Table S1: Demographic, social, and economic characteristics of Shelby County, TN, DeSoto County, MS, and Crittenden County, AR; Figure S1: Annual averages of $PM_{2.5}$ concentrations (in $\mu g/m^3$) and 8 h max daily ozone concentrations (in ppb) in the three urban counties in the Mid-South Region between 1999 and 2019; Figure S2: Age-adjusted all-cause annual average mortality rates in the Mid-South Metropolitan Region and three counties between 1999 and 2019; Figure S3: Age-adjusted all-cause annual average mortality rates attributable to $PM_{2.5}$ exposure in the Mid-South Region and three counties between 1999 and 2019; Figure S4: Age-adjusted all-cause annual average mortality rates attributable to ozone exposure in the Mid-South Metropolitan Region and three counties between 1999 and 2019; Figure S5: Contributions of $PM_{2.5}$-related ΔMRs to all-cause ΔMRs in three urban counties in the Mid-South Region. Notes: The baseline year was 1999; Figure S6: Contributions of ozone-related ΔMRs to all-cause ΔMRs in three urban counties in the Mid-South Region. Notes: The baseline year was 1999.

Author Contributions: Conceptualization, C.J., H.Z., and I.K.; methodology, C.J., H.Z., I.K., and Y.L.; formal analysis, N.B. and H.Z.; data curation, N.B. and H.Z.; writing—original draft preparation, C.J.; writing—review and editing, Y.L. and A.M.N.; visualization, N.B. and C.J.; supervision, C.J.; project administration, C.J.; funding acquisition, C.J. All authors have read and agreed to the published version of the manuscript.

Funding: This research was partially funded by a sub-award (Number: 2-340-0219134-67597L) from the U.S. Environmental Protection Agency's Environmental Justice Thriving Communities Technical Assistance Center (EJ TCTAC) Program (Cooperative Agreement 13746193).

Data Availability Statement: The mortality data were downloaded from CDC's WONDER Database at https://wonder.cdc.gov/ (accessed on 23 May 2023). The air quality data were downloaded from

EPA's Air Quality System at https://aqs.epa.gov/aqsweb/airdata/download_files.html (accessed on 22 May 2023). The temperature data were downloaded from NOAA's Local Climatological Data (LCD) at https://www.ncdc.noaa.gov/cdo-web/datatools/lcd (accessed on 22 May 2023). Details of the data have been described in the Methods section of this manuscript.

Conflicts of Interest: The authors declare no conflicts of interest.

References

1. Manisalidis, I.; Stavropoulou, E.; Stavropoulos, A.; Bezirtzoglou, E. Environmental and Health Impacts of Air Pollution: A Review. *Front. Public Health* **2020**, *8*, 14. [CrossRef] [PubMed]
2. USEPA. *Our Nation's Air: Trends Through 2022*; U.S. Environmental Protection Agency: Washington, DC, USA, 2023.
3. Pope, C.A.; Ezzati, M.; Dockery, D.W. Fine-Particulate Air Pollution and Life Expectancy in the United States. *N. Engl. J. Med.* **2009**, *360*, 376–386. [CrossRef]
4. Correia, A.W.; Pope, C.A., 3rd; Dockery, D.W.; Wang, Y.; Ezzati, M.; Dominici, F. Effect of air pollution control on life expectancy in the United States: An analysis of 545 U.S. counties for the period from 2000 to 2007. *Epidemiology* **2013**, *24*, 23–31. [CrossRef] [PubMed]
5. Cromar, K.R.; Gladson, L.A.; Ewart, G. Trends in Excess Morbidity and Mortality Associated with Air Pollution above American Thoracic Society-Recommended Standards, 2008–2017. *Ann. Am. Thorac. Soc.* **2019**, *16*, 836–845. [CrossRef] [PubMed]
6. Wong, E.Y.; Gohlke, J.; Griffith, W.C.; Farrow, S.; Faustman, E.M. Assessing the health benefits of air pollution reduction for children. *Environ. Health Perspect.* **2004**, *112*, 226–232. [CrossRef] [PubMed]
7. Gauderman, W.J.; Urman, R.; Avol, E.; Berhane, K.; McConnell, R.; Rappaport, E.; Chang, R.; Lurmann, F.; Gilliland, F. Association of improved air quality with lung development in children. *N. Engl. J. Med.* **2015**, *372*, 905–913. [CrossRef]
8. Garcia, E.; Berhane, K.T.; Islam, T.; McConnell, R.; Urman, R.; Chen, Z.; Gilliland, F.D. Association of Changes in Air Quality with Incident Asthma in Children in California, 1993–2014. *JAMA J. Am. Med. Assoc.* **2019**, *321*, 1906–1915. [CrossRef]
9. Schraufnagel, D.E.; Balmes, J.R.; de Matteis, S.; Hoffman, B.; Kim, W.J.; Perez-Padilla, R.; Rice, M.; Sood, A.; Vanker, A.; Wuebbles, D.J. Health benefits of air pollution reduction. *Ann. Am. Thorac. Soc.* **2019**, *16*, 1478–1487. [CrossRef] [PubMed]
10. Ye, T.; Guo, S.; Xie, Y.; Chen, Z.; Abramson, M.J.; Heyworth, J.; Hales, S.; Woodward, A.; Bell, M.; Guo, Y.; et al. Health and related economic benefits associated with reduction in air pollution during COVID-19 outbreak in 367 cities in China. *Ecotoxicol. Environ. Saf.* **2021**, *222*, 112481. [CrossRef]
11. USEPA. *Climate Change Indicators: Heat Waves*; U.S. Environmental Protection Agency: Washington, DC, USA, 2023.
12. Wu, S.; Mickley, L.J.; Leibensperger, E.M.; Jacob, D.J.; Rind, D.; Streets, D.G. Effects of 2000–2050 global change on ozone air quality in the United States. *J. Geophys. Res. Atmos.* **2008**, *113*, 112481. [CrossRef]
13. Jacob, D.J.; Winner, D.A. Effect of climate change on air quality. *Atmos. Environ.* **2009**, *43*, 51–63. [CrossRef]
14. Hedegaard, G.B.; Christensen, J.H.; Brandt, J. The relative importance of impacts from climate change vs. emissions change on air pollution levels in the 21st century. *Atmos. Chem. Phys.* **2013**, *13*, 3569–3585. [CrossRef]
15. Fu, T.-M.; Tian, H. Climate Change Penalty to Ozone Air Quality: Review of Current Understandings and Knowledge Gaps. *Curr. Pollut. Rep.* **2019**, *5*, 159–171. [CrossRef]
16. Shen, L.; Mickley, L.J.; Murray, L.T. Influence of 2000–2050 climate change on particulate matter in the United States: Results from a new statistical model. *Atmos. Chem. Phys.* **2017**, *17*, 4355–4367. [CrossRef]
17. Shi, L.; Liu, P.; Zanobetti, A.; Schwartz, J. Climate Penalty: Climate-driven increases in ozone and PM2.5 levels and mortality. *Environ. Epidemiol.* **2019**, *3*, 365. [CrossRef]
18. Chen, K.; Vicedo-Cabrera, A.M.; Dubrow, R. Projections of Ambient Temperature- and Air Pollution-Related Mortality Burden Under Combined Climate Change and Population Aging Scenarios: A Review. *Curr. Environ. Health Rep.* **2020**, *7*, 243–255. [CrossRef] [PubMed]
19. Limaye, V.S.; Vargo, J.; Harkey, M.; Holloway, T.; Patz, J.A. Climate change and heat-related excess mortality in the eastern USA. *EcoHealth* **2018**, *15*, 485–496. [CrossRef] [PubMed]
20. Neumann, J.E.; Amend, M.; Anenberg, S.; Kinney, P.L.; Sarofim, M.; Martinich, J.; Lukens, J.; Xu, J.W.; Roman, H. Estimating PM2.5-related premature mortality and morbidity associated with future wildfire emissions in the western US. *Environ. Res. Lett.* **2021**, *16*, 035019. [CrossRef]
21. Wilson, A.; Reich, B.J.; Nolte, C.G.; Spero, T.L.; Hubbell, B.; Rappold, A.G. Climate change impacts on projections of excess mortality at 2030 using spatially varying ozone-temperature risk surfaces. *J. Expo. Sci. Environ. Epidemiol.* **2017**, *27*, 118–124. [CrossRef]
22. Zhang, Y.; Smith, S.J.; Bowden, J.H.; Adelman, Z.; West, J.J. Co-benefits of global, domestic, and sectoral greenhouse gas mitigation for US air quality and human health in 2050. *Environ. Res. Lett.* **2017**, *12*, 114033. [CrossRef]

23. Sacks, J.D.; Fann, N.; Gumy, S.; Kim, I.; Ruggeri, G.; Mudu, P. Quantifying the public health benefits of reducing air pollution: Critically assessing the features and capabilities of WHO's AirQ+ and U.S. EPA's environmental benefits mapping and analysis program-community edition (BenMAP-CE). *Atmosphere* **2020**, *11*, 516. [CrossRef]
24. Curtin, S.; Spencer, M.R. *Trends in Death Rates in Urban and Rural Areas: United States, 1999–2019*; National Center for Health Statistics: Hyattsville, MD, USA, 2021.
25. Commane, R.; Schiferl, L.D. Climate mitigation policies for cities must consider air quality impacts. *Chem* **2022**, *8*, 910–923. [CrossRef]
26. Castillo, M.D.; Anenberg, S.C.; Chafe, Z.A.; Huxley, R.; Johnson, L.S.; Kheirbek, I.; Malik, M.; Marshall, J.D.; Naidoo, S.; Nelson, M.L.; et al. Quantifying the Health Benefits of Urban Climate Mitigation Actions: Current State of the Epidemiological Evidence and Application in Health Impact Assessments. *Front. Sustain. Cities* **2021**, *3*, 768227. [CrossRef]
27. CDC. *CDC WONDER Online Databases*; U.S. Centers for Disease Control and Prevention: Atlanta, GA, USA. Available online: https://wonder.cdc.gov/ (accessed on 23 May 2023).
28. Chen, J.; Hoek, G. Long-term exposure to PM and all-cause and cause-specific mortality: A systematic review and meta-analysis. *Environ. Int.* **2020**, *143*, 105974. [CrossRef]
29. Turner, M.C.; Jerrett, M.; Pope III, C.A.; Krewski, D.; Gapstur, S.M.; Diver, W.R.; Beckerman, B.S.; Marshall, J.D.; Su, J.; Crouse, D.L. Long-term ozone exposure and mortality in a large prospective study. *Am. J. Respir. Crit. Care Med.* **2016**, *193*, 1134–1142. [CrossRef] [PubMed]
30. Zanobetti, A.; Schwartz, J. Temperature and mortality in nine US cities. *Epidemiology* **2008**, *19*, 563. [CrossRef]
31. Voorhees, A.S.; Fann, N.; Fulcher, C.; Dolwick, P.; Hubbell, B.; Bierwagen, B.; Morefield, P. Climate change-related temperature impacts on warm season heat mortality: A proof-of-concept methodology using BenMAP. *Environ. Sci. Technol.* **2011**, *45*, 1450–1457. [CrossRef] [PubMed]
32. CDC. Age Adjustment. Available online: https://www.cdc.gov/nchs/hus/sources-definitions/age-adjustment.htm (accessed on 22 January, 2025).
33. CDC. Compressed Mortality File. Available online: https://wonder.cdc.gov/wonder/help/cmf.html#Age-Adjusted%20Rates (accessed on 10 February, 2025).
34. Wyatt, L.H.; Peterson, G.C.L.; Wade, T.J.; Neas, L.M.; Rappold, A.G. The contribution of improved air quality to reduced cardiovascular mortality: Declines in socioeconomic differences over time. *Environ. Int.* **2020**, *136*, 105430. [CrossRef] [PubMed]
35. USEPA. *Air Quality System*; U.S. Environmental Protection Agency: Washington, DC, USA. Available online: https://aqs.epa.gov/aqsweb/airdata/download_files.html (accessed on 22 May 2023).
36. Jia, C.; Fu, X.; Bartelli, D.; Smith, L. Insignificant Impact of the "Stay-At-Home" Order on Ambient Air Quality in the Memphis Metropolitan Area, U.S.A. *Atmosphere* **2020**, *11*, 630. [CrossRef]
37. NOAA. *Data Tools: Local Climatological Data (LCD)*; National Oceanic and Atmospheric Administration: Washington, DC, USA. Available online: https://www.ncdc.noaa.gov/cdo-web/datatools/lcd (accessed on 22 May 2023).
38. Madrigano, J.; Jack, D.; Anderson, G.B.; Bell, M.L.; Kinney, P.L. Temperature, ozone, and mortality in urban and non-urban counties in the northeastern United States. *Environ. Health* **2015**, *14*. [CrossRef]
39. Nordio, F.; Zanobetti, A.; Colicino, E.; Kloog, I.; Schwartz, J. Changing patterns of the temperature-mortality association by time and location in the US, and implications for climate change. *Environ. Int.* **2015**, *81*, 80–86. [CrossRef] [PubMed]
40. USEPA. *Estimating PM2.5- and Ozone-Attributable Health Benefits*; U.S. Environmental Protection Agency: Research Triangle Park, NC, USA, 2023.
41. WHO. *WHO Global Air Quality Guidelines: Particulate Matter (PM2.5 and PM10), Ozone, Nitrogen Dioxide, Sulfur Dioxide and Carbon Monoxide*; World Health Organization: Geneva, Switzerland, 2021.
42. Anderson, B.G.; Bell, M.L. Weather-Related Mortality: How Heat, Cold, and Heat Waves Affect Mortality in the United States. *Epidemiology* **2009**, *20*, 205–213. [CrossRef] [PubMed]
43. Li, Y.; Ren, T.; Kinney, P.L.; Joyner, A.; Zhang, W. Projecting future climate change impacts on heat-related mortality in large urban areas in China. *Environ. Res.* **2018**, *163*, 171–185. [CrossRef] [PubMed]
44. Li, Y.; Akkus, C.; Yu, X.; Joyner, A.; Kmet, J.; Sweat, D.; Jia, C. Heatwave events and mortality outcomes in Memphis, Tennessee: Testing effect modification by socioeconomic status and urbanicity. *Int. J. Environ. Res. Public Health* **2019**, *16*, 4568. [CrossRef]
45. GBD 2017 Risk Factor Collaborators. Global, regional, and national comparative risk assessment of 84 behavioural, environmental and occupational, and metabolic risks or clusters of risks for 195 countries and territories, 1990–2017: A systematic analysis for the Global Burden of Disease Study 2017. *Lancet* **2018**, *392*, 1923–1994. [CrossRef]
46. Mokdad, A.H.; Marks, J.S.; Stroup, D.F.; Gerberding, J.L. Actual causes of death in the United States, 2000. *JAMA* **2004**, *291*, 1238–1245. [CrossRef]
47. The US Burden of Disease Collaborators. The State of US Health, 1990-2016: Burden of Diseases, Injuries, and Risk Factors Among US States. *JAMA* **2018**, *319*, 1444–1472. [CrossRef]

48. Thun, M.J.; Jemal, A. How much of the decrease in cancer death rates in the United States is attributable to reductions in tobacco smoking? *Tob. Control* **2006**, *15*, 345–347. [CrossRef] [PubMed]
49. Bowe, B.; Xie, Y.; Yan, Y.; Al-Aly, Z. Burden of Cause-Specific Mortality Associated With PM2.5 Air Pollution in the United States. *JAMA Netw. Open* **2019**, *2*, e1915834. [CrossRef]
50. Wang, J.; Xing, J.; Mathur, R.; Pleim, J.E.; Wang, S.; Hogrefe, C.; Gan, C.-M.; Wong, D.C.; Hao, J. Historical Trends in PM2.5-Related Premature Mortality during 1990-2010 across the Northern Hemisphere. *Environ. Health Perspect.* **2017**, *125*, 400–408. [CrossRef] [PubMed]
51. Zhang, Y.; Jason West, J.; Mathur, R.; Xing, J.; Hogrefe, C.; Roselle, S.J.; Bash, J.O.; Pleim, J.E.; Gan, C.M.; Wong, D.C. Long-term trends in the ambient PM2.5- and O3-related mortality burdens in the United States under emission reductions from 1990 to 2010. *Atmos. Chem. Phys.* **2018**, *18*, 15003–15016. [CrossRef] [PubMed]
52. Castillo, M.D.; Kinney, P.L.; Southerland, V.; Arno, C.A.; Crawford, K.; van Donkelaar, A.; Hammer, M.; Martin, R.V.; Anenberg, S.C. Estimating Intra-Urban Inequities in PM2.5-Attributable Health Impacts: A Case Study for Washington, DC. *GeoHealth* **2021**, *5*, e2021GH000431. [CrossRef] [PubMed]
53. Chalbot, M.C.G.; Jones, T.A.; Kavouras, I.G. Trends of non-accidental, cardiovascular, stroke and lung cancer mortality in Arkansas are associated with ambient PM2.5 reductions. *Int. J. Environ. Res. Public Health* **2014**, *11*, 7442–7455. [CrossRef] [PubMed]
54. Vodonos, A.; Schwartz, J. Estimation of excess mortality due to long-term exposure to PM2.5 in continental United States using a high-spatiotemporal resolution model. *Environ. Res.* **2021**, *196*, 110904. [CrossRef] [PubMed]
55. Bobb, J.F.; Peng, R.D.; Bell, M.L.; Dominici, F. Heat-related mortality and adaptation to heat in the United States. *Environ. Health Perspect.* **2014**, *122*, 811–816. [CrossRef]
56. Requia, W.J.; Coull, B.A.; Koutrakis, P. Where air quality has been impacted by weather changes in the United States over the last 30 years? *Atmos. Environ.* **2020**, *224*, 117360. [CrossRef]
57. Bell, M.L.; Goldberg, R.; Hogrefe, C.; Kinney, P.L.; Knowlton, K.; Lynn, B.; Rosenthal, J.; Rosenzweig, C.; Patz, J.A. Climate change, ambient ozone, and health in 50 US cities. *Clim. Change* **2007**, *82*, 61–76. [CrossRef]
58. Yin, L.; Bai, B.; Zhang, B.; Zhu, Q.; Di, Q.; Requia, W.J.; Schwartz, J.D.; Shi, L.; Liu, P. Climate Penalty on Air Pollution Abated by Anthropogenic Emission Reductions in the United States. *Res. Sq.* **2023**, preprint. [CrossRef]
59. Sheridan, S.C.; Kalkstein, A.J. Seasonal variability in heat-related mortality across the United States. *Nat. Hazards* **2010**, *55*, 291–305. [CrossRef]
60. Wang, H.; Naghavi, M.; Allen, C.; Barber, R.M.; Bhutta, Z.A.; Carter, A.; Casey, D.C.; Charlson, F.J.; Chen, A.Z.; Coates, M.M.; et al. Global, regional, and national life expectancy, all-cause mortality, and cause-specific mortality for 249 causes of death, 1980–2015: A systematic analysis for the Global Burden of Disease Study 2015. *Lancet* **2016**, *388*, 1459–1544. [CrossRef] [PubMed]
61. Barreca, A.; Clay, K.; Deschenes, O.; Greenstone, M.; Shapiro, J.S. Adapting to climate change: The remarkable decline in the US temperature-mortality relationship over the Twentieth Century. *J. Political Econ.* **2016**, *124*, 105–159. [CrossRef]
62. Stafoggia, M.; Michelozzi, P.; Schneider, A.; Armstrong, B.; Scortichini, M.; Rai, M.; Achilleos, S.; Alahmad, B.; Analitis, A.; Åström, C.; et al. Joint effect of heat and air pollution on mortality in 620 cities of 36 countries. *Environ. Int.* **2023**, *181*, 108258. [CrossRef]
63. Ma, Y.; Zang, E.; Opara, I.; Lu, Y.; Krumholz, H.M.; Chen, K. Racial/ethnic disparities in PM2.5-attributable cardiovascular mortality burden in the United States. *Nat. Hum. Behav.* **2023**, *7*, 2074–2083. [CrossRef] [PubMed]
64. Garcia, M.C.; Faul, M.; Massetti, G.; Thomas, C.C.; Hong, Y.; Bauer, U.E.; Iademarco, M.F. Reducing potentially excess deaths from the five leading causes of death in the rural United States. *MMWR Surveill. Summ.* **2017**, *66*, 1–7. [CrossRef] [PubMed]
65. Li, Y.; Odame, E.A.; Silver, K.; Zheng, S. Comparing Urban and Rural Vulnerability to Heat-Related Mortality: A Systematic Review and Meta-analysis. *J. Glob. Epidemiol. Environ. Health* **2017**, *1*, 9–15. [CrossRef]

climate

MDPI

Article

Views of Health Professionals About Climate and Health in Sierra Leone: A Cross-Sectional Study

Isaac S. Sesay and Konstantinos C. Makris *

Cyprus International Institute for Environmental and Public Health, School of Health Sciences, Cyprus University of Technology, Limassol 3041, Cyprus; im.sesay@edu.cut.ac.cy

* Correspondence: konstantinos.makris@cut.ac.cy; Tel.: +357-25002398

Abstract: Climate change presents one of the biggest global threats to society, while the impact of its manifestations on human health has been poorly characterized and quantified, especially in middle- and low-income countries. The perceptual views of health professionals about the climate and health nexus are critical for the effective implementation of climate policies. The Sierra Leone health professionals are no exception to this, and no such data exist for their country. To this extent, we distributed a cross-sectional survey to understand the perceptual views and beliefs of health professionals in Sierra Leone about the climate and health nexus. A validated international questionnaire on the topic was electronically administered to 265 participants. A descriptive analysis of the survey responses was conducted. Results showed that almost all of the respondents (97%) felt that climate change is an important issue; more than half (68%) of them were very worried about climate change, and 28% were somewhat worried. About half of respondents believed that human activities mostly caused climate change, while 40% of health professionals felt this was equally caused by human activities and natural changes in the environment. The need to engage health professionals with the public and policymakers to bring the health effects of climate change to their attention was particularly highlighted; however, most respondents (81%) stated that numerous barriers impede them from doing so. The most widely reported barriers and needs were the need for training to communicate effectively about climate change and health (96%) and guidance on creating sustainable workplaces (94%), followed by the need for lifelong training and education programs on climate and health, and the lack of time (73%). These survey findings would be valuable to policymakers in Sierra Leone and the broader African regions towards mitigating and adapting to climate change threats to human health.

Keywords: health professionals; climate change; public health; policy advocacy; barriers; perceptions; views

1. Introduction

Sierra Leone is a country in West Africa with a population of ~7.65 million [1]. The country's climate is tropical with two distinct seasons—the rainy season (May to November) and the dry season (December to April). Average temperatures range between 25 and 27 °C and between 22 and 25 °C during the dry and rainy seasons, respectively [2]. In the World Risk Index 2021, Sierra Leone is ranked 47 among 181 countries with high scores for vulnerability to, susceptibility to, and absence of coping capacities for climate change [3]. Furthermore, the country is among the most vulnerable African countries in terms of the increasing frequency of climate change impacts; the Intergovernmental Panel on Climate Change (IPCC) ranked Sierra Leone the third most vulnerable to climate change in the world, after Bangladesh and Guinea Bissau [3].

Climate manifestations heavily impact Sierra Leone and its people. In August 2017, a 6-kilometer mudslide and flooding occurred in the country's capital city, Freetown, and its outskirts—the worst natural disaster in modern history; over 1000 people were reported dead or missing, over 11,000 were displaced, and over 300 buildings were submerged [3,4]. In 2018, ~2.2 million people were adversely impacted by destructive cyclones, resulting in severe flooding, high death tolls, and forced migration out of their countries (Malawi, Mozambique, and Zimbabwe) [5].

Evidence suggests that the current climate change crisis could undermine public health gains made during the past 2–3 decades in tackling major causes of mortality and morbidity burden of disease. The changing weather and abnormal environmental patterns affect proximal agents of disease, such as the transmission patterns of various pathogens in the environment. The number of months with environmentally suitable conditions for malaria (*Plasmodium falciparum*) transmission has increased by 39% from the 1950s to 2019 in highland areas of low human development index countries [6]. Similar increases in *Vibrio cholerae* transmission rates were observed in coastal areas between 2003 and 2019 with 98% of these coastal areas located in low human development index countries [6]. Studies suggest that climate change directly influences the etiology of vector-borne diseases, particularly malaria and dengue fever [7]. Prediction models for the period 2020–2040 suggest that increased air temperatures are anticipated to adversely affect Africa's malaria burden [8]. Approximately 34% of the global disability-adjusted life years (DALYs) presented by the adverse effects of climate change are in sub-Saharan Africa [9].

Climate change may be associated with non-communicable diseases and their risk factors as well. According to Charlson et al. [10], climate change manifestations were associated with psychological distress and deteriorating mental health, i.e., anxiety, post-traumatic stress, and depression [11,12]. A systematic review found that exposures to heat and air pollution were linked with severe birth effects such as pre-term birth and low birth weight [13]. Climate change stressors bring about hostile pregnancy results, influencing the development of offspring [14]. Additionally, climate change is associated with respiratory difficulties, deteriorating asthma, and other lung diseases [15]. Studies have shown that climate change can significantly increase chronic malnutrition among under-five children and stunting cases, especially in the most vulnerable countries [16,17].

Climate change presents considerable difficulties for global food security. This affects farming routines, agricultural productivity, and the dietary quality of food produced. Emerging and classic plant pests are affected by climate change manifestations, becoming increasingly important in food production [18,19]. Heatwaves, humidity, droughts, uncontrollable wildfires, and torrential flooding [10] would potentially affect food security. The Intergovernmental Panel on Climate Change (IPCC) reports predicted declining agricultural productivity in sub-Saharan Africa from 21% to 9% by 2080 because of climate change. The increasing high temperature will reduce the production of food eaten by many up to 50%. Approximately two-thirds of the continent's agricultural productive lands are expected to be lost by 2025 due to a lack of rainfall and drought [20]. Globally, around 700 million and 832 million people were food insecure in 2019 and 2020, respectively, and this number is projected to reach over a billion by 2050 due to climate change [19].

It is well appreciated that health professionals hold major responsibility in assessing and managing the human health risks associated with climatic manifestations. The health of patients, families, communities, and the population at large is already impacted as a result of changing weather patterns [21]. The public looks up to health professionals for information regarding their health, including information about climate change impacts on individual health [22]. However, evidence supporting health professionals' preparedness and literacy regarding their effective engagement with climate change remains unclear across many parts of the world [23,24].

In Sierra Leone, to the best of our knowledge, no study has been published about the perceptual views of health professionals on climate change and the human health nexus. Sierra Leoneans trust health professionals for information regarding their health and other related issues. The planetary health theoretical framework and its practical applications throughout the globe were underpinned here to describe linkages and systems jointly operating between proximate environmental agents of human disease, global ecological drivers such as climate change, and human health outcomes [25]. Hence, this is the first-ever survey on the perceptual views of health professionals in Sierra Leone about the climate and health nexus, which serves as a crucial baseline assessment to design and implement climate policies. Thus, this study's objectives were to (i) document the perceptual views of

Sierra Leone health professionals about the climate and health nexus and (ii) document the extent to which they support action to address the climate crisis.

2. Materials and Methods

The study population was a random cross-section of health professionals in Sierra Leone, including medical doctors, nurses, public health professionals (individuals with public health qualifications), midwives, and other allied health professionals. The validated international health survey on health professionals' views about climate change and health was used here [26], with written permission from Prof. J. Kotcher. The survey was administered in English (as this is the official language in the country) and online, using REDCap from July to November 2023. To facilitate cultural adaptation, the questionnaire was pre-tested with twenty selected health professionals in Sierra Leone. A total of 265 volunteers responded partially to the survey and 215 (81%) complete responses were recorded. Data were managed using the REDCap electronic data capture tools. The questionnaire used a Likert scale which can be found in Supplementary Materials. The Sierra Leone Ethics and Scientific Review Committee granted ethical approval (SLESRC #019/04/2023) to the study.

The survey was distributed via emails and via select professional WhatsApp platforms (Health Promotion Network, Malaria Fighters, and One Health Comms Team) of the Ministry of Health and Sanitation (MoHS) in Sierra Leone. The inclusion criteria were (i) >18 years of age and (ii) actively working in Sierra Leone as a health professional. A synopsis of the research was provided to a representative of the MoHS who distributed the survey link to the various WhatsApp groups of the Ministry of Health. Those who responded positively received the survey electronic link with instructions. Follow-up messages on WhatsApp and emails were sent as a reminder within two weeks of the initial email/WhatsApp dispatch.

The survey dataset was divided and analyzed by sub-sections with unweighted data: understanding key facts on the issue and engagement; risk perceptions of the changing climate as a threat to health; engagement with the public and policymaker on the issue; professional bodies' policies related to the issue; and barriers to engagement and helpful resources in addressing the issue [26]. The incomplete questionnaires (19%) were not included in the main analysis [27]. Descriptive statistics included means and standard deviations (SDs) for continuous variables or sample sizes and percentages for categorical variables. Statistical analysis was carried out using the R app version 4.2.2 for Windows.

3. Results

About 65% of survey participants were men, while 54% of participants were aged between 30 and 44 years old, 30% were between 45 and 59 years old, 10% were between 18 and 29, and 6% were >60 years of age (Table 1). Almost half (46%) of participants described their primary occupation as public health professionals, whereas 25% were medical doctors, with nurses comprising 16%, midwives 5%, and other allied health professionals 9%.

Table 1. Study population characteristics.

	n%		
Category of Health Professionals	Medical Doctors	53	25
	Nurses	34	16
	Midwives	10	5
	Public Health Practitioners	99	46
	Others (e.g., Pharmacist, District Social Mobilization Coordinator, Lab Technician)	19	9
Age (in Years)	18–29	22	10
	30–44	116	54
	45–59	64	30
	60+	13	6.0
Gender	Male	140	65
	Female	75	35

3.1. Understanding Human-Caused Climate Crisis and Its Importance

The vast majority (98%) of participating health professionals in Sierra Leone responded that climate change is happening (Figures S2 and S3). A total of 92% of respondents felt sure that climate change is happening (38% were extremely sure; 54% were very sure). Responding to the causes of the climate crisis, 46% of respondents believed that human activities mostly caused climate change, while a considerable portion of respondents (39%) felt that climate change was equally caused by human activities and natural changes in the environment; only 8% reported climate crisis is entirely caused by human activities (Figure 1).

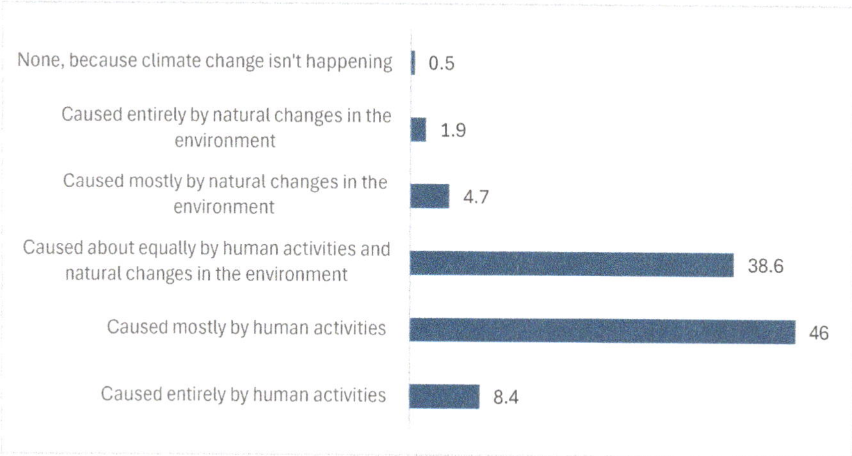

Figure 1. Views of health professionals in Sierra Leone about the causes of climate change (Q: Assuming climate change is happening, do you think it is?).

The majority (60.5%) reported that the climate crisis is very important to them personally, 23.3% reported it is extremely important, and 13.5% indicated it is somewhat important (Figure 2). About 68% were very worried about climate change, 28.4% were somewhat worried, and 3.7% were not too worried (Figure S7).

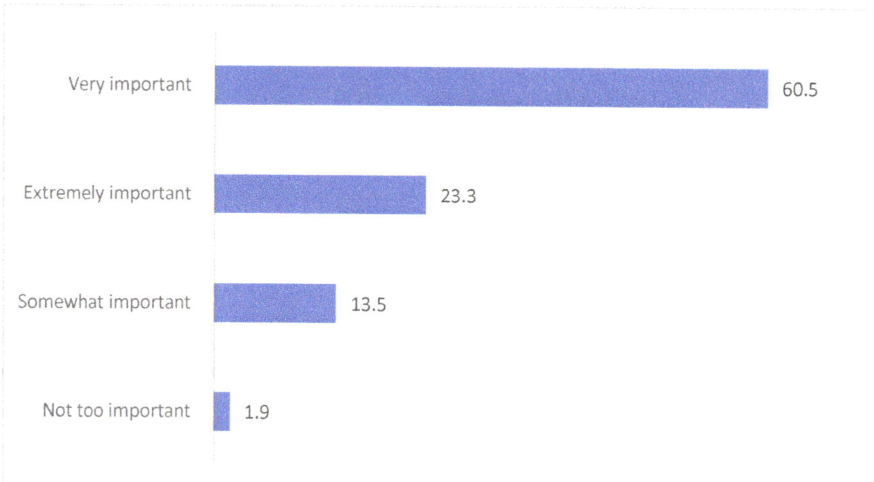

Figure 2. Views of health professionals in Sierra Leone about the importance of climate change (Q: How important is the issue of climate change to you personally?).

3.2. Risk Perceptions and the Impact of the Climate Crisis on Human Health

About 72% of respondents reported the climate change crisis would cause harm to them personally, their patients (54%), people in their community (84%), people in Sierra Leone (88%), and future generations (87%) at a moderate amount or a great deal (Figure S8). When asked about specific health effects, most of the survey respondents stated that the climatic crisis has already negatively affected the health of people in Sierra Leone, either in a moderate amount or a great deal (Figure 3). Approximately 52% of participants reported heat-related illnesses, 71% reported physical or mental harm from storms and floods, 58% reported physical or mental harm from forest fires or bushfires, 72% reported vector-borne infectious diseases, 75% reported water-borne and food-borne diseases, 51% reported housing loss for people displaced by extreme weather events, 75% reported hunger and malnutrition as a result of rising food prices, and 75% reported higher poverty risk due to economic hardship and resulting health problems.

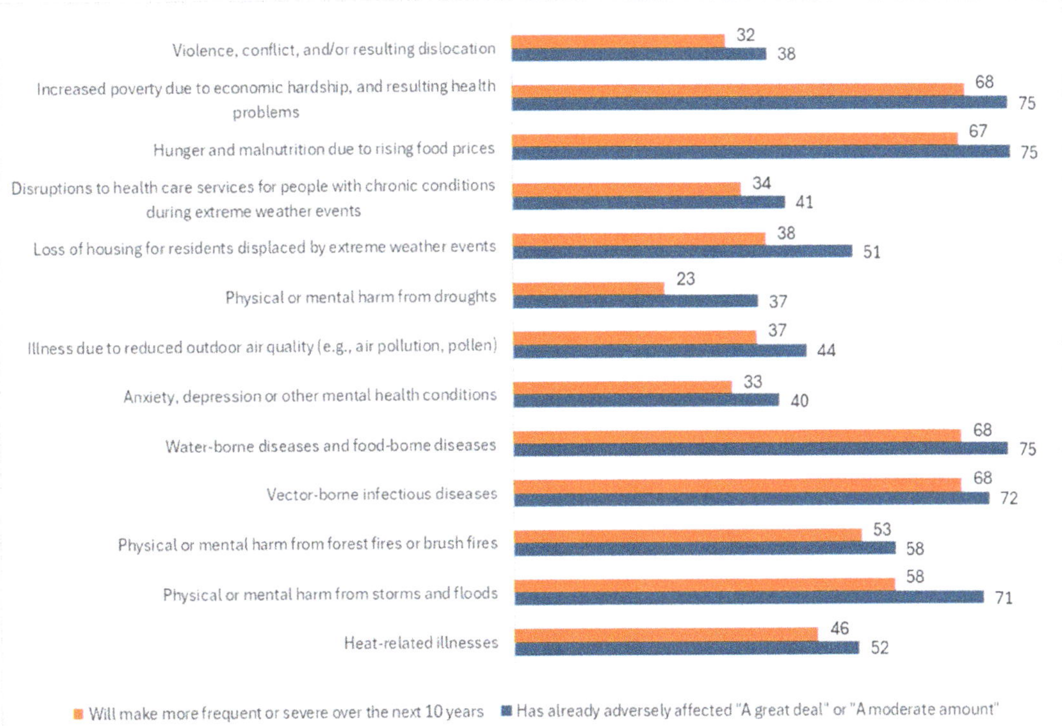

Figure 3. Views of health professionals in Sierra Leone about the adverse effect of climate change on human health (Qs: How much, if at all, has climate change already adversely affected these health issues in Sierra Leone? Over the next 10 years, do you think climate change will make these health issues in Sierra Leone more frequent or severe, less frequent or severe, or do you think they will remain largely unchanged?).

About half of the survey respondents acknowledged other adverse health consequences due to the changing weather patterns, including anxiety, depression, or other mental health conditions (40%), illness due to reduced outdoor air quality (e.g., air pollution or pollen) (44%), physical or mental harm from droughts (37%), disruptions to health care services for people with chronic conditions during extreme weather events (41%), and violence, conflict, and resulting dislocation (38%).

Participants indicated that climate change will greatly influence human health outcomes more frequently or severely in Sierra Leone for the next decade with the most common being vector-borne infectious diseases (68%), water- and food-borne diseases (68%), increased poverty due to economic hardship and resulting health problems (68%), hunger and malnutrition due to rising food prices (67%), physical or mental harm from storms and floods (58%), physical or mental harm from forest fires or bush fires (53%), and health-related illnesses (46%) (Figure 3).

3.3. Engagement with the Public and Policymakers

The vast majority of study participants (88% strongly or somewhat agreed) felt a responsibility to bring the health effects of the changing climate to the attention of the public (Figure S16) and policymakers (Figure S17). Half of them strongly agreed (46.5%) and 41.4% somewhat agreed with actively encouraging national leaders to strengthen national commitment towards lowering emissions to achieve the Paris Climate Agreement. Similarly, 44.7% of respondents strongly agreed and 43.3% somewhat agreed with actively encouraging all world leaders to strengthen national commitment towards lowering emissions (Figure S21). Almost all participants (96%) believed that health organizations should advocate for increased investments in the health sector (Figure S1). Responding on their willingness to participate in a global advocacy campaign to encourage world leaders to prioritize their commitments, 56% said they are personally willing to participate, 36% are possibly willing to participate, but they need more information on the issue, 7% of participants think they would support such a campaign, but they might not personally participate, and 0.5% of respondents felt that they would not participate, nor support such a campaign (Figure 4).

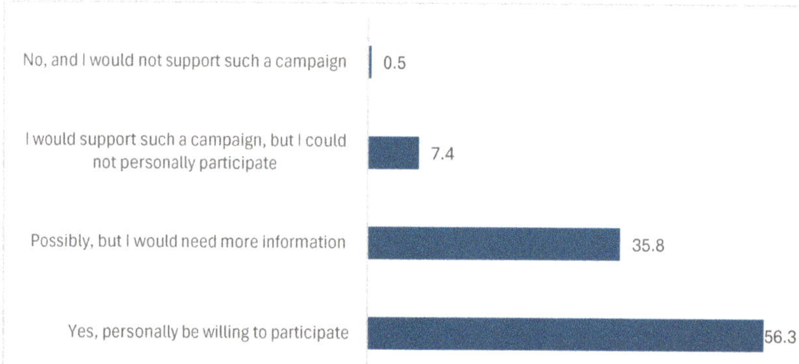

Figure 4. Views of health professionals in Sierra Leone about their willingness to participate in climate change global advocacy campaigns (Q: Would you personally be willing to participate in a global advocacy campaign by health professionals to encourage all world leaders to strengthen their commitment to achieving the goal of the Paris Climate Agreement?).

3.4. Policies of Health Professional Society/Association

Responses from survey participants on the question about policies of their professional society/association indicated that 13% of survey participants strongly agreed, 37% somewhat agreed, and 36% neither agreed nor disagreed that the organizations they belong to should cut any ties that they may have with companies producing fossil fuel (Figure S18). This includes divestment of any fossil fuel reserves or bonds. Furthermore, of the total survey participants, 21% strongly agreed, 38% somewhat agreed, and 31% neither agreed nor disagreed that their professional entities or associations should offer opportunities for members to partake in online sessions like seminars and conferences held by their society or association to decrease emissions (Figure S19).

3.5. Barriers to Public Engagement and Useful Resources

A total of 73% and 50% indicated insufficient time and lack of knowledge, respectively, influencing their willingness (from a moderate amount or a great deal) to engage and communicate with the public about the climate crisis (Figure 5); other reasons included beliefs that such efforts would be futile (47%), controversy of the topic (33%), no support from peers (26%), and too risky for them professionally or personally (13%). Most survey respondents (83%) stated having little formal education on climate change and health (Figure S23).

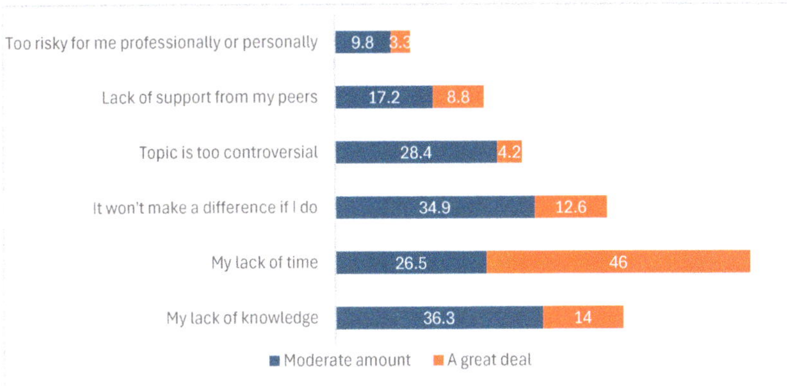

Figure 5. Views of health professionals in Sierra Leone about the barriers to engaging the public on climate change and health issues (Q: To what extent do these factors reduce your willingness to communicate with the public about climate change and health?).

Participants expressed different views on the resources to address climate change and health (Figure 6): training for communicating the science of climate and health (70% very helpful, 26% moderately helpful), direction or guidance on how to ensure a sustainable workplace (67% very helpful, 27% moderately helpful), continuous professional education on climate and health (58% very helpful, 31% moderately helpful), early notification mechanism (action alerts) to engage and advocate with policymakers (53% very helpful), the issuing of relevant policy statements by the MoHS or the Sierra Leone Medical and Dental Association (47% very helpful), and patient education materials (41% very helpful).

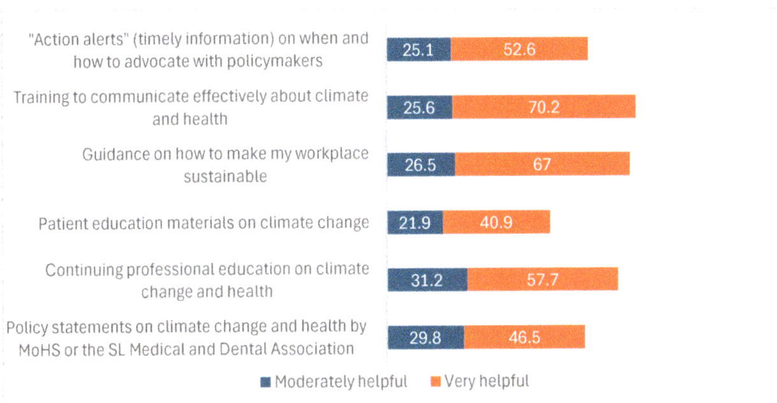

Figure 6. Views of health professionals in Sierra Leone about helpful resources to address the problem of climate change and health (Q: How helpful, if at all, would the following resources be to you?).

4. Discussion

To the best of our knowledge, this is the first survey in Sierra Leone on the views of health professionals about the changing climate and its effects on human health. Health professionals reported that a median of 75% (range of 39–100%) of climate scientists would think that human-caused climate change is happening. Interestingly, about half of participating health professionals in Sierra Leone believed that human activities mostly caused climate change, while about 40% felt this was equally caused by human activities and natural changes in the environment. This result highlights the mixed views of health professionals in Sierra Leone about the causes of climate change.

A large number (96%) of Sierra Leone's healthcare workers were worried that the effects of climate change would impact them personally, their patients, people in their community, people in Sierra Leone, and future generations. This possibly echoes concerns about the country's high vulnerability to climate change and low readiness to mitigate and adapt to climate change. Records of numerous weather events, like extreme heat, rising sea levels, recurring storms, floods, bushfires, coastal erosion, mudslides, etc., threaten Sierra Leone's food security, health and infrastructure [28,29]. The country's economy is dependent on natural resources and fragile ecosystems, which present with low agricultural productivity indices, while they have repeatedly suffered from environmental disasters, leaving many communities vulnerable to hunger and malnutrition [30,31].

Furthermore, communicable and non-communicable diseases (NCDs) have a strong grip on the country. In 2019 alone, 34% of all-cause mortality was attributed to NCD [32,33]. The country's vulnerability to climatic, environmental, and energy-related changes has been associated with increasing rates of asthma and other respiratory diseases [2]. Studies have revealed that increasing climatic conditions are associated with heightened episodes of diarrheal diseases, water-borne diseases, seafood poisoning, dangerous pollutants, and a rise in malaria infections [30]. Malaria is one of the major health challenges in the country responsible for most of the out-patient visits and in-patient admissions, especially among children under five [34]. Consistent with previous studies [25,35], health professionals in Sierra Leone agreed that climate change has already negatively impacted the country and its population in numerous ways, such as affecting food security, hunger, and malnutrition due to rising food prices. Survey participants felt that the climate crisis would increase the incidence of NCDs over the next decade in Sierra Leone [2].

Health professionals in Sierra Leone strongly felt the responsibility to bring climate and health issues to the attention of policymakers and the public/patients, similar to surveys conducted among healthcare professionals in the Caribbean [25,35]. This would include health education/promotion activities, with 56% of the respondents willing to participate in advocacy campaigns, both at the national and global levels [36]. The need to communicate with the public about the climate and health nexus was also brought forward. However, several barriers and needs were recorded among the health professionals that would prevent them from doing so, such as lack of time or knowledge on the topic. Interestingly, about half of them (48%) reported that such efforts would not make a difference. Similarly, health professionals in the Caribbean (53%) mentioned a lack of time, 45% mentioned a lack of knowledge, and 36% said it will not make a difference if they do [25,35]. In April 2024, MoHS launched the country's climate health unit as a continuing commitment to adapting and mitigating the adverse effects of the changing climate on human health and survival [37]. This is a positive move as it will provide interested health professionals with opportunities to collaborate with the unit, engage in advocacy, and educate the public on the issue. The climate health unit in collaboration with health professional associations in the country should lead in the production of resources mentioned by the survey participants to support health professionals in implementing sustainable practices.

5. Conclusions

The views of health professionals highlighted in this research represent a cross-section of health workers in Sierra Leone. Health professionals in the country are worried about the climate crisis and its impact on human health. Regarding health professionals' involvement, provision of resources, and addressing of the barriers that prevent health professionals from being more actively involved with the climate and health nexus, a call for action is needed by regulatory agencies and policymakers. The recently formed unit on climate and health within the Ministry of Health and Sanitation in Sierra Leone should closely collaborate with health professionals to accelerate the implementation of climate and environment strategies for the country and the broader African region.

It is suggested that health professional societies or organizations, non-governmental organizations, and the Government of Sierra Leone should work together to empower health professionals through structured professional training programs in climate and health, while strengthening ties with stakeholders, such as NGOs, international organizations, and the public in terms of enhanced adaptation and mitigation to climate change.

6. Limitations and Future Directions

This study took place during the country's presidential and parliamentary general elections; however, a large proportion of the study sample participated in the study (81%), and responses by health professionals originated from different parts of the country. The online data collection method was a challenge for many participants to complete the survey due to poor or no access to internet data. This made it difficult to obtain data from remote areas of the country, which is a limitation. The data presented in this study were a representation of survey participants from cities and towns across the country where internet access is less problematic.

This baseline assessment of health professionals' views on climate and health should act as the guide for the design and implementation of future health education and health promotion programs for health professionals and also for the general public.

Supplementary Materials: The following supporting information can be downloaded at https://www.mdpi.com/article/10.3390/cli12120216/s1: All associated figures for each item of the questionnaire used in this study.

Author Contributions: Conceptualization, K.C.M.; methodology, K.C.M. and I.S.S.; validation, K.C.M.; formal analysis, I.S.S.; investigation, I.S.S.; resources, K.C.M.; data curation, I.S.S.; writing—original draft preparation, I.S.S.; writing—review and editing, I.S.S. and K.C.M.; supervision, K.C.M.; project administration, K.C.M. All authors have read and agreed to the published version of the manuscript.

Funding: This research received no external funding.

Data Availability Statement: Interested parties can approach the corresponding author for data-sharing options.

Acknowledgments: The authors would like to thank John Kotcher for the permission to use the survey. IS dearly appreciates the assistance of relatives, friends, and colleagues in Sierra Leone. We extend special thanks to Snr. Wani of NMCP for giving us permission to use the MoHS platform to disseminate the survey. Finally, I am greatly indebted to my brothers Hon. Abdulai Daniel Sesay and Solomon Sulay Sesay for their financial and moral support and to my beloved wife and daughter for the joy, smiles, moral support, and constant encouragement they provided me with throughout this project.

Conflicts of Interest: The authors declare no conflicts of interest. The funders had no role in the design of the study; in the collection, analyses, or interpretation of the data; in the writing of the manuscript; or in the decision to publish the results.

References

1. The World Bank. Sierra Leone Programmatic Public Expenditure Reviews 2021: Improving Quality of Public Expenditure in Health. 2021. In *Macroeconomics Trade and Investment Global Practice, World Bank Group*; The World Bank Group: Washington, DC, USA, 2021.
2. UNICEF. *Climate Landscape Analysis for Children (CLAC) Report 2023*; UNICEF: Freetown, Sierra Leone, 2023. Available online: https://www.unicef.org/sierraleone/reports/climate-landscape-analysis-children-clac (accessed on 27 August 2024).

3. African Development Bank. Country Focus Report 2022-Supporting Climate Resilience and a Just Energy Transition in Sierra Leone. African Development Bank Group. Available online: https://www.afdb.org/en/documents/country-focus-report-2022-supporting-climate-resilience-and-just-energy-transition-sierra-leone (accessed on 27 August 2024).
4. Musoke, R.; Chimbaru, A.; Jambai, A.; Njuguna, C.; Kayita, J.; Bunn, J.; Latt, A.; Yao, M.; Yoti, Z.; Yahaya, A.; et al. A Public Health Response to a Mudslide in Freetown, Sierra Leone, 2017: Lessons Learnt. *Disaster Med. Public Health Prep.* **2020**, *14*, 256–264. [CrossRef] [PubMed]
5. Atwoli, L.; Erhabor, G.E.; Gbakima, A.A.; Haileamlak, A.; Kayembe Ntumba, J.-M.; Kigera, J.; Laybourn-Langton, L.; Mash, B.; Muhia, J.; Mulaudzi, F.M.; et al. COP27 Climate Change Conference: Urgent action needed for Africa and the world. *Rheumatol. Adv. Pract.* **2022**, *6*, rkac082. [CrossRef] [PubMed]
6. Romanello, M.; McGushin, A.; Napoli, C.D.; Drummond, P.; Hughes, N.; Jamart, L.; Kennard, H.; Lampard, P.; Rodriguez, B.S.; Arnell, N.; et al. The 2021 report of the Lancet Countdown on health and climate change: Code red for a healthy future. *Lancet* **2021**, *398*, 1619–1662. [CrossRef] [PubMed]
7. Kulkarni, M.A.; Duguay, C.; Ost, K. Charting the evidence for climate change impacts on the global spread of malaria and dengue and adaptive responses: A scoping review of reviews. *Glob. Health* **2022**, *18*, 1. [CrossRef]
8. Li, C.; Managi, S. Global malaria infection risk from climate change. *Environ. Res.* **2022**, *214*, 114028. [CrossRef]
9. Opoku, S.K.; Leal Filho, W.; Hubert, F.; Adejumo, O. Climate Change and Health Preparedness in Africa: Analysing Trends in Six African Countries. *Int. J. Environ. Res. Public Health* **2021**, *18*, 4672. [CrossRef]
10. Charlson, F.; Ali, S.; Benmarhnia, T.; Pearl, M.; Massazza, A.; Augustinavicius, J.; Scott, J.G. Climate Change and Mental Health: A Scoping Review. *Int. J. Environ. Res. Public Health* **2021**, *18*, 4486. [CrossRef]
11. Palinkas, L.A.; Wong, M. Global climate change and mental health. *Curr. Opin. Psychol.* **2020**, *32*, 12–16. [CrossRef]
12. Sharpe, I.; Davison, C.M. Climate change, climate-related disasters and mental disorder in low- and middle-income countries: A scoping review. *BMJ Open* **2021**, *11*, e051908. [CrossRef]
13. Agache, I.; Sampath, V.; Aguilera, J.; Akdis, C.A.; Akdis, M.; Barry, M.; Bouagnon, A.; Chinthrajah, S.; Collins, W.; Dulitzki, C.; et al. Climate change and global health: A call to more research and more action. *Allergy* **2022**, *77*, 1389–1407. [CrossRef]
14. Olson, D.M.; Metz, G.A.S. Climate change is a major stressor causing poor pregnancy outcomes and child development. *F1000Research* **2020**, *9*, 1222. [CrossRef] [PubMed]
15. Nilsson, M.; Sie, A.; Muindi, K.; Bunker, A.; Ingole, V.; Ebi, K.L. Weather, climate, and climate change research to protect human health in sub-Saharan Africa and South Asia. *Glob. Health Action* **2021**, *14* (Suppl. 1), 1984014. [CrossRef] [PubMed]
16. Mirzabaev, A.; Bezner Kerr, R.; Hasegawa, T.; Pradhan, P.; Wreford, A.; Cristina Tirado von der Pahlen, M.; Gurney-Smith, H. Severe climate change risks to food security and nutrition. *Clim. Risk Manag.* **2023**, *39*, 100473. [CrossRef]
17. Sorgho, R.; Quiñonez, C.A.M.; Louis, V.R.; Winkler, V.; Dambach, P.; Sauerborn, R.; Horstick, O. Climate Change Policies in 16 West African Countries: A Systematic Review of Adaptation with a Focus on Agriculture, Food Security, and Nutrition. *Int. J. Environ. Res. Public Health* **2020**, *17*, 8897. [CrossRef]
18. European Food Safety Authority (EFSA); Maggiore, A.; Afonso, A.; Barrucci, F.; Sanctis, G.D. Climate change as a driver of emerging risks for food and feed safety, plant, animal health and nutritional quality. *EFSA Support. Publ.* **2020**, *17*, 1881E. [CrossRef]
19. Muzerengi, F.; Gandidzanwa, C.P.; Chirubvu, L. Impacts of climate change on household food security in Matande communal lands, Mwenezi district in Zimbabwe. *Jamba* **2023**, *15*, 1499. [CrossRef]
20. Masipa, T.S. The impact of climate change on food security in South Africa: Current realities and challenges ahead. *Jamba* **2017**, *9*, 411. [CrossRef]
21. McKinnon, S.; Breakey, S.; Fanuele, J.R.; Kelly, D.E.; Eddy, E.Z.; Tarbet, A.; Nicholas, P.K.; Ros, A.M.V. Roles of health professionals in addressing health consequences of climate change in interprofessional education: A scoping review. *J. Clim. Change Health* **2022**, *5*, 100086. [CrossRef]
22. Kreslake, J.M.; Sarfaty, M.; Roser-Renouf, C.; Leiserowitz, A.A.; Maibach, E.W. The Critical Roles of Health Professionals in Climate Change Prevention and Preparedness. *Am. J. Public Health* **2018**, *108*, S68–S69. [CrossRef]
23. Yang, L.; Liu, C.; Hess, J.; Phung, D.; Huang, C. Health professionals in a changing climate: Protocol for a scoping review. *BMJ Open* **2019**, *9*, e024451. [CrossRef]
24. Dupraz, J.; Burnand, B. Role of Health Professionals Regarding the Impact of Climate Change on Health—An Exploratory Review. *Int. J. Environ. Res. Public Health* **2021**, *18*, 3222. [CrossRef] [PubMed]
25. Myers, S.S. Planetary health: Protecting human health on a rapidly changing planet. *Lancet* **2017**, *390*, 2860–2868. [CrossRef] [PubMed]
26. Kotcher, J.; Maibach, E.; Miller, J.; Campbell, E.; Alqodmani, L.; Maiero, M.; Wyns, A. Views of health professionals on climate change and health: A multinational survey study. *Lancet Planet. Health* **2021**, *5*, e316–e323. [CrossRef] [PubMed]
27. Aste, M.; Boninsegna, M.; Freno, A.; Trentin, E. Techniques for dealing with incomplete data: A tutorial and survey. *Pattern Anal. Appl.* **2015**, *18*, 1–29. [CrossRef]
28. Margaret, E. AD725: Sierra Leoneans Demand Urgent Climate Action from Government and Other Stakeholders. Available online: https://www.afrobarometer.org/publication/ad725-sierra-leoneans-demand-urgent-climate-action-from-government-and-other-stakeholders/ (accessed on 12 March 2024).

29. The World Bank. Sustainable Cities: Urban Areas and Climate Change in Sierra Leone | PreventionWeb. Available online: https://www.preventionweb.net/publication/sustainable-cities-urban-areas-and-climate-change-sierra-leone (accessed on 12 March 2024).
30. Government of Sierra Leone. *National Adaptation Plan (NAP) 2021*; Government of Sierra Leone: Freetown, Sierra Leone, 2021.
31. Government of Sierra Leone. *National Climate Change Policy, Final 2021*; Government of Sierra Leone: Freetown, Sierra Leone, 2021.
32. WHO. Country Disease Outlook. Sierra Leone. World Health Organisation, African Region, August 2023. Available online: https://www.afro.who.int/sites/default/files/2023-08/Sierra%20Leone.pdf (accessed on 7 July 2024).
33. Institute for Health Metrics and Evaluation; Human Development Network; The World Bank. *The Global Burden of Disease: Generating Evidence, Guiding Policy—Sub-Saharan Africa Regional Edition*; IHME: Seattle, WA, USA, 2013. Available online: https://www.healthdata.org/sites/default/files/files/data_for_download/2013/WorldBank_SubSaharanAfrica/IHME_GBD_WorldBank_SubSaharanAfrica_FullReport.pdf (accessed on 19 March 2024).
34. National Malaria Control Programme (NMCP) [Sierra Leone]; Statistics Sierra Leone; University of Sierra Leone; Catholic Relief Services; Utica International. *Sierra Leone Malaria Indicator Survey 2021*; NMCP, SSL, CRS, and Utica International: Freetown, Sierra Leone, 2021. Available online: https://malariasurveys.org/documents/SL%20MIS%20final%20report-4.3.2022.pdf (accessed on 10 January 2024).
35. De Freitas, L.; Bahadursingh, S.; Basdeo, D.; Kotcher, J.; Hospedales, J. Caribbean health professional views on climate change and health. *J. Clim. Change Health* **2023**, *12*, 100248. [CrossRef]
36. Maibach, E.; Sarfaty, M.; Gould, R.; Damle, N.; Armstrong, F. A Call to Action by Health Professionals. In *Health of People, Health of Planet and Our Responsibility: Climate Change, Air Pollution and Health*; Al-Delaimy, W.K., Ramanathan, V., Sánchez Sorondo, M., Eds.; Springer International Publishing: Cham, Switzerland, 2020; pp. 395–405. ISBN 978-3-030-31125-4.
37. MoHS. The Ministry of Health launches Sierra Leone's First Climate-Health Unit as the Country Pledges to go Green for Health. Ministry of Health and Sanitation. Available online: https://mohs.gov.sl/the-ministry-of-health-launches-sierra-leones-first-climate-health-unit-as-the-country-pledges-to-go-green-for-health/ (accessed on 29 August 2024).

climate

MDPI

Article

Climatology and Long-Term Trends in Population Exposure to Urban Heat Stress Considering Variable Demographic and Thermo–Physiological Attributes

Christos Giannaros [1,2,*], Elissavet Galanaki [3] and Ilias Agathangelidis [1]

[1] Department of Physics, National and Kapodistrian University of Athens, 15784 Athens, Greece; iliasaga@phys.uoa.gr
[2] Institute for Environmental Research and Sustainable Development, National Observatory of Athens, 15236 Athens, Greece
[3] Research Support Directorate, National Observatory of Athens, 15236 Athens, Greece; galanaki@noa.gr
* Correspondence: cgiannaros@phys.uoa.gr

Abstract: Previous studies assessing population exposure to heat stress have focused primarily on environmental heat loads without accounting for variations in human thermo–physiological responses to heat. A novel 30-year (1991–2020) human thermal bioclimate dataset, consisting of hourly mPET (modified physiologically equivalent temperature) values for diverse populations, was employed in the present study to assist in addressing this gap. Focusing on the Athens urban area (AUA), Greece, the climatology and long-term trends in acclimatization-based strong heat stress (accliSHS) experienced by average male and female adult and senior individuals during the warm period of the year (April–October) were investigated. Results showed that an average adult (senior) in AUA experienced, on average, approximately 13 (18) additional days with at least 1 h accliSHS in 2020 compared with 1991. The increasing rates per year were particularly pronounced for days with $\geq$6 h accliSHS, indicating a rise in the daily duration of heat stress in AUA from 1991 to 2020. Combining the variations in climate and demographics in AUA during the examined 30-year period, the long-term trends in $\geq$1 h accliSHS exposure for the study population types were further examined. This analysis revealed that seniors' exposure to $\geq$1 h accliSHS in AUA increased by up to +153,000 person-days $\times$ year^{-1} from 1991 to 2020. Increasing population aging was the main driver of this outcome, highlighting the urgent need for heat–health action planning in Greece.

Keywords: climatology; trend analysis; human thermal bioclimate; heat stress; modified physiologically equivalent temperature; Athens urban area; Greece

1. Introduction

Anthropogenic-driven global warming and the consequent intensification of weather and climate extremes induce severe impacts on humans [1]. One of the most important climate-related threats is related to the thermal environment. Population exposure to elevated heat conditions endangers human livability and survivability, as hot weather and heatwaves are associated with adverse birth outcomes [2,3], worsened mental health [4] and sleep quality [5], limited ability to exercise, play, and work [6–9], and increased morbidity and mortality, primarily due to cardiorespiratory diseases [10,11]. These heat-related issues in turn cause a substantial economic burden on healthcare systems and various labor sectors, especially those heavily relying on an outdoor workforce [12,13].

The spatiotemporal assessment of heat stress exposure patterns in the recent past [14–22] and future [23–32] can provide essential guidance for targeted adaptation and action aimed at alleviating the above impacts, such as the development of early warning systems. However, previous relevant studies have not accounted for exposure to heat stress within a human–biometeorological and health-related context. Single meteorological variables (e.g., air temperature [19]) or simple composite indices (e.g., heat index (HI) [31]) were primarily used in these studies, even though the comprehensive assessment of heat stress requires considering the collective impact of all heat-related environmental drivers (i.e., temperature, humidity, wind speed, and radiation loads). Most importantly, it requires solving the human energy balance equation, integrating anthropometric data (e.g., age), clothing, and activity in addition to the above environmental variables [33–36]. Thus, even with the employment of more advanced direct indices such as outdoor wet-bulb globe temperature (WBGT; e.g., [20]), the existing literature on the topic has ignored the thermo–physiological basis of the human thermal environment.

In this article, to assist in addressing this research gap, we expand the concept of heat stress exposure assessment to account for the thermo–physiological responses of different population groups. To that end, we employed a novel 30-year (1991–2020) human thermal bioclimate dataset consisting of hourly mPET (modified physiologically equivalent temperature) values to characterize heat stress levels for diverse population types [36]. The study area was the Athens urban area (AUA) in Greece, situated in the Eastern Mediterranean. Initially, we analyzed the long-term trends in acclimatization-based strong heat stress (accliSHS) experienced by average male and female adults and senior individuals in five regional units in AUA during the warm period of the year (April–October). Emphasis was placed on accliSHS duration, which is of paramount importance in relation to adverse health outcomes [37–39]. Then, we computed accliSHS exposure and assessed its long-term trends for each examined demographic type and regional unit by combining the mPET estimates with population data. Contributions arising from changes in climate and population sizes were determined; these are discussed, emphasizing the differences between the targeted demographic groups and regional units in AUA.

2. Materials and Methods

2.1. Study Urban Area

AUA is situated in south Greece (Figure 1), characterized by a Mediterranean climate with hot summers. It consists of five regional units, namely North Athens, West Athens; Central Athens, Piraeus, and South Athens (Figure 1), which accommodate around three million inhabitants in total [40]. Central and West Athens feature compact built-up areas characterized by narrow streets and flat-roofed, light-colored buildings. These areas are also marked by a lack of green spaces. This urban setting description holds for much of Piraeus as well, whereas regions with a more dispersed built-up environment and an abundance of greenery are primarily located in South and North Athens [40,41]. These urban characteristics contribute to the development of prominent overheating, particularly in Central and West Athens and Piraeus [41]. The urban heat island (UHI) effect in these areas is particularly intense during the night (>4 °C; [42]), while high spatial variability characterizes the entire AUA in terms of responsiveness to heat stress [40].

Moreover, in recent decades, the climatic features of AUA have been significantly altered due to anthropogenic climate change. Specifically, increasing trends in the duration, frequency, and intensity of heat extremes have been evident, notably since 1990 [43–46], with a lengthening of the "hot extremes season" also apparent [43,45]. These changes contribute to the establishment of more stressful human bioclimatic conditions in the area [44,45,47–49]. Thus, significant challenges in relation to public health arise in AUA, especially when considering that the population exposed to these stressful conditions is increasingly aging [50].

2.2. Human Bioclimate and Population Data

The human bioclimate data used in the current study consisted of hourly, population-weighted mPET index values for average male and female adult and senior individuals. They spanned from 1991 to 2020 in the five regional units in AUA and focused on the warm period (April–October).

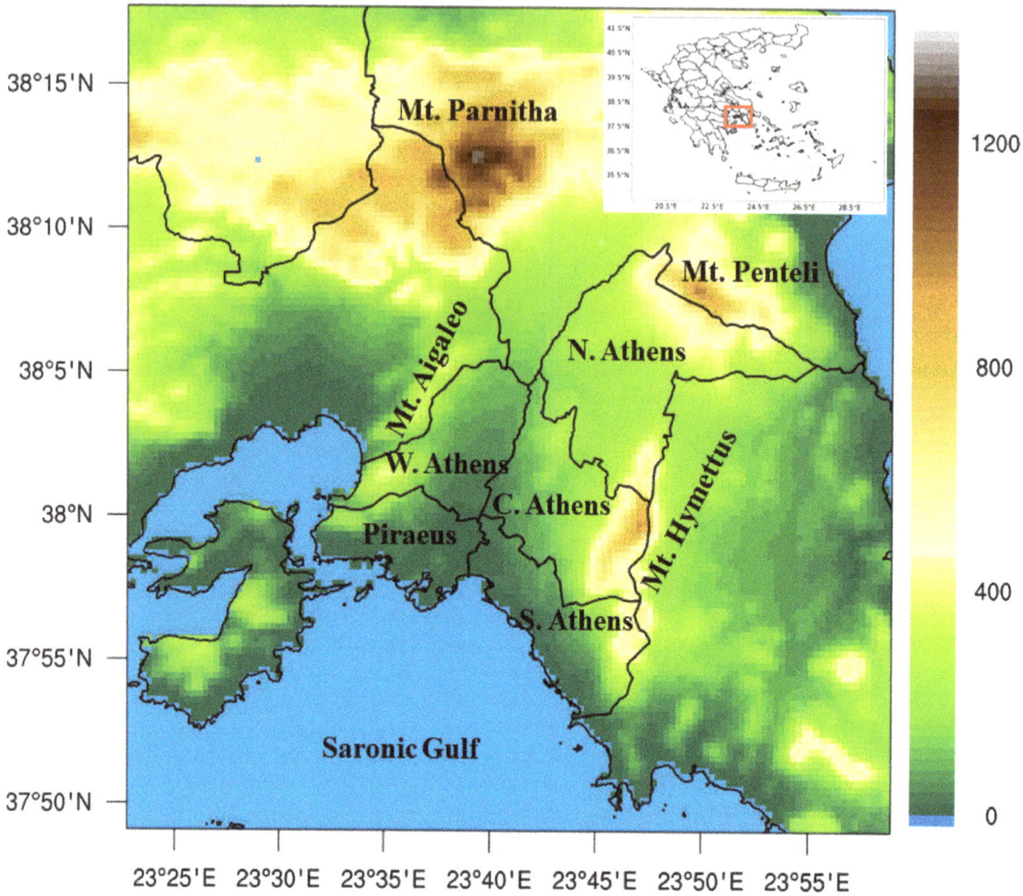

Figure 1. AUA topography (elevation in meters) based on SRTM (Shuttle Radar Topography Mission) data [40] with identification of main geomorphological features. The black boundaries indicate the five AUA regional units (N. Athens: North Athens; W. Athens: West Athens; C. Athens: Central Athens; Piraeus; S. Athens: South Athens), while the red square in the nested map of Greece shows the wider study area.

Briefly, mPET is an advanced human-biometeorological index that is intimately related to human thermal perception and stress. It is computed based on human energy balance modeling that incorporates both the environmental (air and radiant temperature, humidity, and wind speed) and personal/thermo–physiological (sex, age, height, weight, metabolic activity, and clothing coverage) factors that affect the human thermal environment [51,52]. Furthermore, the mPET thermoregulation model can be applied for different demographic and thermo-physiological attributes that correspond to the characteristics of average individuals within population sub-groups. Thus, compared with other thermal stress metrics, mPET assesses heat stress in a thermo–physiologically significant way while allowing consideration of population-wide variability in heat stress [36,39]. This means that given the same environmental conditions in each AUA regional unit, different mPET values and corresponding heat stress levels were estimated for each average individual within each targeted population type. The dataset used to extract the mPET data for the study area and populations is publicly available (https://zenodo.org/records/10893914; accessed on 1 September 2024) and has been described in more detail by Giannaros et al. [36].

The population data for each AUA regional unit were retrieved from the Hellenic Statistical Authority and Eurostat. They correspond to the 1991, 2001, 2011, and 2021 censuses, grouped by sex and age. Thus, to estimate group-specific populations for each day from 1 January 1991 to 31 December 2020, linear interpolation was applied to the censuses population counts for each targeted demographic group and AUA regional unit. Considering the relatively limited data points and the impact of the immigration outflow following the 2009 financial crisis in Greece, linear interpolation was considered to provide more robust and reliable estimates than alternative methods such as exponential or multivariate approaches [53]

2.3. Exposure Estimations and Trend Analysis

The methods used in the existing literature to define heat stress exposure are characterized by significant diversity. For example, Tuholske et al. used fixed thresholds to identify extreme heat episodes [17], whereas Dong et al. applied percentile-based criteria to define humid heatwaves [22]. Here, we focused on days with strong heat stress (SHS), following previous epidemiological (e.g., [39,54,55]) and thermo–physiological (e.g., [56,57]) evidence showing that heat-related health risks are associated with SHS exposure.

Specifically, we considered accliSHS days to account for population- and location-specific short-term acclimatization effects. The definition of accliSHS was based on variable mPET thresholds derived from adjusting the fixed SHS mPET threshold (i.e., 35 °C; [58]) according to the thermal stress conditions prevailing in the 30 days previous, specific to each population and regional unit. This was carried out through the application of a Gaussian filter with 30 significant weights, reflecting a thermo–physiologically relevant short-term (30 days) adaptation to SHS exposure [39]. This process used a formula based on the HeRATE (Health-Related Assessment of the Thermal Environment) approach [59], applied to the population-and location-specific mPET estimates (see Section 2.2). Further details, including the applied formula, can be found in Giannaros et al. [39]. Accounting for different exposure hours within a given day, we computed the annual number of days when population- and location-specific mPET values exceeded the accliSHS mPET thresholds for at least 1 h and 6 h during the warm period of the year (April–October), for each year from 1991 to 2020. Following previous studies [17,41], long-term (1991–2020) trends in the annual number of days with $\geq$1 h and $\geq$6 accliSHS for each examined population type and regional unit were assessed using an ordinary least squares (OLS) regression model. The statistical significance of trends was computed based on Student's t-test at 95% significance level (p-value ≤ 0.05) [45,60]. Further, changes in population- and location-based multi-year numbers of days with $\geq$1 h and $\geq$ 6 accliSHS were computed. Concerning accliSHS exposure, the location- and population-specific $\geq$ 1 h accliSHS exposure per year was quantified in person-days $\times$ year^{-1}, based on Tuholske et al. [17]. In particular, for each year between 1991 and 2020 and for each AUA regional unit, we multiplied the annual number of days with $\geq$1 h accliSHS and the population size of each targeted demographic type. We then applied OLS regression to assess the long-term (1991–2020) trends in population- and location-specific $\geq$1 h accliSHS exposure. Finally, we referred to Tuholske et al. [17] to estimate the relative contribution shares (%) in the exposure trends arising from climate- and demographic-related changes. In more detail, the climate-related contributions were assessed in person-days $\times$ year^{-1} by multiplying the annual number of days with $\geq$1 h accliSHS by the corresponding population size fixed in 1991, while the contributions of demographics were assessed in person-days $\times$ year^{-1} by multiplying the annual number of days with $\geq$1h accliSHS by the corresponding annual population change since 1991. The rates of changes in these contributions were estimated by applying OLS regression. These rates were subsequently used to compute an index ($[-1, 1]$) indicating the relative shares (%) of the $\geq$1 h accliSHS exposure trends from changes in climate and population size from 1991 to 2020 (100% dominated by climate-related changes versus 100% dominated by demographic-related changes) [17].

3. Results

3.1. Trends in Annual accliSHS Days

Figure 2 illustrates the rate of change in numbers of days annually with at least one 1 h accliSHS, expressed in days × year^{-1}, for each examined population type in AUA from 1991 to 2020. Overall, an increasing trend was evident, ranging from +0.22 (for an average male adult in West Athens) to +0.78 (for an average male senior in South Athens) days × year^{-1} with ≥1 h accliSHS, across AUA. These results were statistically significant at a 95% confidence level in all regional units, expect for the average male and female adult individuals in West Athens. For all population types, South and Central Athens demonstrated the highest increasing trends (>+0.5 days × year^{-1}), while North Athens and Piraeus were characterized by increasing trends that spanned from +0.37 to +0.55 days × year^{-1} (Figure 2).

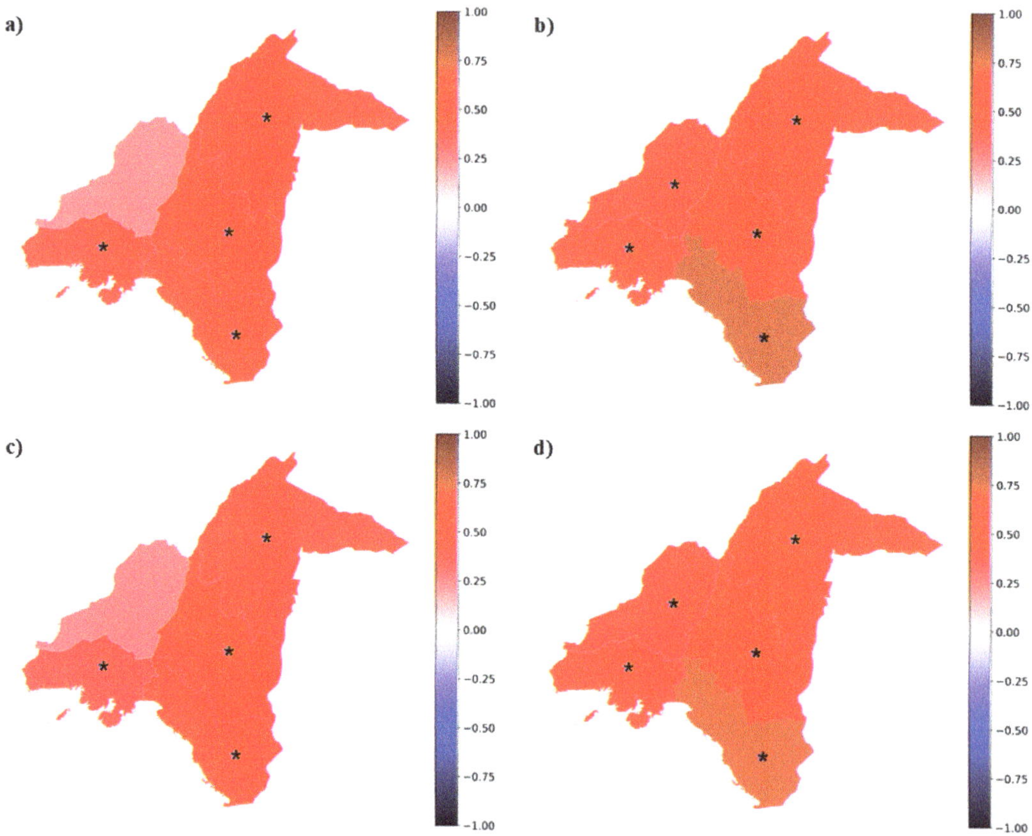

Figure 2. Trends (1991–2020) in numbers of annual days with ≥1 h accliSHS in days × year^{-1} for a (**a**) male adult, (**b**) male senior, (**c**) female adult, and (**d**) female senior individual in AUA. Asterisks indicate statistical significance at the 95% confidence level.

Regarding age-related differences, the average seniors were characterized by overall higher increasing trends in the number of days per year with ≥1 h accliSHS compared with average adults, throughout AUA. For example, the increasing trend for the average male adult in South Athens was equal to +0.67 days × year^{-1}, whereas it was equal to +0.78 days × year^{-1} for the average male senior in the same regional unit. Similarly, the average female senior in North Athens (+0.55 days × year^{-1}) experienced a higher increase in ≥1 h accliSHS compared with the average female adult (+0.42 days × year^{-1}) by +0.13 days/year^{-1}. Sex-related differences were less evident, as the trends in increase were approximately the same for the average male and female individuals, both adults and seniors, in all AUA regional units (Figure 2).

The trends described above denote that, on average, the annual days with ≥1 h accliSHS for an average adult and senior in AUA increased by ~13 and ~18 days, respectively, within the 30-year study period (1991–2020). Specifically, the increase in the average annual number of days with ≥1 h accliSHS from early 1990s to late 2020s ranged from 4 days for an average female adult in West Athens to 12 days for the same population type in South Athens, as shown in Table 1, corresponding to percentage increases ranging from 3.2% to 11.8%. Similar findings were evident for an average male adult, whereas the average number of days per year with accliSHS for at least 1 h increased by 7.7% (7.6%) for an average male (female) senior in West Athens and by 21% (19.2%) in South Athens from the early 1990s to the late 2010s (Table 1).

Table 1. Five-year average numbers of days with ≥1 h accliSHS for the examined population types in AUA regional units in early 1990s (1999–1995) and late 2010s (2016–2020).

	Demographic Type			
AUA Regional Unit	**Male Adult**	**Male Senior**	**Female Adult**	**Female Senior**
North Athens	early 1990s: 94 late 2020s: 103	early 1990s: 77 late 2020s: 87	early 1990s: 99 late 2020s: 107	early 1990s: 79 late 2020s: 90
West Athens	early 1990s: 121 late 2020s: 125	early 1990s: 104 late 2020s: 112	early 1990s: 124 late 2020s: 128	early 1990s: 106 late 2020s: 114
Central Athens	early 1990s: 109 late 2020s: 119	early 1990s: 90 late 2020s: 105	early 1990s: 112 late 2020s: 122	early 1990s: 93 late 2020s: 107
Piraeus	early 1990s: 100 late 2020s: 107	early 1990s: 83 late 2020s: 94	early 1990s: 105 late 2020s: 111	early 1990s: 85 late 2020s: 95
South Athens	early 1990s: 97 late 2020s: 111	early 1990s: 81 late 2020s: 98	early 1990s: 102 late 2020s: 114	early 1990s: 83 late 2020s: 99

The patterns of increase in the trends outlined earlier (Figure 1) were similarly observed when focusing on annual days with ≥6 h accliSHS, as shown in Figure A1. As such, a significant contribution to the rise in the annual number of days with ≥1 h accliSHS seems to have arisen from the increased numbers of days with more than 6 h accliSHS. This indicates that AUA experienced an increase in the daily duration of accliSHS, in addition to the increase in the number of annual days with at least 1 h accliSHS, during the examined 30-year period (1991–2020). The related evidence of this became more apparent when comparing the percentage increases in the average numbers of days with ≥1 h and ≥6 h accliSHS, as shown in Tables 1 and A1, respectively. Indicatively, these increases ranged from 5.7% (female adult) to 13.2% (male senior) for ≥1 h accliSHS in Pireaus, whereas they ranged from 42.3% (average female adult) to 87.7% (average male senior) for ≥6 h accliSHS in the same regional unit (Tables 1 and A1).

3.2. Trends in Annual accliSHS Exposure

Figures 3, 4, A2 and A3 show the 1991–2020 trends in ≥1 h accliSHS exposure and in climate- and demographic-related changes for the examined demographic types in the five AUA regional units. The warming referenced in the first column of plots refers to the increasing trend in annual ≥ 1 h accliSHS person-days × year^{-1}, similar to the growing numbers of annual ≥ 1 h accliSHS days per year documented in Section 3.1. The population sizes stated in the second column of plots may be associated with an increasing or decreasing trend, expressed in person-days × year^{-1} (Figures 3, 4, A2 and A3). The warming and population size trends both contribute to the population- and location-specific ≥ 1 h accliSHS exposure trends, illustrated in person-days × year^{-1} in the third column of plots in Figures 3, 4, A2 and A3. For adults (Figures 3 and A2), the demographic changes exhibit parabolic shapes due to the population increase from 1991 to 2000 and the subsequent population decrease since 2000, which was particularly pronounced after 2010. As such, the population size trends in Figures 3 and A2 indicate the general direction of the demographic shifts from 1991 to 2020, indicating an overall decrease in the AUA adult population, except in North Athens, during the examined 30-year period. This overall decreasing trajectory in population size was more noticeable in Central Athens, as indicated by the steeper trend slopes ($-462{,}000$ person-days $\times$ year^{-1} for male adults and $-620{,}000$ person-days $\times$ year^{-1} for female adults). As a result, the $\geq$1 h accliSHS exposure of adults in this regional unit was reduced ($-253{,}000$ person-days $\times$ year^{-1} for male adults and $-396{,}000$ person-days $\times$ year^{-1} for female adults), despite the significant warming the Central Athens experienced (+209,000 person-days $\times$ year^{-1} for male adults and +224,000 person-days $\times$ year^{-1} for female adults). On the other hand, the overall decline in the adult population in the other AUA regional units after 2000 was less evident. This fact, combined with warming in these areas (see also Section 3.1), resulted primarily in increases in $\geq$1 h accliSHS exposure for adults, ranging from +21,000 person-days $\times$ year^{-1} (West Athens) to +139,000 person-days $\times$ year^{-1} (North Athens) for female adults and from +23,000 person-days $\times$ year^{-1} (Piraeus) to +75,000 person-days $\times$ year^{-1} (North Athens) for male adults (Figures 3 and A2).

For seniors, the results of the present analysis are clearer. All AUA regional units experienced warming (see also Section 3.1). Additionally, all examined areas were characterized by an increasingly elderly population, including both males and females, although this increase was less evident in Central Athens. Combined, these trends led to significant increases in seniors' exposure to $\geq$1 h accliSHS across the AUA. These values ranged from +64,000 person-days $\times$ year^{-1} (Piraeus) to +109,000 person-days $\times$ year^{-1} (North Athens) for male seniors, while the highest increase in exposure was evident for female seniors, surpassing +100,000 person-days $\times$ year^{-1} in most regional units and reaching up to +153,000 person-days $\times$ year^{-1} in North Athens. This outcome was primarily driven by greater increases in the elderly female population compared with elderly males. Indicatively, the exposure of female and male adults in South Athens to at least 1 h accliSHS grew by +136,000 person-days $\times$ year^{-1} and +97,000 person-days $\times$ year-1, respectively. Warming contributed almost equally to these trends for both sexes (+19,000 person-days $\times$ year^{-1} for females and +25,000 person-days $\times$ year^{-1} for males), whereas population size changes contributed by +111,000 person-days $\times$ year^{-1} and +78,000 person-days $\times$ year^{-1} for females and males, respectively (Figures 4 and A3).

The above contributions are summarized in Figure 5, which illustrates the relative contribution shares (%) in the 30-year (1991–2020) trends in exposure to $\geq$1 h accliSHS for the examined populations in AUA. Specifically, the decreasing trends in exposure for male and female adults in Central Athens (Figures 3 and A2) were driven by the population decline by ~37.7% and ~46.9%, respectively (Figure 5a,c). In contrast, the increase in exposure for male adults in North and South Athens and Piraeus and for female adults in North, West, and South Athens was attributed primarily to warming (Figures 3 and A2), with relative contribution shares ranging from ~20.9% in Piraeus to ~92.2% in North Athens for female adults and from ~8.1% in North Athens to ~78.3% in South Athens for male adults (Figure 5a,c). For seniors, the increase in the size of their population contributed more than 60% to the exposure of both males and females in North, West, and South Athens and Piraeus to at least 1 h accliSHS. The highest value in these relative contribution shares (~82.9%) was found for female seniors in West Athens (Figures 4, 5b,d and A3). On the other hand, warming primarily drove the exposure of seniors in Central Athens to $\geq$1 h accliSHS (Figure 5b,d), as an outcome of the less pronounced increase in the elderly population in this area (Figures 4 and A3). Indicatively, warming contributed by ~28.8% to the increase in $\geq$1 h accliSHS exposure for male seniors in Central Athens (Figure 5b,d).

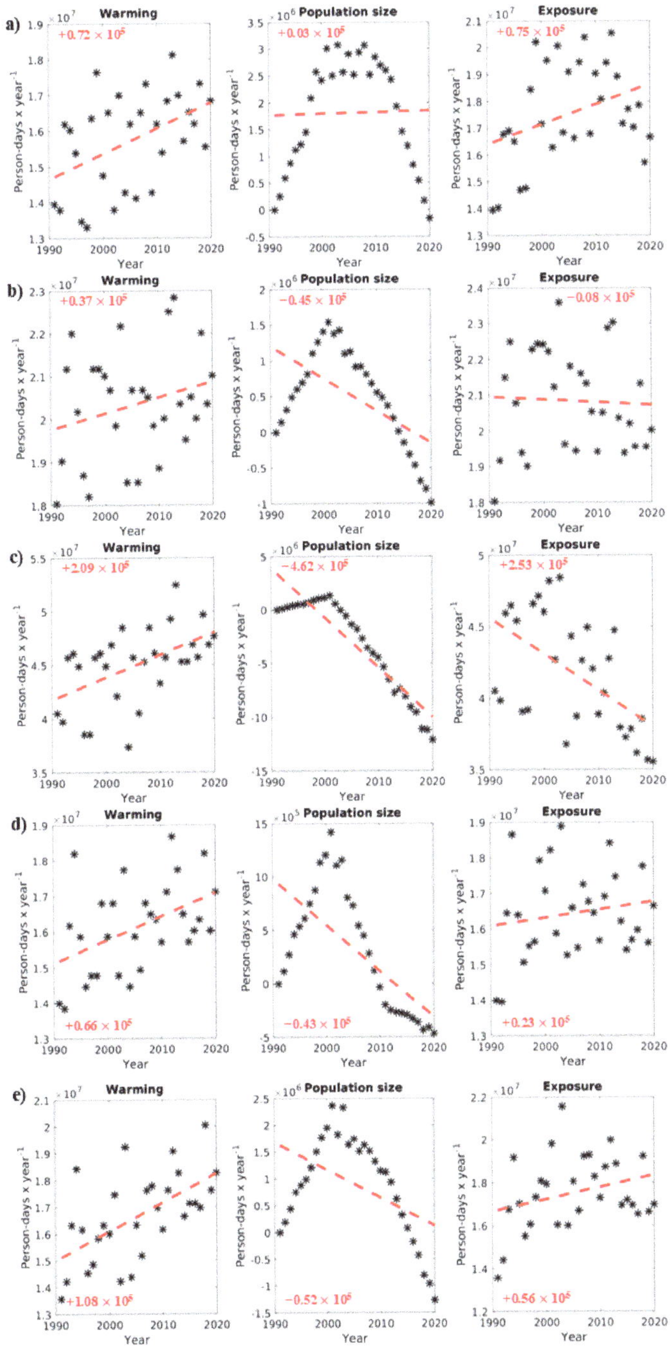

Figure 3. Warming and population size evolution for male adults (person-days × year^{-1}) contributing to their exposure (person-days × year^{-1}) to ≥1 h accliSHS, from 1991 to 2020 (*) in (**a**) North Athens, (**b**) West Athens, (**c**) Central Athens, (**d**) Piraeus, and (**e**) South Athens. The red text and dotted lines in the plots indicate the trends in changes in the presented variables in person-days × year^{-1}.

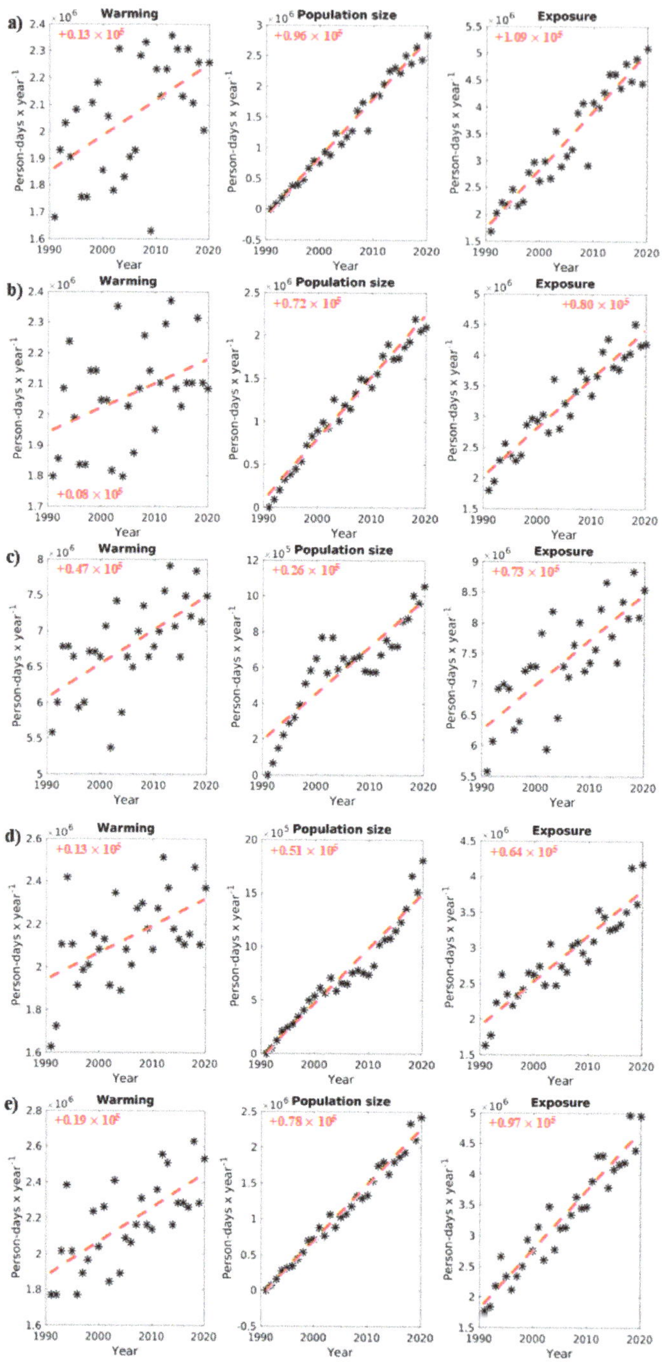

Figure 4. Warming and population size evolution for male seniors (person-days $\times$ year^{-1}) contributing to their exposure (person-days $\times$ year^{-1}) to $\geq$1 h accliSHS, from 1991 to 2020 (*) in (**a**) North Athens, (**b**) West Athens, (**c**) Central Athens, (**d**) Piraeus, and (**e**) South Athens. The red text and dotted lines in the plots indicate the trends in changes in the presented variables in person-days $\times$ year^{-1}.

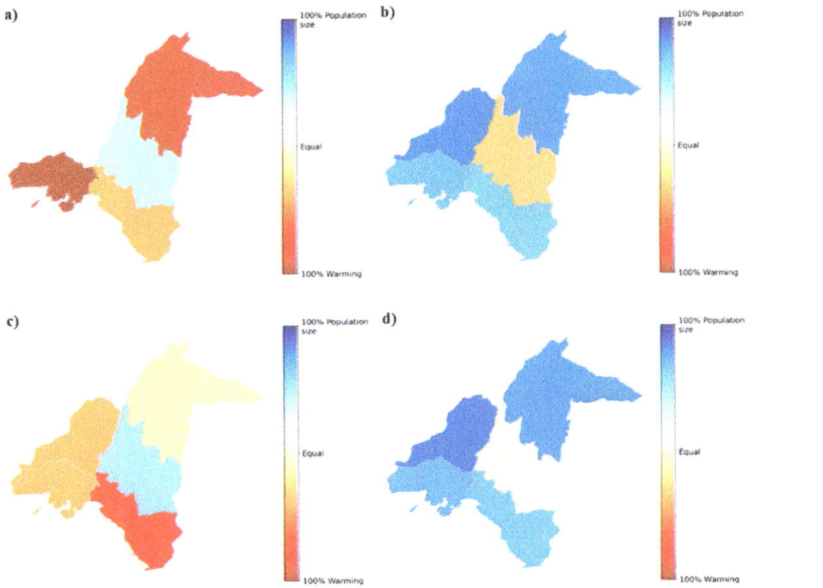

Figure 5. The relative contribution shares (%) of climate- and demographic-related changes in ≥1 h accliSHS exposure trends (1991–2020) for (**a**) male adults, (**b**) male seniors, (**c**) female adults, and (**d**) female seniors in AUA.

4. Discussion

In the present study, we assessed the climatology and long-term (1991–2020) trends in population exposure to heat stress, considering population-wide variability in heat stress and short-term acclimatization effects. This was achieved by employing a novel human thermal bioclimate dataset containing hourly values of an advanced human-biometeorological index, namely mPET, for diverse demographic types [36]. We specifically focused on acclimatization-based SHS experienced by average male and female adult and senior individuals living in AUA, Greece, which is of particular interest with respect to heat-related issues due to the dangerous confluence of accelerated warming [61] and increased population ageing [50] in the region. The definition of aclliSHS was based on variable mPET thresholds, thus accounting for population- and location-specific short-term thermo–physiological adaptation to heat, which is critical with respect to the occurrence of heat-related health effects [39].

The first-stage climatological analysis showed that days with at least 1 h aclliSHS constitute a common feature during the warm period of the year (April–October) in the study area. This was expected because of the hot summers in the Mediterranean climate of AUA [40]. An average adult in particular, may have been exposed to around 100 days with ≥1 h accliSHS in the early 1990s (Table 1). For an average senior, the number of days with ≥1 h accliSHS were lower overall (~85 in early 1990s; Table 1), because of the reduced physical activity assumed in the mPET computations for this population [36,39], consistent with the metabolic activity rates of elderly people under hot environmental conditions [62,63]. However, it is interesting that higher increasing trends (1991–2020) in days with ≥1 h accliSHS were documented for the average senior compared with the average adult across AUA (Figure 1). Given the significant warming and other thermal-related changes between 1991 and 2020 in the greater study area (e.g., [45,46,64,65]), this outcome indicates that thermo–physiologically it is more difficult for seniors to cope with the increasing heat stress conditions induced by climate change. Indeed, the human body's capacity to dissipate heat decreases as age increases [62].

Overall, the upward trends observed in our study in relation to the annual numbers of days with at least 1 h accliSHS in AUA are in accordance with previous similar studies in the area [43,47–49]. Regional differences in these trends are driven by geomorphological features and background atmospheric circulations that affect the thermal environment in AUA [40]. Furthermore, high rates of increase were particularly evident for days when mPET values for the examined demographic types exceeded the accliSHS threshold by more than 6 h, indicating a rise from 1991 to 2020 in the daily duration of heat stress in AUA, in agreement with previous research [49]. We focused on $\geq$ 6 h accliSHS because prolonged exposure to strong heat stress during the day is an important determinant for adverse health outcomes [39]. Thus, the increase greater than 8% in the average annual number of days with $\geq$6 h accliSHS (over 40% for an average senior; Table A1) is a particularly concerning outcome.

The significance of this outcome is heightened when it is considered alongside the increase in the elderly population from 1991 to 2020 (Figures 3, 4, A2 and A3). This combination resulted in increases in seniors' exposure to $\geq$1 h accliSHS across AUA that reached up to +153,000 person-days $\times$ year^{-1} for female seniors in North Athens (Figure A3). The relevant outcomes for adults were more varied (Figures 3 and 4), driven by population decline in Central Athens and primarily by warming in the rest of the examined AUA regional units (Figure 5). These varying contributions to the trends in $\geq$1 h accliSHS exposure reflect the urban and population expansion in suburban areas, particularly in North Athens [41], as well as the immigration wave associated with the outbreak of the financial crisis in Greece in 2009 [66].

5. Concluding Remarks

In this study, we expanded on existing heat stress exposure research by incorporating thermo–physiological principles through the use of an advanced human-biometeorological index, namely mPET. This approach allowed us to account for the variability in human thermoregulation responses, including those related to short-term acclimatization. As a result, we were able to provide a more nuanced and realistic assessment of the heat stress experienced by different population sub-groups. We demonstrated this novel technique in AUA, Greece, focusing on accliSHS. This research revealed increases in annual accliSHS days in the study area from 1991 to 2020 (i.e., warming). These trends, which were particularly pronounced for prolonged accliSHS exposure exceeding 6 h, pose significant thermo–physiological challenges for senior individuals. Furthermore, combined with the increase in size of the elderly population in AUA, these trends have led to a substantial and concerning increase in seniors' accliSHS exposure over the past 30 years (1991–2020). Changes in population size linked to socioeconomic factors, such as urban expansion in northern AUA suburbs and the 2009 financial crisis in Greece, also contributed to the accliSHS exposure trends for adults, which showed both increases and decreases from 1991 to 2020. These findings underscore the multi-faceted nature of heat stress exposure, influenced by both climate change and demographic shifts [15,17,19,21,22]. Given the significant health risks associated with heat stress exposure, the current study's findings also underscore the urgent need for heat–health action planning [67–69], as it is likely that continuation or intensification of the presented increasing trends in accliSHS in AUA, particularly for seniors, will amplify the burden on Greece's healthcare system.

Author Contributions: Conceptualization, C.G. and E.G.; methodology, C.G, E.G. and I.A.; software, C.G., E.G. and I.A.; validation, C.G., E.G. and I.A.; formal analysis, C.G., E.G. and I.A.; investigation, C.G., E.G. and I.A.; resources, C.G., E.G. and I.A.; data curation, C.G., E.G. and I.A.; writing—original draft preparation, C.G.; writing—review and editing, C.G., E.G. and I.A.; visualization, E.G.; supervision, C.G.; funding acquisition, C.G. All authors have read and agreed to the published version of the manuscript.

Funding: This research is funded by the Hellenic Foundation for Research and Innovation (H.F.R.I.) under the "3d Call for H.F.R.I. Research Projects to support Post-Doctoral Researchers" (Project Number: 06885).

Data Availability Statement: The human-biometeorological data used in the present study are publicly available at the Zenodo repository: https://zenodo.org/records/10893914 (accessed on 1 September 2024). The population data used in the current study were retrieved from the Hellenic Statistical Authority (https://www.statistics.gr/en/; accessed on 1 September 2024) and Eurostat (https://ec.europa.eu/eurostat; accessed on 1 September 2024).

Conflicts of Interest: The authors declare no conflicts of interest.

Appendix A

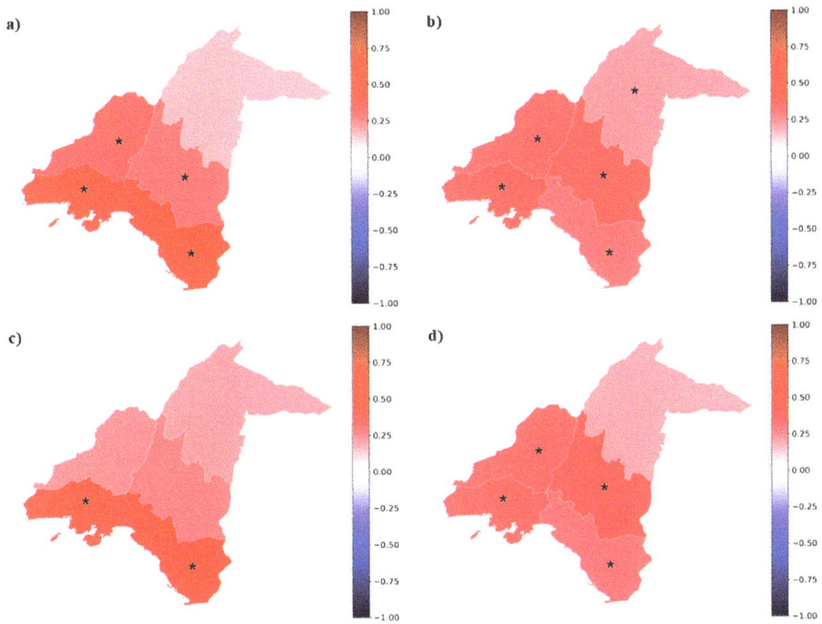

Figure A1. Trends (1991–2020) in numbers of annual days with ≥6 h accliSHS in days × year^{-1} for a (**a**) male adult, (**b**) male senior, (**c**) female adult, and (**d**) female senior individual in AUA. Asterisks indicate statistical significance at the 95% confidence level.

Table A1. Five-year average numbers of days with ≥6 h accliSHS for the examined population types in AUA regional units in early 1990s (1999–1995) and late 2010s (2016–2020).

	Demographic Type			
AUA Regional Unit	**Male Adult**	**Male Senior**	**Female Adult**	**Female Senior**
North Athens	early 1990s: 20 late 2020s: 23	early 1990s: 11 late 2020s: 16	early 1990s: 22 late 2020s: 27	early 1990s: 12 late 2020s: 17
West Athens	early 1990s: 33 late 2020s: 38	early 1990s: 19 late 2020s: 27	early 1990s: 38 late 2020s: 41	early 1990s: 20 late 2020s: 28
Central Athens	early 1990s: 27 late 2020s: 33	early 1990s: 15 late 2020s: 23	early 1990s: 30 late 2020s: 35	early 1990s: 16 late 2020s: 25
Piraeus	early 1990s: 23 late 2020s: 34	early 1990s: 11 late 2020s: 21	early 1990s: 26 late 2020s: 37	early 1990s: 13 late 2020s: 22
South Athens	early 1990s: 21 late 2020s: 34	early 1990s: 13 late 2020s: 21	early 1990s: 24 late 2020s: 37	early 1990s: 13 late 2020s: 21

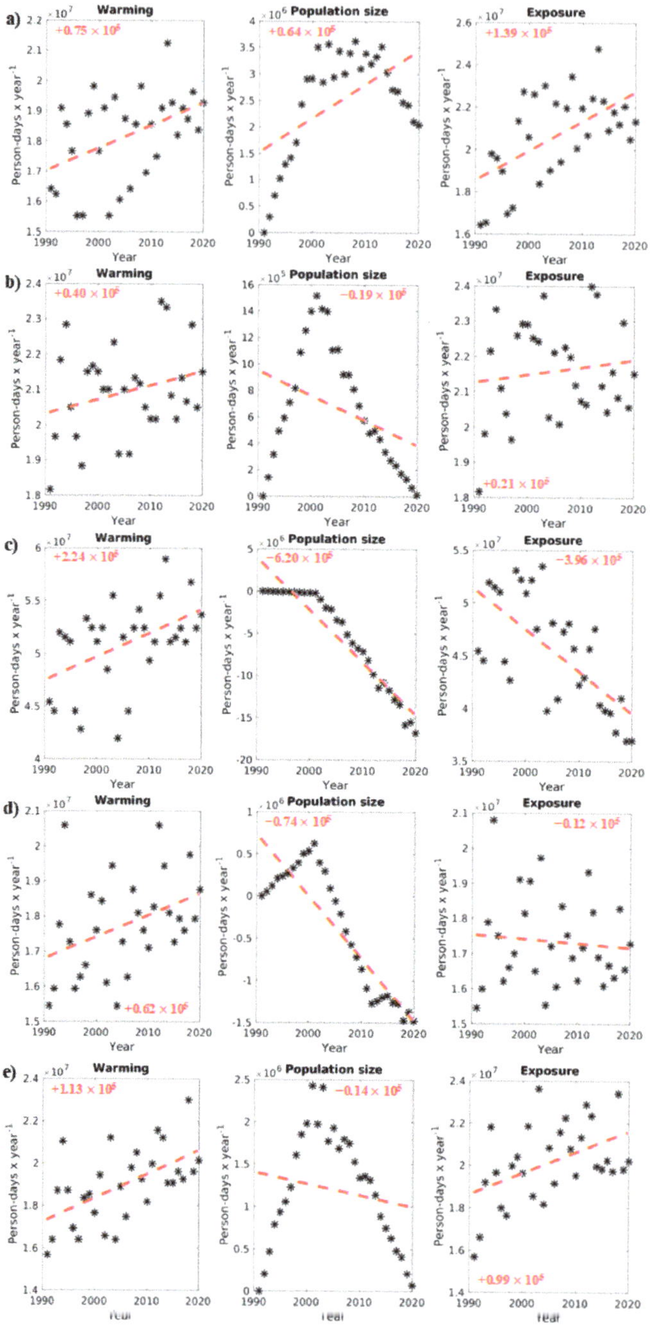

Figure A2. Warming and population size evolution for female adults (person-days $\times$ year^{-1}) contributing to their exposure (person-days $\times$ year^{-1}) to $\geq$1 h accliSHS from 1991 to 2020 (*) in (**a**) North Athens, (**b**) West Athens, (**c**) Central Athens, (**d**) Piraeus, and (**e**) South Athens. The red text and dotted lines in the plots indicate the trends in changes in the presented variables in person-days $\times$ year^{-1}.

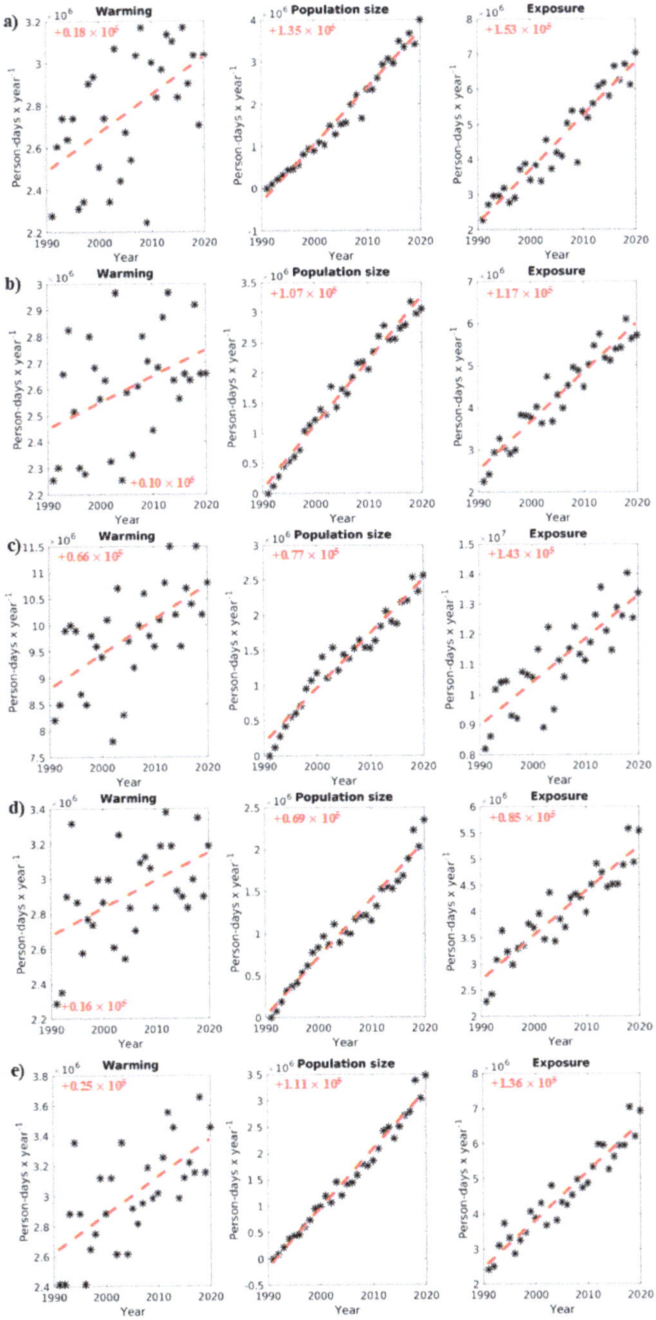

Figure A3. Warming and population size evolution for female seniors (person-days $\times$ year^{-1}) contributing to their exposure (person-days $\times$ year^{-1}) to $\geq$1 h accliSHS from 1991 to 2020 (*) in (**a**) North Athens, (**b**) West Athens, (**c**) Central Athens, (**d**) Piraeus, and (**e**) South Athens. The red text and dotted lines in the plots indicate the trends in changes in the presented variables in person-days $\times$ year^{-1}.

References

1. Seneviratne, S.I.; Zhang, X.; Adnan, M.; Badi, W.; Dereczynski, C.; Di Luca, A.; Ghosh, S.; Iskandar, I.; Kossin, J.; Lewis, S.; et al. Weather and Climate Extreme Events in a Changing Climate. In *Climate Change 2021: The Physical Science Basis*; Contribution of Working Group I to the Sixth Assessment Report of the Intergovernmental Panel on Climate, Change; Masson-Delmotte, V., Zhai, P., Pirani, A., Connors, S.L., Péan, C., Berger, S., Caud, N., Chen, Y., Goldfarb, L., Gomis, M.I., et al., Eds.; Cambridge University Press: Cambridge, UK; New York, NY, USA, 2023; pp. 1513–1766. [CrossRef]
2. Zhang, Y.; Yu, C.; Wang, L. Temperature Exposure during Pregnancy and Birth Outcomes: An Updated Systematic Review of Epidemiological Evidence. *Environ. Pollut.* **2017**, *225*, 700–712. [CrossRef] [PubMed]
3. McElroy, S.; Ilango, S.; Dimitrova, A.; Gershunov, A.; Benmarhnia, T. Extreme Heat, Preterm Birth, and Stillbirth: A Global Analysis across 14 Lower-Middle Income Countries. *Environ. Int.* **2022**, *158*, 106902. [CrossRef] [PubMed]
4. Liu, J.; Varghese, B.M.; Hansen, A.; Xiang, J.; Zhang, Y.; Dear, K.; Gourley, M.; Driscoll, T.; Morgan, G.; Capon, A.; et al. Is There an Association between Hot Weather and Poor Mental Health Outcomes? A Systematic Review and Meta-Analysis. *Environ. Int.* **2021**, *153*, 106533. [CrossRef]
5. Minor, K.; Bjerre-Nielsen, A.; Jonasdottir, S.S.; Lehmann, S.; Obradovich, N. Rising Temperatures Erode Human Sleep Globally. *One Earth* **2022**, *5*, 534–549. [CrossRef]
6. Bernard, P.; Chevance, G.; Kingsbury, C.; Baillot, A.; Romain, A.-J.; Molinier, V.; Gadais, T.; Dancause, K.N. Climate Change, Physical Activity and Sport: A Systematic Review. *Sports Med.* **2021**, *51*, 1041–1059. [CrossRef]
7. Wallenberg, N.; Lindberg, F.; Thorsson, S.; Jungmalm, J.; Fröberg, A.; Raustorp, A.; Rayner, D. The Effects of Warm Weather on Children's Outdoor Heat Stress and Physical Activity in a Preschool Yard in Gothenburg, Sweden. *Int. J. Biometeorol.* **2023**, *67*, 1927–1940. [CrossRef]
8. Nazarian, N.; Liu, S.; Kohler, M.; Lee, J.K.W.; Miller, C.; Chow, W.T.L.; Alhadad, S.B.; Martilli, A.; Quintana, M.; Sunden, L.; et al. Project Coolbit: Can Your Watch Predict Heat Stress and Thermal Comfort Sensation? *Environ. Res. Lett.* **2021**, *16*, 34031. [CrossRef]
9. Flouris, A.D.; Dinas, P.C.; Ioannou, L.G.; Nybo, L.; Havenith, G.; Kenny, G.P.; Kjellstrom, T. Workers' Health and Productivity under Occupational Heat Strain: A Systematic Review and Meta-Analysis. *Lancet Planet. Health* **2018**, *2*, e521–e531. [CrossRef]
10. Cheng, J.; Xu, Z.; Bambrick, H.; Prescott, V.; Wang, N.; Zhang, Y.; Su, H.; Tong, S.; Hu, W. Cardiorespiratory Effects of Heatwaves: A Systematic Review and Meta-Analysis of Global Epidemiological Evidence. *Environ. Res.* **2019**, *177*, 108610. [CrossRef]
11. Parliari, D.; Cheristanidis, S.; Giannaros, C.; Keppas, S.C.; Papadogiannaki, S.; de'Donato, F.; Sarras, C.; Melas, D. Short-Term Effects of Apparent Temperature on Cause-Specific Mortality in the Urban Area of Thessaloniki, Greece. *Atmosphere* **2022**, *13*, 852. [CrossRef]
12. Wondmagegn, B.Y.; Xiang, J.; Williams, S.; Pisaniello, D.; Bi, P. What Do We Know about the Healthcare Costs of Extreme Heat Exposure? A Comprehensive Literature Review. *Sci. Total Environ.* **2019**, *657*, 608–618. [CrossRef] [PubMed]
13. Parsons, L.A.; Masuda, Y.J.; Kroeger, T.; Shindell, D.; Wolff, N.H.; Spector, J.T. Global Labor Loss due to Humid Heat Exposure Underestimated for Outdoor Workers. *Environ. Res. Lett.* **2022**, *17*, 14050. [CrossRef]
14. De Perez, E.C.; Van Aalst, M.; Bischiniotis, K.; Mason, S.; Nissan, H.; Pappenberger, F.; Stephens, E.; Zsoter, E.; Van Den Hurk, B. Global Predictability of Temperature Extremes. *Environ. Res. Lett.* **2018**, *13*, 054017. [CrossRef]
15. Chambers, J. Global and Cross-Country Analysis of Exposure of Vulnerable Populations to Heatwaves from 1980 to 2018. *Clim. Chang.* **2020**, *163*, 539–558. [CrossRef]
16. Li, L.; Zha, Y. Population Exposure to Extreme Heat in China: Frequency, Intensity, Duration and Temporal Trends. *Sustain. Cities Soc.* **2020**, *60*, 102282. [CrossRef]
17. Tuholske, C.; Caylor, K.; Funk, C.; Verdin, A.; Sweeney, S.; Grace, K.; Peterson, P.; Evans, T. Global Urban Population Exposure to Extreme Heat. *Proc. Natl. Acad. Sci. USA* **2021**, *118*, e2024792118. [CrossRef] [PubMed]
18. Alizadeh, M.R.; Abatzoglou, J.T.; Adamowski, J.F.; Prestemon, J.P.; Chittoori, B.; Akbari Asanjan, A.; Sadegh, M. Increasing Heat-Stress Inequality in a Warming Climate. *Earth's Future* **2022**, *10*, e2021EF002488. [CrossRef]
19. Romanello, M.; di Napoli, C.; Green, C.; Kennard, H.; Lampard, P.; Scamman, D.; Walawender, M.; Ali, Z.; Ameli, N.; Ayeb-Karlsson, S.; et al. The 2023 Report of the Lancet Countdown on Health and Climate Change: The Imperative for a Health-Centred Response in a World Facing Irreversible Harms. *Lancet* **2023**, *402*, 2346–2394. [CrossRef]
20. Kyaw, A.K.; Hamed, M.M.; Kamruzzaman, M.; Shahid, S. Spatiotemporal Changes in Population Exposure to Heat Stress in South Asia. *Sustain. Cities Soc.* **2023**, *93*, 104544. [CrossRef]
21. Dong, D.; Tao, H.; Zhang, Z. Historic Evolution of Population Exposure to Heatwaves in Xinjiang Uygur Autonomous Region, China. *Sci. Rep.* **2023**, *13*, 7401. [CrossRef]
22. Dong, J.; Hu, T.; Liu, Y.; Peng, J. Urban-Rural Differences in the Local Human Exposure to Humid Heatwaves in the Northern Hemisphere. *Sustain. Cities Soc.* **2024**, *107*, 105415. [CrossRef]
23. Liu, Z.; Anderson, B.; Yan, K.; Dong, W.; Liao, H.; Shi, P. Global and Regional Changes in Exposure to Extreme Heat and the Relative Contributions of Climate and Population Change. *Sci. Rep.* **2017**, *7*, 43909. [CrossRef] [PubMed]
24. Coffel, E.D.; Horton, R.M.; De Sherbinin, A. Temperature and Humidity Based Projections of a Rapid Rise in Global Heat Stress Exposure during the 21st Century. *Environ. Res. Lett.* **2018**, *13*, 014001. [CrossRef] [PubMed]
25. Jones, B.; Tebaldi, C.; O'Neill, B.C.; Oleson, K.; Gao, J. Avoiding Population Exposure to Heat-Related Extremes: Demographic Change vs Climate Change. *Clim. Chang.* **2018**, *146*, 423–437. [CrossRef]
26. Rohat, G.; Flacke, J.; Dosio, A.; Pedde, S.; Dao, H.; van Maarseveen, M. Influence of Changes in Socioeconomic and Climatic Conditions on Future Heat-Related Health Challenges in Europe. *Glob. Planet. Chang.* **2019**, *172*, 45–59. [CrossRef]

27. Broadbent, A.M.; Krayenhoff, E.S.; Georgescu, M. The Motley Drivers of Heat and Cold Exposure in 21st Century US Cities. *Proc. Natl. Acad. Sci. USA* **2020**, *117*, 21108–21117. [CrossRef]
28. Iyakaremye, V.; Zeng, G.; Yang, X.; Zhang, G.; Ullah, I.; Gahigi, A.; Vuguziga, F.; Asfaw, T.G.; Ayugi, B. Increased High-Temperature Extremes and Associated Population Exposure in Africa by the Mid-21st Century. *Sci. Total Environ.* **2021**, *790*, 148162. [CrossRef]
29. Chen, H.; He, W.; Sun, J.; Chen, L. Increases of Extreme Heat-Humidity Days Endanger Future Populations Living in China. *Environ. Res. Lett.* **2022**, *17*, 064013. [CrossRef]
30. Freychet, N.; Hegerl, G.C.; Lord, N.S.; Lo, Y.T.E.; Mitchell, D.; Collins, M. Robust Increase in Population Exposure to Heat Stress with Increasing Global Warming. *Environ. Res. Lett.* **2022**, *17*, 064049. [CrossRef]
31. Lei, Y.; Wang, Z.; Zhang, X.; Che, H.; Yue, X.; Tian, C.; Zhong, J.; Guo, L.; Li, L.; Zhou, H. Avoided Population Exposure to Extreme Heat under Two Scenarios of Global Carbon Neutrality by 2050 and 2060. *Environ. Res. Lett.* **2022**, *17*, 094041. [CrossRef]
32. Nishant, N.; Ji, F.; Guo, Y.; Herold, N.; Green, D.; Di Virgilio, G.; Beyer, K.; Riley, M.L.; Perkins-Kirkpatrick, S. Future Population Exposure to Australian Heatwaves. *Environ. Res. Lett.* **2022**, *17*, 064030. [CrossRef]
33. Vanos, J.K.; Warland, J.S.; Gillespie, T.J.; Kenny, N.A. Review of the Physiology of Human Thermal Comfort While Exercising in Urban Landscapes and Implications for Bioclimatic Design. *Int. J. Biometeorol.* **2010**, *54*, 319–334. [CrossRef] [PubMed]
34. McGregor, G.R.; Vanos, J.K. Heat: A Primer for Public Health Researchers. *Public Health* **2018**, *161*, 138–146. [CrossRef] [PubMed]
35. Matzarakis, A. Curiosities about Thermal Indices Estimation and Application. *Atmosphere* **2021**, *12*, 721. [CrossRef]
36. Giannaros, C.; Agathangelidis, I.; Galanaki, E.; Cartalis, C.; Kotroni, V.; Lagouvardos, K.; Giannaros, T.M.; Matzarakis, A. Hourly Values of an Advanced Human-Biometeorological Index for Diverse Populations from 1991 to 2020 in Greece. *Sci. Data* **2024**, *11*, 76. [CrossRef]
37. Matzarakis, A. A Note on the Assessment of the Effect of Atmospheric Factors and Components on Humans. *Atmosphere* **2020**, *11*, 1283. [CrossRef]
38. Vanos, J.K.; Baldwin, J.W.; Jay, O.; Ebi, K.L. Simplicity Lacks Robustness When Projecting Heat-Health Outcomes in a Changing Climate. *Nat. Commun.* **2020**, *11*, 6079. [CrossRef]
39. Giannaros, C.; Economou, T.; Parliari, D.; Galanaki, E.; Kotroni, V.; Lagouvardos, K.; Matzarakis, A. A Thermo-Physiologically Consistent Approach for Studying the Heat-Health Nexus with Hierarchical Generalized Additive Modelling: Application in Athens Urban Area (Greece). *Urban Clim.* **2024**, *58*, 102206. [CrossRef]
40. Giannaros, C.; Agathangelidis, I.; Papavasileiou, G.; Galanaki, E.; Kotroni, V.; Lagouvardos, K.; Giannaros, T.M.; Cartalis, C.; Matzarakis, A. The Extreme Heat Wave of July–August 2021 in the Athens Urban Area (Greece): Atmospheric and Human-Biometeorological Analysis Exploiting Ultra-High Resolution Numerical Modeling and the Local Climate Zone Framework. *Sci. Total Environ.* **2023**, *857*, 159300. [CrossRef]
41. Agathangelidis, I.; Cartalis, C.; Santamouris, M. Integrating Urban Form, Function, and Energy Fluxes in a Heat Exposure Indicator in View of Intra-Urban Heat Island Assessment and Climate Change Adaptation. *Climate* **2019**, *7*, 75. [CrossRef]
42. Giannaros, T.M.; Melas, D.; Daglis, I.A.; Keramitsoglou, I.; Kourtidis, K. Numerical Study of the Urban Heat Island over Athens (Greece) with the WRF Model. *Atmos. Environ.* **2013**, *73*, 103–111. [CrossRef]
43. Founda, D.; Pierros, F.; Katavoutas, G.; Keramitsoglou, I. Observed Trends in Thermal Stress at European Cities with Different Background Climates. *Atmosphere* **2019**, *10*, 436. [CrossRef]
44. Katavoutas, G.; Founda, D. Response of Urban Heat Stress to Heatwaves in Athens (1960–2017). *Atmosphere* **2019**, *10*, 483. [CrossRef]
45. Founda, D.; Katavoutas, G.; Pierros, F.; Mihalopoulos, N. Centennial Changes in Heat Waves Characteristics in Athens (Greece) from Multiple Definitions Based on Climatic and Bioclimatic Indices. *Glob. Planet. Chang.* **2022**, *212*, 103807. [CrossRef]
46. Galanaki, E.; Giannaros, C.; Kotroni, V.; Lagouvardos, K.; Papavasileiou, G. Spatio-Temporal Analysis of Heatwaves Characteristics in Greece from 1950 to 2020. *Climate* **2023**, *11*, 5. [CrossRef]
47. Matzarakis, A.; Nastos, P.T. Human-Biometeorological Assessment of Heat Waves in Athens. *Theor. Appl. Climatol.* **2011**, *105*, 99–106. [CrossRef]
48. Giannaros, T.M.; Kotroni, V.; Lagouvardos, K.; Matzarakis, A. Climatology and Trends of the Euro-Mediterranean Thermal Bioclimate. *Int. J. Climatol.* **2018**, *38*, 3290–3308. [CrossRef]
49. Katavoutas, G.; Founda, D. Intensification of Thermal Risk in Mediterranean Climates: Evidence from the Comparison of Rational and Simple Indices. *Int. J. Biometeorol.* **2019**, *63*, 1251–1264. [CrossRef]
50. Lamnisos, D.; Giannakou, K.; Jakovljevic, M. (Michael) Demographic Forecasting of Population Aging in Greece and Cyprus: One Big Challenge for the Mediterranean Health and Social System Long-Term Sustainability. *Health Res. Policy Syst.* **2021**, *19*, 21. [CrossRef]
51. Chen, Y.; Matzarakis, A. Modified Physiologically Equivalent Temperature—Basics and Applications for Western European Climate. *Theor. Appl. Climatol.* **2018**, *132*, 1275–1289. [CrossRef]
52. Chen, Y.-C.; Chen, W.-N.; Chou, C.C.-K.; Matzarakis, A. Concepts and New Implements for Modified Physiologically Equivalent Temperature. *Atmosphere* **2020**, *11*, 694. [CrossRef]
53. Fukuda, K. Interpolation and Forecasting of Population Census Data. *J. Popul. Res.* **2010**, *27*, 1–13. [CrossRef]
54. Matzarakis, A.; Muthers, S.; Koch, E. Human Biometeorological Evaluation of Heat-Related Mortality in Vienna. *Theor. Appl. Climatol.* **2011**, *105*, 1–10. [CrossRef]
55. Kuchcik, M. Mortality and Thermal Environment (UTCI) in Poland—Long-Term, Multi-City Study. *Int. J. Biometeorol.* **2021**, *65*, 1529–1541. [CrossRef] [PubMed]

56. Bouchama, A.; Abuyassin, B.; Lehe, C.; Laitano, O.; Jay, O.; O'Connor, F.G.; Leon, L.R. Classic and Exertional Heatstroke. *Nat. Rev. Dis. Prim.* **2022**, *8*, 8. [CrossRef]
57. Cramer, M.N.; Jay, O. Biophysical Aspects of Human Thermoregulation during Heat Stress. *Auton. Neurosci. Basic Clin.* **2016**, *196*, 3–13. [CrossRef]
58. Matzarakis, A.; Mayer, H.; Iziomon, M.G. Applications of a Universal Thermal Index: Physiological Equivalent Temperature. *Int. J. Biometeorol.* **1999**, *43*, 76–84. [CrossRef]
59. Koppe, C.; Jendritzky, G. Inclusion of Short-Term Adaptation to Thermal Stresses in a Heat Load Warning Procedure. *Meteorol. Z.* **2005**, *14*, 271–278. [CrossRef]
60. Giannaros, C.; Kotroni, V.; Lagouvardos, K.; Oikonomou, C.; Haralambous, H.; Papagiannaki, K. Hydrometeorological and Socio-Economic Impact Assessment of Stream Flooding in Southeast Mediterranean: The Case of Rafina Catchment (Attica, Greece). *Water* **2020**, *12*, 2426. [CrossRef]
61. Urdiales-Flores, D.; Zittis, G.; Hadjinicolaou, P.; Osipov, S.; Klingmüller, K.; Mihalopoulos, N.; Kanakidou, M.; Economou, T.; Lelieveld, J. Drivers of Accelerated Warming in Mediterranean Climate-Type Regions. *NPJ Clim. Atmos. Sci.* **2023**, *6*, 97. [CrossRef]
62. Matzarakis, A.; Laschewski, G.; Muthers, S. The Heat Health Warning System in Germany—Application and Warnings for 2005 to 2019. *Atmosphere* **2020**, *11*, 170. [CrossRef]
63. Tony Wolf, S.; Cottle, R.M.; Fisher, K.G.; Vecellio, D.J.; Larry Kenney, W. Heat Stress Vulnerability and Critical Environmental Limits for Older Adults. *Commun. Earth Environ.* **2023**, *4*, 486. [CrossRef]
64. Galanaki, E.; Emmanouil, G.; Lagouvardos, K.; Kotroni, V. Long-Term Patterns and Trends of Shortwave Global Irradiance over the Euro-Mediterranean Region. *Atmosphere* **2021**, *12*, 1431. [CrossRef]
65. Lagouvardos, K.; Dafis, S.; Kotroni, V.; Kyros, G.; Giannaros, C. Exploring Recent (1991–2020) Trends of Essential Climate Variables in Greece. *Atmosphere* **2024**, *15*, 1104. [CrossRef]
66. Zyrichidou, I.; Balis, D.; Koukouli, M.E.; Drosoglou, T.; Bais, A.; Gratsea, M.; Gerasopoulos, E.; Liora, N.; Poupkou, A.; Giannaros, C.; et al. Adverse Results of the Economic Crisis: A Study on the Emergence of Enhanced Formaldehyde (HCHO) Levels Seen from Satellites over Greek Urban Sites. *Atmos. Res.* **2019**, *224*, 42–51. [CrossRef]
67. Martinez, G.S.; Kendrovski, V.; Salazar, M.A.; de'Donato, F.; Boeckmann, M. Heat-Health Action Planning in the WHO European Region: Status and Policy Implications. *Environ. Res.* **2022**, *214*, 113709. [CrossRef] [PubMed]
68. Matzarakis, A. Communication Aspects about Heat in an Era of Global Warming –The Lessons Learnt by Germany and Beyond. *Atmosphere* **2022**, *13*, 226. [CrossRef]
69. Winklmayr, C.; Matthies-Wiesler, F.; Muthers, S.; Buchien, S.; Kuch, B.; an der Heiden, M.; Mücke, H.-G. Heat in Germany: Health Risks and Preventive Measures. *J. Health Monit.* **2023**, *8*, 3–32. [CrossRef]

climate

MDPI

Article

Exploring the Effects of Greenhouse Gases and Particulate Emissions on Quality of Life: A Country-Level Empirical Study

Dongli Zhang [1], Wullianallur Raghupathi [1,*] and Viju Raghupathi [2]

[1] Gabelli School of Business, Fordham University, 140 W. 62nd Street, New York, NY 10023, USA; dzhang@fordham.edu

[2] Koppelman School of Business, Brooklyn College, City University of New York, Brooklyn, NY 11210, USA; vraghupathi@brooklyn.cuny.edu

* Correspondence: raghupathi@fordham.edu

Abstract: This study explores the relationship between greenhouse gases (GHGs) and particulate emissions and quality of life. The aim is to understand how emissions affect quality of life globally—across countries, regions, and the global population. Statistical methods were used to examine the impact of various emissions' indicators on different aspects of quality of life. The study highlights the urgent need for climate change action and encourages policymakers to take strategic steps. Climate change adversely affects numerous aspects of daily life, leading to significant consequences that must be addressed through policy changes and global governance recommendations. Key findings include that higher CO_2 and methane emissions and air pollution negatively impact quality of life. CO_2 emissions are positively associated with electricity while air pollution is positively associated with GDP and negatively with unemployment. Air pollution has an adverse effect on all three aspects of the children's welfare dimension of quality of life. These results provide timely and convincing insights for policy- and decision-making aimed at mitigating the impact of emissions on quality of life.

Keywords: access to electricity; air pollution; climate change; CO_2 emission; freshwater withdrawal; GDP; greenhouse gas emission (GHG); immunization rate; methane emission; mortality rate; quality of life; school enrollment; unemployment

1. Introduction

The purpose of this exploratory work is to examine the association between greenhouse gas and other emissions (GHGs) and quality of life (QoL) at the macro-level. When we talk about the macro-level, we are referring to the impact on QoL that arises out of the adverse effects of emissions at the country level [1–7]. Prior research has mostly focused on narrative or large case studies and hardly any empirical studies have been conducted. In this paper, we develop a framework for conceptualizing and studying the association between emission factors and QoL indicators at the country level. Our research is motivated by various propositions that climate change factors such as CO_2 emissions, PM2.5 air pollution, methane emissions, annual freshwater withdrawals, and others have the potential to adversely impact QoL indicators such as access to electricity, GDP growth, mortality rates, unemployment, school enrollment, and the percentage of measles immunization at the country level [8–13]. By identifying the specific associations, policymakers may implement targeted solutions to prevent or reduce emissions (and thereby climate change) and mitigate its adverse effects on QoL [3,6–9]. This study addresses relationships empirically using statistical analysis. Although studies in past decades have examined and generally confirmed the negative association between emissions and QoL, very few have conducted empirical studies; most are anecdotal or case studies or descriptive in nature. Hence, the causal or associative relation between the two is not well understood. Unlike many of the others, our study is data-driven and provides empirical evidence upon which country leaders and policymakers can base decisions to allocate their limited resources to prevent or mitigate climate change by reducing emissions to improve QoL.

In this study, GGE climate change is characterized by several world development indicators (WDIs) that collectively specify the climate change in a country (https://datatopics.worldbank.org/world-development-indicators/, accessed on 2 October 2024). The World Bank has already pre-classified several key indicators as characterizing climate change and, therefore, is an established and recognized source of climate change variables. Quality of life is represented by a set of variables from historical research [2] and the data drawn from the WDI database. We believe that important insights can be derived from studying potential associations between GHG and other emissions and quality of life. This study makes several contributions. First, we introduce the contemporary literature on climate change, emissions, and QoL. Second, we develop an initial framework of the relationship between emissions and QoL. Third, we empirically examine the relationship with a large dataset, so the generalizability of the conclusions is possible. Fourth, by studying the macro-effects, countries can develop preventive and mitigation strategies for climate change to promote a higher quality of life.

The balance of the paper is organized as follows: we first provide background information with respect to climate change and emissions and quality of life. We then outline the conceptual framework of the association between emissions and QoL and develop the hypotheses. Fourth, we describe the methods and results. Fifth, we discuss the results of our statistical analysis and implications. We then identify the scope and limitations. This is followed by a discussion of the key contributions, conclusions, and directions for future research.

2. Research Background

Global climate change is a serious threat to quality of life as it affects many aspects of human well-being [1–5]. This research examines country-level data to examine how one aspect of climate variability, namely, GHG and other emissions, correlates with quality of life variables. By leveraging data analytics, the goal is to gain a better understanding of how GHGs impact living standards. These insights are vital for policymakers and researchers in creating strategies to improve resiliency and quality of life amidst environmental changes [3,6–9]. The investigation focused on the effects of indicators such as CO_2 and methane emissions, PM2.5 air pollution, annual freshwater withdrawals, and others on quality of life variables including mortality rates, immunization against measles, school enrollment, GDP growth and unemployment, and others [8,10–13]. This analysis of direct impacts helps identify vulnerable populations and critical areas requiring immediate attention. Beyond direct effects, climate change triggers a series of indirect consequences through environmental degradation [14,15]. These include, for example, increased greenhouse gas emissions which affect public health and socio-economic stability [3,16–22]. Understanding these chain reactions is essential for developing holistic and sustained climate resilience strategies [23,24]. The concept of quality of life encompasses many aspects of human experience, including economic, physical, and social well-being [25–31]. Climate change will have a profound impact on these aspects: climate policies, for example, that aim to reduce CO_2 emissions may have an impact on access to electricity due to the use of fossil fuels in the generation of electricity [3]. Climate change could make it more difficult to achieve adequate immunization because of the spread of diseases [10,32–38]. Climate-related food- and water-borne disease risks could increase mortality rates, especially for children under five [39,40]. The number of students enrolled in school may be affected if families prioritize short-term survival over educational opportunities [41]. Climate change could also have a significant impact on food production [3]. The study provides critical insights into how climate change affects quality of life variables across different countries and regions. By studying various socio-economic factors such as health, education, and economic stability, researchers can gain an understanding of the broader economic and social impacts [42–45].

A key assumption is that GHGs indeed have an adverse effect on quality of life. Key to this research, therefore, is the question of whether there is an association between GHGs and quality of life: are GHGs directly associated with quality of life? If so, which variables are significant?

As the study is data-driven and experiments and control groups are not appropriate in this study, we can at best examine association. Additionally, our study is limited to the emission aspect of climate change. The insights gained from this research can be used to shape strategies by governments and non-governmental organizations to increase resilience against climate impacts on quality of life. This research contributes significantly to academic discourse and practical discussion on sustainable development. It offers new perspectives for adapting to climate change and fighting it. Understanding the interdependencies between climate change and quality of life is pivotal for crafting resilient economic, education, environmental, and public health policies while simultaneously reducing climate change [46,47]. Unpacking the association between climate change and quality of life data informs sustainable economic growth models and global climate agreements [48–50]. Identifying which quality of life variables are most sensitive to climate change helps prioritize resource allocation and international aid [37,51].

Climate change refers to long-term alterations in temperature and weather patterns. While these changes can occur naturally, such as through variations in solar activity or significant volcanic eruptions, human activities have been the dominant cause since the 1800s. This is largely due to the burning of fossil fuels, including coal, oil, and gas [3,4,18]. Burning fossil fuels releases greenhouse gasses that function like a blanket around the Earth, trapping heat from the sun and causing temperatures to rise. The primary greenhouse gases driving climate change are carbon dioxide and methane [5,8]. These emissions result from activities like using gasoline to power vehicles or burning coal to heat buildings, clearing land, and deforestation [23,36,49]. The primary sectors responsible for greenhouse gases include energy, industry, transportation, buildings, agriculture, and land use [1,3,9].

The indirect impacts from climate change extend to an individual's quality of life along the lines of economic stability and social, physical, and mental well-being. Quality of life (QoL) is defined by the World Health Organization (WHO) as an individual's perception of their position in life, taking into account the cultural and value systems that one is embedded in, as well as the goals, standards, expectations, and concerns (https://www.who.int/toolkits/whoqol accessed on 2 October 2024). The concept of QoL addresses the overall well-being, inclusive of positive and negative aspects at a point in time [52–55]. Indicators to study QoL include those that are relevant to material living conditions (such as food, clothing, and shelter) as well as to quality of life (such as environment, education, community, health, governance, life satisfaction, safety, and work–life balance) [56–58]. Traditionally, quality of life was assessed using GDP. However, in recent years, this view has been considered myopic, as it does not cover aspects of a person's current and future living conditions [59–62]. While GDP and economic growth remain key for well-being, the goals and aspirations of people are equally important in considering the overall quality of life for a sustainable society [56,63–66].

Climate change has a detrimental socio-economic impact on the quality of life of the population [4,9,27,43]. Environmental effects such as rising sea levels and saltwater intrusion have forced communities to relocate, while droughts have exposed people to famine. Extreme weather conditions have exacerbated the incidence of food scarcity, droughts, and displacement worldwide [15,45,50]. The number of people facing acute food scarcity worldwide has increased from 149 million prior to 2019 (pre-COVID-19) to 333 million in 2023 (post COVID-19) in countries monitored by the World Food Program. Between 2010 and 2020, the death toll from climate change events was 15 times higher in highly vulnerable regions [7]. The negative impacts of climate change on health range from heat-related or vector-borne diseases to water-borne infections, allergies, malnutrition, respiratory issues, and mental health challenges [10,34,37,44,46]. These environmental and health threats cover a wide array of civil, political, economic, social, and cultural rights, including the rights to life, water, food, shelter, health, security, and cultural preservation [4,18,50].

At the national and local levels, the groups most vulnerable to the environmental and health impacts of climate change include poor and minority communities, women, children, people with chronic illnesses and disabilities, and those living in regions that are prone to extreme weather and climates [13,15,17,44,45]. At the global level, there is also a significant disparity in the effects of climate change on the quality of life. Despite being the lowest contributors to climate change, low-income countries suffer the most impacts, while high-income countries which are the highest contributors suffer less severe consequences [36]. This is because of the inequality in the capacity to adapt to the challenges posed by climate change [4]. As an example, the per capita GHG emissions in 2004 in developed nations such as the United States, Canada, and Australia approached 6 metric tons, and those in Japan and Western European countries ranged from 2 to 5 metric tons. In contrast, annual per capita GHG emissions in developing countries overall approximate 0.6 metric tons, and more than 50 developing countries have annual per capita GHG emissions less than 0.2 metric tons [4].

As global temperature increases, rich countries' economies continue to prosper, but the economic growth of poor countries is seriously impaired more than previously estimated [23,24]. The consequences for economic growth in poor countries will be substantial if we continue down a "business-as-usual" path of increasing carbon dioxide concentrations and rapid climate change, with poor countries' mean annual growth rate decreasing from 3.2% to 2.6%. 27 Poor countries are likely to suffer a greater negative effect than rich countries from climate change since they more often experience high temperatures [4,43]. Additionally, their economic growth depends very much on agriculture, natural resource extraction, and other sectors exposed to extreme weather fluctuations. Furthermore, air conditioning, insurance, and other risk management alternatives are less available in poor countries than in rich countries [4]. Considering the major adverse effects of Greenhouse Gas and other emissions on various dimensions of QoL, this study undertakes to shed light on the association between the two and find out the significant variables and linkages. Below, we next outline the conceptual framework and hypotheses.

3. Conceptual Framework and Hypotheses

Based on the background discussion above, we draw broadly from the literature in several disciplines, including the conceptual basis for climate change [3,23], quality of life [61,62,66], and the association between the two [49,51], global development [36,49,53], and the substantive number of publications at the UN, World Bank, IPCC, and others, to develop the conceptual framework for this study shown in Figure 1 below.

The independent variables reflecting climate change include CO_2 emissions, PM2.5 air pollution, annual freshwater withdrawals, and methane emissions. These are drawn from the World Bank's 'climate change' world development indicators [67–71]. The dependent variables characterizing QoL include access to electricity, GDP growth, mortality rates, unemployment, school enrollment, and the percentage of measles immunization. The six dependent variables are categorized into two dimensions: three variables relating to economic growth (access to electricity, GDP growth, and unemployment) and three variables relating to the education and health welfare of children (mortality rate, school enrollment, and immunization rate). These are drawn from the QoL literature and data retrieved from the WDI database [72–77]. Inflation and population are used as control variables. CO_2 emissions are mainly caused by the burning of coal, oil, and natural gas for energy production and transportation. Other significant sources include industrial processes, deforestation, and certain agricultural practices. Increased CO_2 emissions could indicate economic growth and the expansion of manufacturing industries. Therefore, we hypothesize that there may be a positive relationship between CO_2 emissions and the economic growth dimension of quality of life. However, CO_2 emissions will also lead to climate change and adversely affect the children's welfare dimension of quality of life. PM2.5 air pollution has a similar impact on the quality of life as CO_2 emissions. Annual water withdrawals will benefit both the economic growth and children's welfare dimensions of quality of life in the short term. Methane emissions are mainly caused by agricultural activities and the burning of organic materials. While highly industrialized regions can also have high methane emissions due to activities such as fossil fuel extraction and waste management, for the purpose of this analysis, we are theorizing that high methane emissions correlate with the levels of industrialization. Therefore, methane emissions are influenced by a variety of factors, and high emissions do not necessarily correlate with the level of industrialization. In other words, methane emissions are negatively related to both dimensions of quality of life. Table 1 presents the hypotheses tested in this study.

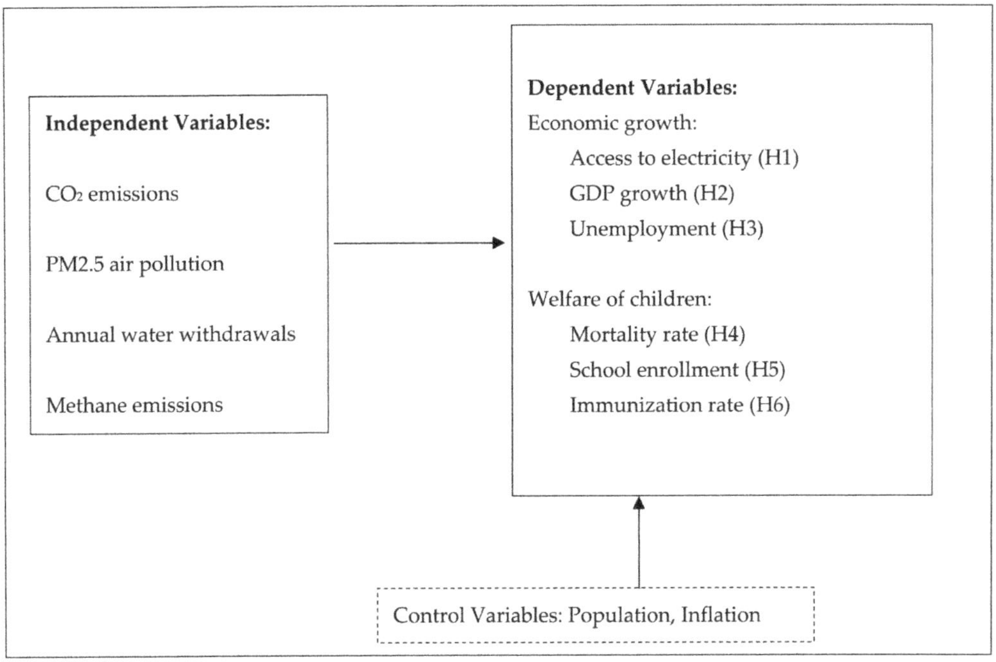

Figure 1. Conceptual framework.

Table 1. Hypotheses of the study.

H1: CO_2 emissions have a positive impact on the economic growth dimension of quality of life.	H1.1: CO_2 emissions are positively related to access to electricity. H1.2: CO_2 emissions are positively related to GDP growth. H1.3: CO_2 emissions are negatively related to unemployment.
H2: CO_2 emissions have a negative impact on the children's welfare dimension of quality of life.	H2.1: CO_2 emissions are positively related to the mortality rate. H2.2: CO_2 emissions are negatively related to school enrollment. H2.3: CO_2 emissions are negatively related to the measles immunization rate.
H3: PM2.5 air pollution has a positive impact on the economic growth dimension of quality of life.	H3.1: PM2.5 air pollution is positively related to access to electricity. H3.2: PM2.5 air pollution is positively related to GDP growth. H3.3: PM2.5 air pollution is negatively related to unemployment.
H4: PM2.5 air pollution has a negative impact on the children's welfare dimension of quality of life.	H4.1: PM2.5 air pollution is positively related to the mortality rate. H4.2: PM2.5 air pollution is negatively related to school enrollment. H4.3: PM2.5 air pollution is negatively related to the measles immunization rate.
H5: Annual water withdrawals have a positive impact on the economic growth dimension of quality of life.	H5.1: Annual water withdrawals are positively related to access to electricity. H5.2: Annual water withdrawals are positively related to GDP growth. H5.3: Annual water withdrawals are negatively related to unemployment.

Table 1. *Cont.*

H6: Annual water withdrawals have a positive impact on the children's welfare dimension of quality of life.	H6.1: Annual water withdrawals are negatively related to the mortality rate. H6.2: Annual water withdrawals are positively related to school enrollment. H6.3: Annual water withdrawals are positively related to the measles immunization rate.
H7: Methane emissions have a negative impact on the economic growth dimension of quality of life.	H7.1: Methane emissions are negatively related to access to electricity. H7.2: Methane emissions are negatively related to GDP growth. H7.3: Methane emissions are positively related to unemployment.
H8: Methane emissions have a negative impact on the children's welfare dimension of quality of life.	H8.1: Methane emissions are positively related to the mortality rate. H8.2: Methane emissions are negatively related to school enrollment. H8.3: Methane emissions are negatively related to the measles immunization rate.

4. Methods

4.1. Data

Data were collected at the country level for the period 2010–2019 for 217 countries. The sample size was 767, providing a robust dataset with sufficient variation for our analysis. Due to the significant differences in data magnitudes, all data were normalized. The collection period spanned ten years, capturing a wide range of economic, social, and environmental variables across different regions and development stages. The process of normalization was crucial to ensure comparability across diverse variables. Normalization involved adjusting the values measured on different scales to a common scale without distorting differences in the ranges of values. This procedure helps in minimizing the impact of scale differences and allows for a more straightforward interpretation of results. Table 2 shows how each variable was measured. This detailed table provides clear definitions and measurements for each variable, ensuring that the data are well understood and accurately interpreted in subsequent analyses. By defining each variable explicitly, we can ensure consistency and reliability in the analysis, allowing for meaningful comparisons and robust conclusions.

Table 2. Measurement of variables.

Variables	Measurement
Population total	Total number of people living in a specific area such as a country
Inflation	Annual percentage change in the cost of goods and services
CO_2 emissions	Average emissions of CO_2 per person measured in metric tons
PM2.5 air pollution	Average concentration of particulate matter smaller than 2.5 microns in the air
Annual water withdrawals	Percentage of country's available water resources used per year
Methane emissions	Total weight of methane gas emissions measured in kt of CO_2 equivalent
Access to electricity	Proportion of population with access to electricity
GDP growth	Yearly percentage increase in national economic output
Mortality rate	Annual deaths of children under five per 1000 live births
Unemployment	Proportion of the labor force that is jobless but seeking employment
School enrollment	Percent of children enrolled in secondary education
Immunization	Proportion of children between 12 and 32 months who have received the measles vaccine

4.2. Results

Table 3 represents the correlation matrix of the variables that were identified in the research framework.

Table 3. Correlation matrix.

	Popu	Infl	CO_2	PM2.5	Water	Metha	Elec	GDP	Mort	Unem	School	Immu
Population	1											
Inflation	0.113 **	1										
CO_2	0.018	−0.188 **	1									
PM2.5	0.150 **	0.179 **	−0.234 **	1								
Water withdrawals	0.029	0.112 **	0.265 **	0.344 **	1							
Methane	0.797 **	0.083 *	0.184 **	−0.032	0.004	1						
Electricity	0.055	−0.223 **	0.460 **	−0.458 **	0.073 *	0.113 **	1					
GDP growth	0.048	0.052	−0.154 **	0.247 **	0.024	−0.032	−0.224 **	1				
Mortality	0.030	0.193 **	−0.523 **	0.604 **	−0.051	−0.059	−0.871 **	0.269 **	1			
Unemployment	−0.078 *	−0.075*	−0.034	−0.181 **	−0.056	−0.061	0.200 **	−0.290 **	−0.228 **	1		
School enrollment	−0.022	−0.220 **	0.598 **	−0.572 **	0.062	0.082 *	0.867 **	−0.292 **	−0.895 **	0.251 **	1	
Immunization	−0.074 *	−0.073 *	0.380 **	−0.334 **	0.080 *	0.007	0.535 **	−0.101 **	−0.652 **	0.206 **	0.594 **	1

** Correlation is significant at the 0.01 level (2-tailed). * Correlation is significant at the 0.05 level (2-tailed).

We tested our hypotheses by conducting regression analyses. The regression model is shown as below:

$$\text{Dependent Variable} = \beta_0 + \beta_1{*}CO_2\text{ emissions} + \beta_2{*}\text{PM2.5 air pollution} + \beta_3{*}\text{Annual water withdrawals} + \beta_4{*}\text{Methane emissions} + \beta_5{*}\text{Population} + \beta_6{*}\text{Inflation} + \text{Year Indicators} + \varepsilon$$

A particularly important control variable in our regression was population. Since data were collected from multiple countries over approximately ten years, the analysis needed to consider the fixed effects of country and time. To account for the fixed effects of country, we included population in the regression model. This allowed us to control for variations across different countries. Additionally, to address the fixed effects of time, we incorporated the year variable into the regression model. The year helped control for any potential time-specific effects that could influence the results. By including these control variables and fixed effects, we aim to isolate the impact of the independent variables on the dependent variables, providing a more accurate and reliable analysis of the data. This approach helps ensure that the results are robust and not biased by unobserved heterogeneity across countries and over time. In summary, our regression analysis framework was designed to rigorously test the relationships between GHG and other emissions' variables (such as CO_2 emissions, PM2.5 air pollution, annual water withdrawals, and methane emissions) and quality of life or economic growth variables (such as access to electricity, GDP growth, mortality rates, unemployment, school enrollment, and the percentage of measles immunization). By accounting for country and time fixed effects, we strive to provide a comprehensive and nuanced understanding of these relationships.

Tables 4 and 5 represent the results of the regression analysis. Standardized regression coefficients and their significance levels are summarized in the tables.

Table 4. The impact of emissions on economic growth dependent variables.

	Dependent Variable		
	Access to Elec.	**GDP Growth**	**Unemployment**
CO_2 emissions	0.312 ***	−0.083 **	−0.101 ***
PM2.5 air pollution	−0.456 ***	0.228 ***	−0.211 ***
Annual water withdrawals	0.154 ***	−0.034	0.051
Methane emissions	−0.136***	−0.061	−0.036
Population	0.234 ***	0.065	−0.010
Inflation	−0.115 ***	−0.003	−0.058
R-Square	0.396	0.095	0.057
p-value of overall model	<0.001	<0.001	<0.01

*** $p < 0.01$. ** $p < 0.05$.

Table 5. The impact of emissions on the welfare of children dependent variables.

	Dependent Variable		
	Mortality Rate	**School Enrollment**	**Measles Immunization**
CO_2 emissions	−0.361 ***	0.454 ***	0.300 ***
PM2.5 air pollution	0.604 ***	−0.529 ***	−0.311 ***
Annual water withdrawals	−0.163***	0.126 ***	0.104 ***
Methane emissions	0.182 ***	−0.150 ***	−0.081
Population	−0.199 ***	0.172 ***	0.026
Inflation	0.043	−0.061 **	0.001
R-Square	0.563	0.588	0.229
p-value of overall model	<0.001	<0.001	<0.001

*** $p < 0.01$. ** $p < 0.05$.

Table 6 presents the results of the hypotheses tested through our regression analyses. Each hypothesis is evaluated based on the direction and significance of the estimated coefficients. The results are categorized as either supported or rejected, with specific notes on the nature of the rejection, such as the sign being opposite to the hypothesis or the result being statistically insignificant.

Table 6. Hypotheses' results based on the data analysis.

Hypothesis		Support/Reject
H1	H1.1	Support
	H1.2	Reject (the opposite sign)
	H1.3	Support
H2	H2.1	Reject (the opposite sign)
	H2.2	Reject (the opposite sign)
	H2.3	Reject (the opposite sign)
H3	H3.1	Reject (the opposite sign)
	H3.2	Support
	H3.3	Support
H4	H4.1	Support
	H4.2	Support
	H4.3	Support
H5	H5.1	Support
	H5.2	Reject (insignificant)
	H5.3	Reject (insignificant)
H6	H6.1	Support
	H6.2	Support
	H6.3	Support
H7	H7.1	Support
	H7.2	Reject (insignificant)
	H7.3	Reject (insignificant)
H8	H8.1	Support
	H8.2	Support
	H8.3	Reject (insignificant)

5. Discussion

The results paint a complicated picture of the association between climate change and human development. Among the key findings, higher CO_2 emission is associated with higher access to electricity. Electricity generation mainly depends on coal and its products, which has led to the increase in CO_2 emissions. By proxy, higher CO_2 emissions imply more demand for electricity thereby resulting in a higher production of electricity utilizing fossil fuels. While electricity itself is good, its production from fossil fuels is indicated in the positive association. On the other hand, higher CO_2 emissions act as a catalyst to the increased use of electricity as society moves towards alternative and clean energies. However, higher CO_2 emission is not associated with higher GDP growth; on the contrary, higher CO_2 is negatively correlated with GDP growth, which means higher CO_2 is associated with lower GDP growth. This paradoxical result needs to be further studied from a reverse causality perspective applying the Granger Causality Test. Historically, causality results confirmed the unidirectional causality running from economic growth to CO_2 emissions but not vice versa. Higher CO_2 emission is associated with lower unemployment. This finding suggests higher CO_2 emission because of rapid industrialization is associated with reduced unemployment. As countries strive for sustainable development and poverty alleviation, emissions are collateral damage resulting in environmental degradation and climate change. A careful balance between economic growth, increased job creation, and higher emissions needs to be achieved. As a contrary indication, higher CO_2 emission is not associated with a higher mortality rate; on the contrary, higher CO_2 emission is associated with a lower mortality rate. A plausible explanation is that CO_2 emission is a consequence of economic growth; simultaneously, health quality improves with higher economic growth. Similarly, higher CO_2 emission is not associated with lower school enrollment; on the contrary, higher CO_2 emission is associated with higher school enrollment. This too is likely due to economic growth. Higher CO_2 emission is not associated with a lower measles immunization rate; on the contrary, higher CO_2 emission is associated with a higher measles immunization rate. Relating to healthcare, economic growth (resulting in increased CO_2 emission) may result in better quality healthcare, with a higher immunization rate being a good indicator of quality healthcare. Overall, the results show that CO_2 emissions have a mixed impact on the economic growth dimension of quality of life. The results also show that CO_2 emissions have a positive impact on the children's welfare dimension of quality of life, which is different from how we hypothesized the relationships.

Addressing air pollution, higher PM2.5 air pollution is not associated with higher access to electricity; on the contrary, higher PM2.5 air pollution is associated with lower access to electricity. This empirical result is quite counterintuitive and presents an interesting research question for future studies. One possible explanation is that lower access to electricity is often a sign of underdevelopment, where the infrastructure for modern, cleaner energy sources may be lacking. In such places, people may rely on cheaper but dirtier fossil fuels. Therefore, when the dataset includes multiple countries at different stages of development, more polluted nations are associated with lower access to electricity. The next two findings are reflecting reality as higher PM2.5 air pollution is associated with higher GDP growth and lower unemployment. One of the key findings of this study relates to the association between air pollution and children's welfare as a proxy for quality of life (health). Higher PM2.5 air pollution is associated with a higher mortality rate, lower school enrollment, and a lower measles immunization rate. It is obvious air pollutions results in premature deaths, may prevent children from attending school (e.g., 'smog' days), and declining immunizations (difficulty in access to immunization due to bad-quality air) [78,79]. In summary, higher PM2.5 air pollution also shows some mixed impacts on the economic growth dimension of quality of life. But in general, the relationships are consistent with what we have hypothesized. PM2.5 air pollution is a more direct measure of climate change and the negative consequences of the change.

Regarding annual freshwater withdrawal and its effects, higher annual freshwater withdrawal is associated with higher access to electricity. This reflects the reality as hydroelectric projects associated with electricity generation result in more consumption (access) of electricity. Contrary to initial expectations, the analysis found no significant correlation between annual freshwater withdrawal and GDP growth. Likewise, the relationship between freshwater withdrawal and unemployment rates is statistically insignificant. This suggests that variations in water usage do not meaningfully affect GDP growth or employment levels. These empirical results do not confirm the hypothesized connections between water usage and key economic indicators, implying that other variables may play a more influential role in these outcomes. Consequently, further research is necessary to explore alternative factors. Concerning the children's dimension of QoL, higher annual freshwater withdrawal is associated with a lower mortality rate. Again, the implication is greater withdrawal means better economic growth resulting in a reduced mortality rate. Likewise, higher annual freshwater withdrawal is associated with higher school enrollment (availability of food, water, and affordability with economic growth). Additionally, higher annual freshwater withdrawal is associated with a higher measles immunization rate. Of course, the rapid depletion of freshwater itself has negative effects so freshwater must be managed efficiently.

Next, in examining the effect of methane emission, higher methane emission is associated with lower access to electricity. It is conceivable that water resources needed for electricity generation are diverted to farming and agricultural activities resulting in higher methane emission but lower electricity production. Further studies are warranted. However, the analysis indicates that there is no significant relationship between methane emissions and GDP growth, nor between methane emissions and unemployment rates. This suggests that methane emissions are not a meaningful indicator of economic growth or employment opportunities. Furthermore, when assessing the welfare dimension related to children's quality of life, higher methane emissions are linked to increased mortality rates and lower school enrollment, but there is no significant connection with measles immunization rates. Methane results in poor air quality by contributing to the formation of ground-level ozone and particulate pollution. Exposure to ozone and particulate pollution damages airways, aggravates lung diseases, causes asthma attacks, increases rates of preterm birth, cardiovascular morbidity, and mortality, and heightens stroke risk. This warrants attention. To summarize, CO_2 and methane emissions do impact some of the economic aspects of QoL. Air pollution is associated with some of the dimensions of economic growth while having a negative impact on the children's welfare component of QoL. These are two of the key findings.

6. Scope and Limitations

While our study is constrained by data availability in terms of the time and number of countries, it has other limitations. For example, the number of countries included is limited, and because the data source is the world development indicators (WDIs) of the World Bank, there is a reporting lag that resulted in numerous missing values. Additionally, being an exploratory study, we considered only a limited set of key variables from the climate change category and key variables from QoL to scope our study. A more comprehensive study could include a wider range of variables from diverse sources, cover more countries, and extend over a longer period. To gain a macro-perspective on the climate change–QoL association, we also incorporated additional control variables. However, there may be additional intervening or compounding variables influencing the adverse relationship with QoL. Furthermore, while our study examines associations and relationships between variables, it does not investigate causality. The study is data-driven and experiments or control groups are not suitable. Lastly, this is an exploratory predictive analytic study; more advanced analytical methods including the Granger Causality Test could be applied in future research.

7. Conclusions and Future Research

Our exploratory study demonstrates that climate change has a negative impact on quality of life. Both climate change and QoL are critical to the health and well-being of individuals and society. To conceptualize these phenomena, it is essential to extend the focus beyond the individual context and incorporate the broader cultural, social, and structural context in which these relationships occur. Furthermore, the ethical, geographical, legal and regulatory, and political contexts must be considered. This requires a multi-dimensional approach, including interdisciplinary research, diverse data sources, and innovative frameworks. Given the ever-increasing disparities in the economy, health, and the environment, new policy and research initiatives are needed to address the adverse effects of climate change on QoL and vice versa.

When identifying and developing action research and policies and strategies, governments and stakeholders must consider the dual role of QoL as both a facilitator of improvement in the human condition and a contributor to climate change, especially through rapid industrialization. Addressing these large-scale disparities requires interventions at the highest levels. Activists, government and non-profit entities, policymakers, researchers, and other stakeholders' policies must work together in this effort: researchers should investigate and share insights and recommendations with policymakers, while policymakers need to communicate the challenges and needs of the economic and human welfare dimensions to researchers. Researchers can leverage variations in national cultural, political, and societal systems to explore how different forms of governance impact the relationship between climate change and QoL. Due to constraints in budgets and time, nations need to prioritize specific aspects of climate change to advance key aspects of QoL. Education plays a crucial role in enabling informed decisions to mitigate the negative impacts of climate change. Additionally, new models and empirical studies are needed to capture the nuances of this relationship and validate the largely anecdotal and case study evidence currently available.

Our study also presents several opportunities for future research. Future research could integrate additional cultural, regional, and social variables—such as population demographics, local cultural and religious beliefs, governance structures, and historical context—for a more comprehensive analysis. Further, incorporating data from other sources could expand the study to include more countries and variables, particularly a more precise classification of income levels, geographic regions, population size, and other factors, thereby broadening the scope of the analysis. Additionally, case studies and action research can complement empirical studies. Methodologically, future research can deploy predictive analytics meta-regression analysis and reverse causal studies to assess the relationships between climate change and QoL variables. Future studies should also move beyond country-level factors to explore the broader social context in which climate change and QoL occur, such as gender inequality, impacts on children and the elderly, and other societal dimensions. This approach will generate insights that can inform effective policies and interventions to address disparities in climate change and QoL. Additionally, ensuring adequate funding, political commitment, technology transfer, and fostering partnerships are essential for more effective emission mitigation and QoL efforts. Although the study of the relationship between emissions and QoL is still in the nascent phase, further empirical research and case studies can help accelerate its maturing process.

Author Contributions: Conceptualization, W.R.; Methodology, W.R., D.Z. and V.R.; Software, W.R. and V.R.; Formal analysis, D.Z. and W.R.; Writing—original draft, W.R. and V.R.; Writing—review & editing, D.Z. and V.R.; Project administration, W.R., D.Z. and V.R. All authors have read and agreed to the published version of the manuscript.

Funding: This research received no external funding.

Data Availability Statement: The data presented in this study are openly available in https://databank.worldbank.org/reports.aspx?source=2&country=ARE (accessed on 2 October 2024).

Conflicts of Interest: The authors declare no conflict of interest.

References

1. Dietz, T.; Shwom, R.L.; Whitley, C.T. Climate change and society. *Annu. Rev. Sociol.* **2020**, *46*, 135–158. [CrossRef]
2. Estoque, R.C.; Togawa, T.; Ooba, M.; Gomi, K.; Nakamura, S.; Hijioka, Y.; Kameyama, Y. A review of quality of life (QOL) assessments and indicators: Towards a "QOL-Climate" assessment framework. *AMBIO* **2018**, *48*, 619–638. [CrossRef] [PubMed]
3. IPCC. Summary for Policymakers. In *Climate Change 2023: Synthesis Report. Contribution of Working Groups I, II and III to the Sixth Assessment Report of the Intergovernmental Panel on Climate Change*; Lee, H., Romero, J., Eds.; IPCC: Geneva, Switzerland, 2023; pp. 1–34. [CrossRef]
4. Levy, B.S.; Patz, J.A. Climate change, human rights, and social justice. *Ann. Glob. Health* **2015**, *81*, 310–322. [CrossRef] [PubMed]
5. Semenza, J.C.; Ploubidis, G.B.; George, L.A. Climate change and climate variability: Personal motivation for adaptation and mitigation. *Environ. Health* **2011**, *10*, 46. [CrossRef]
6. Adger, W.N.; Barnett, J.; Heath, S.; Jarillo, S. Climate change affects multiple dimensions of well-being through impacts, information and policy responses. *Nat. Hum. Behav.* **2022**, *6*, 1465–1473. [CrossRef]
7. Albouy, D.; Graf, W.; Kellogg, R.; Wolff, H. Climate amenities, climate change, and American quality of life. *J. Assoc. Environ. Resour. Econ.* **2016**, *3*, 205–246. [CrossRef]
8. Ding, H.; Nunes, P.A. Modeling the links between biodiversity, ecosystem services and human wellbeing in the context of climate change: Results from an econometric analysis of the European forest ecosystems. *Ecol. Econ.* **2014**, *97*, 60–73. [CrossRef]
9. Fan, Q.; Fisher-Vanden, K.; Klaiber, H.A. Climate change, migration, and regional economic impacts in the United States. *J. Assoc. Environ. Resour. Econ.* **2018**, *5*, 643–671. [CrossRef]
10. Semenza, J.C. Climate change and human health. *Int. J. Environ. Res. Public Health* **2014**, *11*, 7347–7353. [CrossRef]
11. Testa, M.A.; Simonson, D.C. Assessment of quality-of-life outcomes. *N. Engl. J. Med.* **1996**, *334*, 835–840. [CrossRef]

12. Garratt, A.; Schmidt, L.; Mackintosh, A.; Fitzpatrick, R. Quality of life measurement: Bibliographic study of patient assessed health outcome measures. *BMJ* **2002**, *324*, 1417. [CrossRef] [PubMed]
13. van Daalen, K.R.; Romanello, M.; Rocklöv, J.; Semenza, J.C.; Tonne, C.; Markandya, A.; Dasandi, N.; Jankin, S.; Achebak, H.; Ballester, J.; et al. The 2022 Europe report of the Lancet Countdown on health and climate change: Towards a climate resilient future. *Lancet Public Health* **2022**, *7*, e942–e965. [CrossRef] [PubMed]
14. Kannan, R.; James, D.A. Effects of climate change on global biodiversity: A review of key literature. *Trop. Ecol.* **2019**, *50*, 31.
15. Pecl, G.T.; Araújo, M.B.; Bell, J.D.; Blanchard, J.; Bonebrake, T.C.; Chen, I.C.; Clark, T.D.; Colwell, R.K.; Danielsen, F.; Evengard, B.; et al. Biodiversity redistribution under climate change: Impacts on ecosystems and human well-being. *Science* **2017**, *355*, eaai9214. [CrossRef]
16. Bell, M.L.; Davis, D.L.; Cifuentes, L.A.; Krupnick, A.J.; Morgenstern, R.D.; Thurston, G.D. Ancillary human health benefits of improved air quality resulting from climate change mitigation. *Environ. Health* **2008**, *7*, 41. [CrossRef] [PubMed]
17. Doherty, R.M.; Heal, M.R.; O'connor, F.M. Climate change impacts on human health over Europe through its effect on air quality. *Environ. Health* **2017**, *16*, 33–44. [CrossRef]
18. Evans, G.W. Projected Behavioral Impacts of Global Climate Change. *Annu. Rev. Psychol.* **2019**, *70*, 449–474. [CrossRef]
19. Hassan, N.A.; Hashim, Z.; Hashim, J.H. Impact of Climate Change on Air Quality and Public Health in Urban Areas. *Asia Pac. J. Public Health* **2015**, *28* (Suppl. S2), 38S–48S. [CrossRef]
20. Kinney, P.L. Climate change, air quality, and human health. *Am. J. Prev. Med.* **2008**, *35*, 459–467. [CrossRef]
21. Orru, H.; Ebi, K.L.; Forsberg, B. The Interplay of Climate Change and Air Pollution on Health. *Curr. Environ. Health Rep.* **2017**, *4*, 504–513. [CrossRef]
22. Spickett, J.T.; Brown, H.; Rumchev, K. Climate Change and Air Quality: The Potential Impact on Health. *Asia Pac. J. Public Health* **2011**, *23* (Suppl. S2), 37S–45S. [CrossRef] [PubMed]
23. Stern, N. *Stern Review: The Economics of Climate Change*; Cambridge University Press: Cambridge, UK, 2007.
24. Nordhaus, W.D. A review of the Stern review on the economics of climate change. *J. Econ. Lit.* **2007**, *45*, 686–702. [CrossRef]
25. Barcaccia, B. Quality of Life: Everyone Wants It, but What Is It? 4 September 2023. Available online: https://www.forbes.com/sites/iese/2013/09/04/quality-of-life-everyone-wants-it-but-what-is-it/ (accessed on 16 August 2024).
26. Barcaccia, B.; Esposito, G.; Matarese, M.; Bertolaso, M.; Elvira, M.; De Marinis, M.G. Defining Quality of Life: A Wild-Goose Chase? *Eur. J. Psychol.* **2013**, *9*, 185–203. [CrossRef]
27. Gerson, E.M. On "Quality of life". In *American Sociological Review*; American Sociological Association: Washington, DC, USA, 1976; pp. 793–806.
28. Kerce, E.W. *Quality of Life: Meaning, Measurement, and Models*; Navy Personnel Research and Development Center: San Diego, CA, USA, 1992.
29. Nussbaum, M.; Sen, A. (Eds.) *The Quality of Life*; Clarendon Press: Oxford, UK, 1993.
30. Owczarek, K. The concept of quality of life. *Acta Neuropsychol.* **2010**, *8*, 207–213.
31. Schipper, H.; Clinch, J.J.; Olweny, C.L.M. Quality of life studies: Definitions and conceptual issues. In *Quality of Life and Pharmacoeconomics in Clinical Trials*; Spilker, B., Ed.; Lippincott-Raven Publishers: Philadelphia, PA, USA, 1996. [PubMed]
32. Balbus, J.M.; Malina, C. Identifying Vulnerable Subpopulations for Climate Change Health Effects in the United States. *J. Occup. Environ. Med.* **2009**, *51*, 33–37. [CrossRef]
33. Beniston, M. Climatic change: Possible impacts on human health. *Swiss Med. Wkly.* **2002**, *132*, 332–337. [CrossRef]
34. Haines, A.; Patz, J.A. Health effects of climate change. *JAMA* **2004**, *291*, 99–103. [CrossRef]
35. Kovats, R.S.; Haines, A.; Stanwell-Smith, R.; Martens, P.; Menne, B.; Bertollini, R. Climate change and human health in Europe. *BMJ* **1999**, *318*, 1682–1685. [CrossRef]
36. Martens, W.J.; Slooff, R.; Jackson, E.K. Climate change, human health, and sustainable development. *Bull. World Health Organ.* **1997**, *75*, 583.
37. McMichael, A.J. *Climate Change in Australia: Risks to Human Wellbeing and Health*; Austral Special Report; Nautilus Institute: Berkeley, CA, USA, 2009.
38. Wong, C. How climate change is hitting Europe: Three graphics reveal health impacts. *Nature* **2024**, *630*, 800–801. [CrossRef]
39. Green, T.R.; Taniguchi, M.; Kooi, H.; Gurdak, J.J.; Allen, D.M.; Hiscock, K.M.; Treidel, H.; Aureli, A. Beneath the surface of global change: Impacts of climate change on groundwater. *J. Hydrol.* **2011**, *405*, 532–560. [CrossRef]
40. Taylor, R.G.; Scanlon, B.; Döll, P.; Rodell, M.; Van Beek, R.; Wada, Y.; Longuevergne, L.; Leblanc, M.; Famiglietti, J.S.; Edmunds, M.; et al. Ground water and climate change. *Nat. Clim. Chang.* **2013**, *3*, 322–329. [CrossRef]
41. Anderson, A. *Combating Climate Change Through Quality Education*; Brookings Global Economy and Development: Washington, DC, USA, 2010.
42. Alborz, A. The Nature of Quality of Life: A Conceptual Model to Inform Assessment. *J. Policy Pract. Intellect. Disabil.* **2017**, *14*, 15–30. [CrossRef]
43. Newman, R.; Noy, I. The global costs of extreme weather that are attributable to climate change. *Nat. Commun.* **2023**, *14*, 6103. [CrossRef] [PubMed]
44. Thomas, F.; Sabel, C.E.; Morton, K.; Hiscock, R.; Depledge, M.H. Extended impacts of climate change on health and wellbeing. *Environ. Sci. Policy* **2014**, *44*, 271–278. [CrossRef]
45. Tol, R.S.J. The Economic Effects of Climate Change. *J. Econ. Perspect.* **2009**, *23*, 29–51. [CrossRef]

46. Frumkin, H.; McMichael, A.J.; Hess, J.J. Climate Change and the Health of the Public. *Am. J. Prev. Med.* **2008**, *35*, 401–402. [CrossRef]
47. Lorenzoni, I.; Pidgeon, N.F. Public Views on Climate Change: European and USA Perspectives. *Clim. Chang.* **2006**, *77*, 73–95. [CrossRef]
48. Creutzig, F.; Roy, J.; Lamb, W.F.; Azevedo, I.M.; Bruine de Bruin, W.; Dalkmann, H.; Edelenbosch, O.Y.; Geels, F.W.; Grubler, A.; Hepburn, C. Towards demand-side solutions for mitigating climate change. *Nat. Clim. Chang.* **2018**, *8*, 260–263. [CrossRef]
49. Karl, T.R.; Melillo, J.M.; Peterson, T.C. *Global Climate Change Impacts in the United States: A State of Knowledge Report from the US Global Change Research Program*; Cambridge University Press: Cambridge, UK, 2009.
50. Kravchenko, S. Right to Carbon or Right to Life: Human Rights Approaches to Climate Change. *Vt. J. Environ. Law* **2007**, *9*, 513.
51. Lamb, W.F.; Steinberger, J.K. Human well-being and climate change mitigation. *Wiley Interdiscip. Rev. Clim. Chang.* **2017**, *8*, e485. [CrossRef]
52. Aqtam, I.; Ayed, A.; Zaben, K. Quality of Life: Concept Analysis. *Saudi J. Nurs. Health Care* **2023**, *6*, 10–15. [CrossRef]
53. Byravan, S.; Ali, M.S.; Ananthakumar, M.R.; Goyal, N.; Kanudia, A.; Ramamurthi, P.V.; Srinivasan, S.; Paladugula, A.L. Quality of life for all: A sustainable development framework for India's climate policy reduces greenhouse gas emissions. *Energy Sustain. Dev.* **2017**, *39*, 48–58. [CrossRef]
54. Haas, B.K. A Multidisciplinary Concept Analysis of Quality of Life. *West. J. Nurs. Res.* **1999**, *21*, 728–742. [CrossRef]
55. Hörnquist, J.O. The Concept of Quality of Life. *Scand. J. Soc. Med.* **1982**, *10*, 57–61. [CrossRef] [PubMed]
56. OECD. *How's Life? Measuring Well-Being*; Organisation for Economic Co-Operation and Development (OECD): Paris, France, 2011. [CrossRef]
57. WHO Quality of Life Group. What quality of life? World Health Organization quality of life assessment. *World Health Forum* **1996**, *17*, 354–356.
58. WHO. WHOQOL—Measuring Quality of Life. 2012. Available online: https://www.who.int/tools/whoqol (accessed on 2 October 2024).
59. Brock, D. Quality of life measures in health care and medical ethics. In *The Quality of Life*; Oxford University Press: Oxford, UK, 1993; pp. 95–132.
60. Institute of Medicine (US) Division of Health Care Services; Heithoff, K.A.; Lohr, K. (Eds.) Assessing Health-Related Quality of Life Outcomes. In *Effectiveness and Outcomes in Health Care: Proceedings of an Invitational Conference*; National Academies Press: Washington, DC, USA, 1990; Volume 17. Available online: https://www.ncbi.nlm.nih.gov/books/NBK233989/ (accessed on 2 October 2024).
61. Ngan, N.T.; Khoi, H. Factors Influencing on Quality of Life: Model selection by AIC. *Int. J. Psychosoc. Rehabil.* **2020**, *24*, 163–171. [CrossRef]
62. Ventegodt, S.; Merrick, J.; Andersen, N.J. Quality of Life Theory I. The IQOL Theory: An Integrative Theory of the Global Quality of Life Concept. *Sci. World J.* **2003**, *3*, 1030–1040. [CrossRef]
63. Azevedo, G.; Costa, H.; Farias Filho, J. Measuring well-being through OECD Better Life Index: Mapping the gaps. In Proceedings of the International Joint Conference on Industrial Engineering and Operations Management, Rio de Janeiro, Brazil, 8–14 July 2020; pp. 8–11.
64. Hall, J.; Giovannini, E.; Morrone, A.; Ranuzzi, G. *A Framework to Measure the Progress of Societies*; OECD Statistics Directorate Working Paper No. 34; OECD: Paris, France, 2010.
65. Greco, S.; Ishizaka, A.; Resce, G.; Torrisi, G. Measuring well-being by a multidimensional spatial model in OECD Better Life Index framework. *Socio-Econ. Plan. Sci.* **2020**, *70*, 100684. [CrossRef]
66. Kaplan, R.M.; Ries, A.L. Quality of life: Concept and definition. *COPD J. Chronic Obstr. Pulm. Dis.* **2007**, *4*, 263–271. [CrossRef]
67. Black, M.S.; Parry, I.W.; Mylonas, M.V.; Vernon, N.; Zhunussova, K. *The IMF-World Bank Climate Policy Assessment Tool (CPAT): A Model to Help Countries Mitigate Climate Change*; International Monetary Fund: Washington, DC, USA, 2023.
68. Huang, H.F.; Lo, S.L. Review and classify the GHGs-related indicators. *Renew. Sustain. Energy Rev.* **2011**, *15*, 594–602. [CrossRef]
69. Kourkoumpas, D.S.; Benekos, G.; Nikolopoulos, N.; Karellas, S.; Grammelis, P.; Kakaras, E. A review of key environmental and energy performance indicators for the case of renewable energy systems when integrated with storage solutions. *Appl. Energy* **2018**, *231*, 380–398. [CrossRef]
70. Spalevic, V.; Barati, A.A.; Goli, I.; Moghaddam, S.M.; Azadi, H. Do changes in land use and climate change overlap? An analysis of the World Bank Data. In *Land Degradation & Development*; John Wiley & Sons: Hoboken, NJ, USA, 2024.
71. Vasylieva, T.; Lyulyov, O.; Bilan, Y.; Streimikiene, D. Sustainable Economic Development and Greenhouse Gas Emissions: The Dynamic Impact of Renewable Energy Consumption, GDP, and Corruption. *Energies* **2019**, *12*, 3289. [CrossRef]
72. Bérenger, V.; Verdier-Chouchane, A. Multidimensional Measures of Well-Being: Standard of Living and Quality of Life Across Countries. *World Dev.* **2007**, *35*, 1259–1276. [CrossRef]
73. Diener, E.; Suh, E. Measuring quality of life: Economic, Social, and Subjective indicators. *Soc. Indic. Res.* **1997**, *40*, 189–216. [CrossRef]
74. Fayers, P.M.; Hand, D.J. Causal Variables, Indicator Variables and Measurement Scales: An Example from Quality of Life. *J. R. Stat. Soc. Ser. A Stat. Soc.* **2002**, *165*, 233–253. [CrossRef]
75. Gómez, L.E.; Peña, E.; Arias, B.; Verdugo, M.A. Impact of Individual and Organizational Variables on Quality of Life. *Soc. Indic. Res.* **2014**, *125*, 649–664. [CrossRef]

76. Noll, H.H. Social indicators and quality of life research: Background, achievements and current trends. In *Advances in Sociological Knowledge: Over Half a Century*; VS Verlag für Sozialwissenschaften: Wiesbaden, Germany, 2004; Volume 8, pp. 151–181.
77. Ott, J. World Bank World Development Reports. In *Encyclopedia of Quality of Life and Well-Being Research*; Springer: Cham, Switzerland, 2024; pp. 7858–7859.
78. Chong, K.C.; Yeoh, E.K.; Leung, C.C.; Lau, S.Y.F.; Lam, H.C.Y.; Goggins, W.B.; Zhao, S.; Ran, J.; Mohammad, K.N.; Chan, R.W.Y.; et al. Independent effect of weather, air pollutants, and seasonal influenza on risk of tuberculosis hospitalization: An analysis of 22-year hospital admission data. *Sci. Total. Environ.* **2022**, *837*, 155711. [CrossRef]
79. PrayGod, G.; Mukerebe, C.; Magawa, R.; Jeremiah, K.; Török, M.E. Indoor air pollution and delayed measles vaccination increase the risk of severe pneumonia in children: Results from a case-control study in Mwanza, Tanzania. *PLoS ONE* **2016**, *11*, e0160804. [CrossRef]

climate

MDPI

Article

Threat Severity and Threat Susceptibility Are Significantly Correlated with Climate Distress in Australian Mothers

Jennifer L. Barkin [1], James Dimmock [2], Lacee Heenan [2], James Clancy [2], Heather Carr [1] and Madelyn K. Pardon [2,*]

[1] School of Medicine, Mercer University, Macon, GA 31207, USA; barkin_jl@mercer.edu (J.L.B.)
[2] Department of Psychology, James Cook University, Townsville, QLD 4810, Australia
* Correspondence: madelyn.pardon@jcu.edu.au

Abstract: Climate change presents a critical global crisis, characterized by rising temperatures, extreme weather events, and shifting climate patterns. Vulnerable populations bear a disproportionate share of these impacts, with women at heightened risk due to unequal access to resources, decision-making power, and social roles. Postpartum women specifically face further unique challenges as they strive to protect their children, amplifying the psychological toll of climate change. The current study explores climate distress in a sample of 101 postpartum women in Australia (M_{age} = 31.14 years), whose youngest child was (on average) 5 months of age, examining factors associated with their psychological responses to climate threats. Correlational analyses reveal that perceptions of threat severity ($r = 0.621$, $p \leq 0.01$) and susceptibility ($r = 0.695$, $p \leq 0.01$) are strongly linked to climate distress. These findings highlight the need to further investigate the distinct psychological pathways climate-related anxiety operates through in postpartum women. The study underscores the importance of targeted interventions to support this vulnerable population as they face increasing climate-related stressors.

Keywords: climate distress index; Australian mothers; climate change; climate distress

1. Introduction

Climate change is a critical and escalating global crisis, characterized by rising temperatures, increased frequency and severity of extreme weather events, and shifting climate patterns. The Intergovernmental Panel on Climate Change (IPCC) reports that global temperatures have increased by over one-degree Celsius since pre-industrial times, with predictions indicating a further rise if greenhouse gas emissions are not drastically reduced [1]. This warming is contributing to more intense heatwaves, prolonged droughts, and more frequent and severe storms, which in turn lead to devastating impacts on not only natural ecosystems, but also human societies. These changes have far-reaching consequences, impacting food and water security, escalating disease prevalence, worsening social and economic inequalities, and profoundly affecting both individual physical and psychological health [2]. It is important to note that the communities most impacted by climate change, especially in developing countries, contribute the least to greenhouse gas emissions. This discrepancy highlights significant justice and equity concerns, as those least responsible for climate change often bear the brunt of its consequences [1]. Addressing these justice implications is essential to understanding the full scope of climate change's social and ethical impacts.

The impacts of climate change disproportionately affect not only those economically disadvantaged, but also women and children [3–7]. Vulnerable populations often lack the resources to adequately prepare for, respond to, and recover from extreme weather events, making them more susceptible to the adverse effects of climate change [8]. Children are particularly at risk due to their developing bodies and dependence on adults for protection and care [9], making them also at risk of psychological impacts. Women, especially those in low- to middle-income countries, are at a heightened risk, partly due to the unequal distribution of roles, resources, and decision-making power [10]. Mothers, in particular, face unique challenges as they strive to protect and care for their children amidst increasing climate-related threats, amplifying the physical and emotional stress they experience [6,11].

The psychological impact of climate change on women, particularly those of reproductive age, is profound and multifaceted [12,13]. Emotionally, the stress and anxiety associated with climate change can significantly affect mental health, especially for women who are primary caregivers and may already experience heightened levels of stress [4,6,11]. These women often face the dual burden of managing varied responsibilities while coping with the uncertainties and threats posed by a changing climate. New mothers often prioritize long-term considerations due to their dual responsibility for both their own well-being and the future of their children [6,11]. This may heighten their sensitivity to the potential consequences of climate change. As climate change intensifies, women are likely to encounter even greater psychological challenges related to food insecurity, displacement, and the overall safety and well-being of their families. This ongoing stress can lead to increased rates of anxiety, depression, and other mental health issues, highlighting the need for targeted support and interventions to help women navigate the psychological toll of climate change [4,11,14]. Understanding and addressing the unique vulnerabilities of women in the context of climate change is crucial for developing effective adaptation and resilience strategies.

In summary, reproductive-aged women are uniquely susceptible to the impacts of climate change [15], both physiologically and emotionally. The postpartum period is characterized by its own challenges and impacted by many factors including socioeconomic status, fluctuating hormone levels, access to care, mental health history, availability of social support, built environment, and traumatic events—to name a few [4]. Depending on the level and proximity of the exposure, extreme weather events (EWEs) can be quite traumatic. Trauma, in turn, is associated with myriad negative mental health consequences in the perinatal population and beyond [16]. However, we know little about the constellation of factors that are associated with climate distress (and therefore recovery from extreme weather exposure) in pregnant and postpartum women. Until the drivers of climate-related anxiety are understood, we cannot support women appropriately. To address this gap, the current research embarked on an investigation of factors associated with climate distress in a convenience sample of postpartum women in Australia.

2. Materials and Methods

2.1. Design

Participants, recruited through purposive sampling, completed a brief cross-sectional quantitative survey. Demographic information along with data on maternal functioning, depression, climate distress, and social support as reported by the participants were collected within this survey. Ethical approval for the study was granted by the James Cook University Human Ethics Committee (approval number H8916).

2.2. Recruitment

Participants were recruited through online social media platforms (Twitter, Facebook, and Reddit) as well as university networks. They were given a URL to access the Information Sheet, provide their consent, and complete the survey. Participants were provided with detailed information about the study and informed consent procedures that emphasized the voluntary nature of their participation. In cases where the survey questions were distressing, participants were offered access to mental health resources, including referrals to local support services and hotlines. Additionally, participants were informed that they could withdraw from the study at any point without penalty, further mitigating potential stress. These steps were integral to the study design to ethically address the potential for increased distress in this vulnerable group. To be eligible, participants needed to be at least 18 years old, reside in Australia, and have a baby 12 months old or younger. The recruitment was focused on gathering data to investigate the relationship between climate distress and postpartum maternal functioning. The anonymous survey was conducted on the Qualtrics online platform and was available from April 2023 to December 2023. Statistical analyses were carried out using SPSS version 27.

2.3. Data Collection

2.3.1. Sociodemographic Variables

The demographic variables of the study sample included age, relationship status, number of children, age of youngest child, area of residence (rural, metropolitan, remote), and Australian state of residence.

2.3.2. Threat Severity

The perceived severity of climate-related threats was measured using the adapted Risk Behavior Diagnosis Scale (RBDS) [17]. The RBDS, originally designed to measure threat perceptions in the context of health risks, was adapted to reflect climate change threats in this study. Participants indicated their agreement with three statements regarding their perceived severity of climate change threats. Ratings were provided on a 7-point Likert scale, ranging from 1 = *Strongly disagree* to 7 = *Strongly agree*, with higher scores indicating greater perceived severity. Statements such as "*Climate change is a serious threat*" were used to assess this construct. The Cronbach's alpha for the present study sample was 0.955, indicating good reliability.

2.3.3. Threat Susceptibility

Threat susceptibility was also assessed using an adapted version of the Risk Behavior Diagnosis Scale. This construct assessed participants' beliefs about their likelihood of experiencing adverse effects from climate change via three questions on a 7-point Likert scale, ranging from 1 = *Strongly disagree* to 7 = *Strongly agree*. Higher scores indicated greater perceived susceptibility of the threat. Example items included statements such as "*I am susceptible to the negative affects of climate change*". The Cronbach's alpha for the present study sample was 0.947, indicating good reliability.

2.3.4. Self-Efficacy

Self-efficacy, or participants' confidence in their ability to perform behaviors that mitigate climate-related threats, was measured using the adapted Risk Behavior Diagnosis Scale. Participants rated their confidence on a 7-point Likert scale, ranging from 1= *Strongly disagree* to 7= *Strongly agree*. Higher scores indicated higher levels of self-efficacy. Statements such as "*I am able to perform pro-environmental behaviours (for example, saving water, recycling) to prevent climate change*" were used to assess this variable. The Cronbach's alpha for the present study sample was 0.913, indicating good reliability.

2.3.5. Response-Efficacy

Response-efficacy, or the belief in the effectiveness of recommended actions to mitigate climate-related threats, was also measured using the adapted Risk Behavior Diagnosis Scale. Participants indicated their agreement with statements regarding the effectiveness of various responses to climate change questions on a 7-point Likert scale, ranging from 1 = *Strongly disagree* to 7 = *Strongly agree*. Higher scores reflected greater confidence in the efficacy of the recommended responses. Example items included statements such as "*Performing pro-environmental behaviours (for example, saving water, recycling) prevents climate change*". The Cronbach's alpha for the present study sample was 0.923, indicating good reliability.

2.3.6. Consideration of Future Consequences

The Consideration of Future Consequences Scale (CFC) [18] is a 12-item scale that measures immediate and distant outcomes of behavior with questions such as, "*I only act to satisfy immediate concerns, figuring the future will take care of itself"*. The scale is scored on a 5-point Likert Scale from 1 = *Extremely uncharacteristic* to 5 = *Extremely characteristic*. Items 3, 4, 5, 9, 10, 11, and 12 were reverse scored, with all items summed for a total score, with higher scores indicating a greater consideration of future consequences. The Cronbach's alpha for the present study sample was 0.866, indicating good reliability.

2.3.7. Climate Distress

The Climate Distress Index-17 (CDI-17) was completed by all participants with the intent of gauging distress due to climate change/extreme weather (see Figure 1 for full measure). The measure is available in a 16- or 17-item format, with the only difference being that this item—*"The media coverage (t.v., news articles, etc.) of climate change is helpful"* —is excluded from the CDI-16 (but included in the CDI-17). Respondents are asked to rate their experience on items such as *"Future generations will protect the environment"* and *"Climate change has affected my health or safety"* on a 7-point Likert scale, where 0 = *Strongly Disagree* and 6 = *Strongly Agree*. The full measure is shown in Figure 1. To score the CDI-17, items 1-7 must be reverse-coded before the 17 items can be summed. The range for the CDI-17 total score ranges from 0 to 102, with higher scores representing greater levels of climate distress. The Cronbach's alpha for the CDI-17 was 0.83 in the present study, indicating good reliability. Previous studies using the CDI-16 have also indicated good reliability ($\alpha = 0.81$) [11].

Please **circle the number** that best represents your feelings. Please try to answer each question as honestly as possible.

	Strongly Disagree	Disagree	Somewhat Disagree	Neutral	Somewhat Agree	Agree	Strongly Agree
1. There is enough time to stop or slow climate change.	0	1	2	3	4	5	6
2. My community is prepared to deal with the effects of climate change.	0	1	2	3	4	5	6
3. Government officials are doing their best to stop or slow climate change.	0	1	2	3	4	5	6
4. Other people (in society) are doing their best to stop or slow climate change.	0	1	2	3	4	5	6
5.Future generations will protect the environment.	0	1	2	3	4	5	6
6. I am doing my part to protect the environment (recycling, etc.).	0	1	2	3	4	5	6
7. The media coverage (t.v., news articles, etc.) of climate change is helpful.	0	1	2	3	4	5	6
8. I'm concerned that climate change will affect our food supply.	0	1	2	3	4	5	6
9. I'm concerned that climate change will harm nature and animals.	0	1	2	3	4	5	6
10. I'm concerned that climate change will affect my ability to enjoy the outdoors.	0	1	2	3	4	5	6
11. I'm concerned that the earth will not be livable for future generations.	0	1	2	3	4	5	6
12. Climate change has caused financial problems for me/my family.	0	1	2	3	4	5	6
13. I'm concerned that climate change will cause financial problems for me/my family in the future	0	1	2	3	4	5	6
14. Climate change has affected my health or safety.	0	1	2	3	4	5	6
15. Climate change has affected the health or safety of someone I care about.	0	1	2	3	4	5	6
16. I'm concerned that climate change will affect my health or the health of someone I care about in the future.	0	1	2	3	4	5	6
17. I live in an area that is at high risk for extreme weather (heat, wildfires, storms, drought).	0	1	2	3	4	5	6

For questions on the CDI-17, please contact Dr. Jennifer Barkin. Email: barkinj@gmail.com

Figure 1. Climate Distress Index (CDI-17).

2.3.8. Maternal Functioning

The Barkin Index of Maternal Functioning (BIMF) [12,13,19] was used to assess maternal functioning in the postpartum period. The BIMF comprises 20 items, with response options on a 7-point Likert scale, where 0 = *Strongly Disagree* and 6 = *Strongly Agree*. Participants are asked to select the response that best reflects their experience over the past two weeks on statements such as *"I am able to relax and enjoy time with my baby"*. After reverse-coding items 16 and 18, the overall maternal functioning score ranges from 0 to 120, with a score of 120 representing optimal functioning. A Cronbach's alpha of 0.84 for the current study sample indicates good reliability and is similar to that of the original validation study ($\alpha = 0.87$).

2.3.9. Depression

The Center for Epidemiologic Studies Depression Scale (CES-D) [20] was used to assess depressive symptomology. The CES-D comprises 20 items, with response options on a 5-point Likert Scale, where 0 = *Rarely or none of the time (less than 1 day)* and 4 = *Most or all of the time (5–7 days)*. The overall score ranges from 0 to 60, with higher scores representing the presence of greater depressive symptomology. The Cronbach's alpha for the CES-D in the present sample was 0.92, indicating good reliability.

2.3.10. Social Support

The Multidimensional Scale of Perceived Social Support (MSPSS) [21,22] was implemented to assess participants' perceived level of social support with family, friends, and significant others. The MPSS comprises 12 items divided into three sub-scales: Family, Friends, and Significant Other. Items are assessed on a 7-point Likert Scale ranging from 1 = *Very strongly disagree* to 7 = *Very strongly agree*. Respondents are asked to indicate how they feel about statements such as: *"There is a special person who is around when I am in need"*. Scores were averaged for the total scale and for each subscale, with higher scores indicating greater levels of perceived social support. The Cronbach's alpha for the MSPSS was 0.94 for the total scale, 0.94 for the Friends subscale, 0.91 for Family subscale and 0.95 for the Significant Other subscale, indicating strong reliability.

2.3.11. Analytic Approach

In analyzing the data, we employed a variety of statistical methods to understand the intricacies of climate distress among the new motherhood populations. Descriptive statistics were first calculated to summarize the sample's demographic information, including age, relationship status, number of children, and area of residence. In examining the construct validity of the CDI, the relationships between climate distress and consideration of future consequences, maternal functioning, depression, perceived social support, self-efficacy, response-efficacy, threat susceptibility, threat severity, and demographic factors such as age and the age of the youngest child were also explored using Pearson correlation analyses. Additionally, a one-way ANOVA was conducted to assess the impact of relationship status on climate distress and location variables, including state and area classifications.

3. Results

3.1. Sample Characteristics

The study sample consisted of 101 Australian mothers, with an average age of 31.14 years ($SD = 5.51$) (see Table 1 for state-based distribution). A majority of 85 participants (84.2%) were partnered, while the remainder were single mothers. The average age of the participants' youngest child was 5.93 months ($SD = 3.17$). Out of the total sample, forty mothers (39.6%) had two or more children, whereas the rest had only one child. Sixty-seven participants lived in metropolitan areas, twenty-three in rural areas, six in remote areas, and five were uncertain about their area's classification.

Table 1. Participant location based on Australian state.

State	*n*	%
Queensland	42	41.6
New South Wales	15	14.9
South Australia	15	14.9
Northern Territory	13	12.9
Victoria	7	6.9
Western Australia	7	6.9
Tasmania	1	1
Australian Capital Territory	1	1

3.2. Correlates of Climate Distress

Climate Distress (CDI) was significantly, positively, and strongly correlated with threat severity ($r = 0.621$, $p \leq 0.01$) and threat susceptibility ($r = 0.695$, $p \leq 0.01$) (Table 2). Specifically, levels of climate distress increased with higher ratings of threat severity and threat susceptibility. The CDI was also significantly (but weakly) correlated with the Consideration of Future Consequences Scale score ($r = 0.235$, $p \leq 0.05$). Uncorrelated variables included maternal functioning, depression, perceived social support, self-efficacy (related to climate change behaviors), and response-efficacy (also related to climate change behaviors). There were no significant correlations between CDI scores on demographic variables (e.g., age and age of youngest child). See Table 2 for all correlations between all variables of interest and the CDI.

Table 2. Correlation coefficients between all variables of interest and CDI (n=101).

	BIMF	CES-D	MSPSS Total	CFC	Threat Severity	Threat Susceptibility	Self-Efficacy	Response-Efficacy
Climate Distress	−0.068	−0.093	0.004	0.235 *	0.621 **	0.695 **	0.054	0.063

Note: * $p < 0.05$; ** $p < 0.001$.

A one-way ANOVA was also conducted to explore whether relationship status had any impact on CDI scores. Results found a significant difference between those who reported they were single ($M = 64.27$, $SD = 13.57$), partnered ($M = 60.93$, $SD = 14.35$), and other ($M = 33.50$, $SD = 19.09$) on CDI scores ($F_{(2,98)} = 4.09$, $p = 0.20$), with those who were single experiencing more climate distress. There were no significant differences on CDI scores between location variables (state and area classifications).

4. Discussion

This study offers critical insights into the psychological impacts of climate change on a sample of geographically diverse postpartum women, a demographic that is particularly vulnerable yet often overlooked in climate change research. Specifically, the current study sought to examine the factors associated with climate distress in a sample of Australian postpartum women. This study aimed to underscore the importance of acknowledging and addressing climate-related psychological stressors in mothers, as they hold profound implications for maternal mental health and, by extension, family wellbeing. The absence of work to date between climate distress and other mental health variables such as maternal functioning, depression, and perceived social support highlights the complexity of climate-related psychological responses and suggests the need for further exploration into the unique pathways through which climate distress impacts this population.

The significant and strong correlations between climate distress and threat severity ($r = 0.621$, $p < 0.001$) and threat susceptibility ($r = 0.695$, $p < 0.01$) provide critical insight regarding factors associated with climate distress. These results support the existing literature that links perception of environmental threats with increased psychological distress (e.g., Clayton, 2020 [23]). They also indicate that the mother's perception of both her own vulnerability to the effects of climate change and personal degree of risk (severity) influence distress levels. These two strong, significant relationships are intuitive and the direction of the relationships (i.e., level of distress increases as threat susceptibility increases) are as you would expect—providing evidence of the construct validity of the CDI-17.

The positive relationship between the consideration of future consequences (CFC) and the CDI-17 can be attributed to several factors related to this population's unique life circumstances. New mothers are typically highly attuned to future-oriented thinking, as they are not only responsible for their own well-being but also for the future well-being of their children [19]. This forward-looking perspective may make this population more sensitive to the long-term impacts of climate change, thus increasing their climate distress. Mothers often assume the role of primary decision-makers in managing family health, and this heightened responsibility contributes to their increased sensitivity to health-related concerns [4]. New mothers may perceive climate change as a direct threat to the future security and stability of their families, which could contribute to their increased levels of anxiety and concern in this context overall. Furthermore, in Australia, a country that is particularly vulnerable to the effects of climate change, such as extreme weather events, the immediacy and severity of these threats may further intensify climate-related distress among mothers who are deeply concerned about the future.

The non-significant associations between climate distress and other variables such as maternal functioning, depression, and perceived social support warrants further investigation. These findings do not undermine the construct validity of the CDI-17, but rather suggest that climate distress may operate through distinct psychological pathways where perceived vulnerability to threat and perceived degree of threat are more influential than other factors. One possible explanation is that climate distress is a unique form of anxiety that is specifically tied to environmental concerns and is not necessarily linked to general mental health issues [24]. However, the authors are cautious in forming any conclusions based on one study, involving participants from only one country—cultural differences may influence results and the relationship between climate change and maternal mental health is not yet well explored or understood. Alternatively, it may be that existing mental health measures do not adequately capture the nuances of climate-related anxiety. The current findings highlight the need for more comprehensive programming for maternal mental health that integrate environmental stressors, considering that new mothers face the dual burden of parenting and navigating a changing climate.

Strengths and Limitations

In this study, a uniquely vulnerable subgroup of the population regarding climate change (new mothers) was studied. Often referred to as "the last therapeutic orphan", mothers (and their needs/perspectives) are often neglected in favor of an almost myopic focus on the health and wellbeing of children [4]. In that regard, it addresses an under-explored and unnavigated topic that directly impacts the health of the family unit as a whole. Stated differently, how will we protect the "CEO of family health" [25] from the more devastating effects of extreme weather? Additional strengths include the number of variables/indicators of the mother's perspective that were collected and the (within-country) geographic diversity of this sample of Australian women. Limitations include the cross-sectional nature of the study design and that there was no ascertainment of the women's level of recent exposure to extreme weather events.

5. Conclusions

In our sample of Australian mothers, two factors appear to be most directly correlated with increased climate distress: (1) the mother's perception that climate change is a sincere and real threat to her personally, and (2) the belief that the negative effects of climate change are proximal and potentially devastating. Simply put, on average, the mothers that believed that climate change was a clear and present danger were more stressed about it than those who did not have similar beliefs. The present conclusion hints at the possibility that while, in reality, all populations will be impacted by climate change, those that have greater awareness of the threats will suffer more emotionally and earlier. In this case, accurate threat appraisal may also indicate greater vulnerability to related mental health issues.

Author Contributions: Conceptualization, M.K.P. and J.L.B.; methodology, M.K.P., J.L.B. and J.D.; formal analysis, M.K.P., L.H. and J.C.; data curation, J.C. and L.H.; writing—original draft preparation, J.L.B. and M.K.P.; writing—review and editing, J.D., J.C., L.H. and H.C.; supervision, M.K.P. and J.D.; project administration, M.K.P. All authors have read and agreed to the published version of the manuscript.

Funding: This research received no external funding.

Data Availability Statement: Due to the sensitive nature of this research, no supporting data is available.

Conflicts of Interest: The authors declare no conflict of interest.

References

1. Intergovernmental Panel on Climate Change (IPCC). *Climate Change 2022: Impacts, Adaptation, and Vulnerability. Contribution of Working Group II to the Sixth Assessment Report of the Intergovernmental Panel on Climate Change*; Pörtner, H.-O., Roberts, D.C., Tignor, M., Poloczanska, E.S., Mintenbeck, K., Alegría, A., Craig, M., Langsdorf, S., Löschke, S., Möller, V., et al., Eds.; Cambridge University Press: Cambridge, UK, 2022. Available online: https://www.ipcc.ch/report/ar6/wg2/ (accessed on 2 May 2024).
2. Loucks, D.P. Impacts of climate change on economies, ecosystems, energy, environments, and human equity: A systems perspective. In *The Impacts of Climate Change*; Letcher, T.M., Ed.; Elsevier: Amsterdam, The Netherlands, 2021; pp. 19–50. [CrossRef]
3. Al Gasseer, N.; Dresden, E.; Keeney, G.B.; Warren, N. Status of women and infants in complex humanitarian emergencies. *J. Midwifery Women's Health* **2004**, *49*, 7–13. [CrossRef] [PubMed]
4. Barkin, J.L.; Philipsborn, R.P.; Curry, C.L.; Upadhyay, S.; Geller, P.A.; Pardon, M.; Dimmock, J.; Bridges, C.C.; Sikes, C.A.; Kondracki, A.J.; et al. Climate Change is an Emerging Threat to Perinatal Mental Health. *J. Am. Psychiatr. Nurses Assoc.* **2022**, *30*, 683–689. [CrossRef] [PubMed]
5. Intergovernmental Panel on Climate Change (IPCC). *Climate Change 2014: Mitigation of Climate Change. Contribution of Working Group III to the Fifth Assessment Report of the Intergovernmental*; Pachauri, R.K., Meyer, L.A., Eds.; Cambridge University Press: Cambridge, UK, 2022. Available online: https://www.ipcc.ch/report/ar5/syr/ (accessed on 1 November 2022).
6. Lykins, A.D.; Bonich, M.; Sundaraja, C.; Cosh, S. Climate change anxiety positively predicts antenatal distress in expectant female parents. *J. Anxiety Disord.* **2024**, *101*, 102801. [CrossRef] [PubMed]
7. Sato, M.; Nakamura, Y.; Atogami, F.; Horiguchi, R.; Tamaki, R.; Yoshizawa, T.; Oshitani, H. Immediate needs and concerns among pregnant women during and after typhoon Haiyan (Yolanda). *PLoS Curr.* **2016**, *8*. [CrossRef] [PubMed]
8. Ebi, K.L.; Vanos, J.; Baldwin, J.W.; Bell, J.E.; Hondula, D.M.; Errett, N.A.; Hayes, K.; Reid, C.E.; Saha, S.; Spector, J.; et al. Extreme Weather and Climate Change: Population Health and Health System Implications. *Annu. Rev. Public Health* **2021**, *42*, 293–315. [CrossRef] [PubMed]
9. Bolek, M.; Dobrzeniecki, K.; Błaszczyszyn, K.; Hopej, N.; Muc, K.; Turek, K. The impact of climate change on mental health of children and adolescents. *Qual. Sport* **2024**, *16*, 52387. [CrossRef]
10. Sorensen, C.; Murray, V.; Lemery, J.; Balbus, J. Climate change and women's health: Impacts and policy directions. *PLoS Med.* **2018**, *15*, e1002603. [CrossRef] [PubMed]
11. Pardon, M.K.; Dimmock, J.; Chande, R.; Kondracki, A.; Reddick, B.; Davis, A.; Athan, A.; Buoli, M.; Barkin, J.L. Mental health impacts of climate change and extreme weather events on mothers. *Eur. J. Psychotraumatology* **2024**, *15*, 2296818. [CrossRef] [PubMed]
12. Barkin, J.L.; Wisner, K.L.; Bromberger, J.T.; Beach, S.R.; Terry, M.A.; Wisniewski, S.R. Development of the Barkin Index of Maternal Functioning. *J. Women's Health* **2010**, *19*, 2239–2246. [CrossRef] [PubMed]
13. Barkin, J.L.; Wisner, K.L.; Wisniewski, S.R. The Psychometric Properties of the Barkin Index of Maternal Functioning. *J. Obstet. Gynecol. Neonatal Nurs.* **2014**, *43*, 792–802. [CrossRef] [PubMed]
14. Veenema, R.J.; Hoepner, L.A.; Geer, L.A. Climate change-related environmental exposures and perinatal and maternal health outcomes in the U.S. *Int. J. Environ. Res. Public Health* **2023**, *20*, 1662. [CrossRef] [PubMed]
15. Harrington, J. Mapped: How Climate Change Disproportionately Affects Women's Health. Carbon Brief. 25 July 2023. Available online: https://www.carbonbrief.org/mapped-how-climate-change-disproportionately-affects-womens-health/ (accessed on 10 July 2024).

16. Sachdeva, J.; Yang, S.N.; Gopalan, P.; Worley, L.L.; Mittal, L.; Shirvani, N.; Spada, M.; Albertini, E.; Shenai, N.; Simas, T.A.M.; et al. Trauma Informed Care in the Obstetric Setting and Role of the Perinatal Psychiatrist: A Comprehensive Review of the Literature. *J. Acad. Consult.-Liaison Psychiatry* **2022**, *63*, 485–496. [CrossRef] [PubMed]
17. Witte, K.; Cameron, K.A.; McKeon, J.K.; Berkowitz, J.M. Predicting risk behaviors: Development and validation of a diagnostic scale. *J. Health Commun.* **1996**, *1*, 317–342. [CrossRef] [PubMed]
18. Strathman, A.; Gleicher, F.; Boninger, D.S.; Edwards, C.S. The consideration of future consequences: Weighing immediate and distant outcomes of behavior. *J. Personal. Soc. Psychol.* **1994**, *66*, 742–752. [CrossRef]
19. Barkin, J.L.; Willis, G.B.; Hawkins, K.C.; Stanfill-Thomas, T.; Beals, L.; Bloch, J.R. Semantic Assessment of the Barkin Index of Maternal Functioning in a Medically Underserved Obstetric Population. *Perspect. Psychiatr. Care* **2017**, *53*, 95–103. [CrossRef] [PubMed]
20. Radloff, L.S. The CES-D Scale: A Self-Report Depression Scale for Research in the General Population. *Appl. Psychol. Meas.* **1977**, *1*, 385–401. [CrossRef]
21. Zimet, G.D.; Dahlem, N.W.; Zimet, S.G.; Farley, G.K. The Multidimensional Scale of Perceived Social Support. *J. Persinal. Assess.* **1988**, *52*, 30–41. [CrossRef]
22. Zimet, G.D.; Powell, S.S.; Farley, G.K.; Werkman, S.; Berkoff, K.A. Psychometric Characteristics of the Multidimensional Scale of Perceived Social Support. *J. Pers. Assess.* **1990**, *55*, 610–617. [CrossRef]
23. Clayton, S. Climate anxiety: Psychological responses to climate change. *J. Anxiety Disord.* **2020**, *74*, 102263. [CrossRef] [PubMed]
24. Pihkala, P. Anxiety and the Ecological Crisis: An Analysis of Eco-Anxiety and Climate Anxiety. *Sustainability* **2020**, *12*, 7836. [CrossRef]
25. Kaufman, K. Climate Change is Impacting Family, Maternal, and Mental Health: What Can We Do About it? The Hill. 24 July 2024. Available online: https://thehill.com/policy/healthcare/4617753-climate-change-family-maternal-mental-health/ (accessed on 15 July 2024).

climate

MDPI

Article

Set When the Sun Rises, Rise When the Sun Sets: Climate Impact on Health, Safety, and Wellbeing of Smallholder Farmers in Vietnam

Miranda Dally [1,2,*], Thuy Thi Thu Tran [3], Thanh Le Nhat Nguyen [4], Quynh Nguyen [3], Lee S. Newman [1,2,5,6], Mike Van Dyke [1,2], Marcela Tamayo-Ortiz [7], James Crooks [5,8], Lyndsay Krisher [1,2] and Megan Cherewick [9]

1 Center for Health, Work & Environment, Colorado School of Public Health, University of Colorado Anschutz Medical Campus, Aurora, CO 80045, USA
2 Department of Environmental and Occupational Health, Colorado School of Public Health, University of Colorado Anschutz Medical Campus, Aurora, CO 80045, USA
3 Department of Occupational Health and Safety, Hanoi University of Public Health, Hanoi 70000, Vietnam
4 Vietnamese Chamber of Commerce and Industry, Ho Chi Minh City 038986, Vietnam
5 Department of Epidemiology, Colorado School of Public Health, University of Colorado Anschutz Medical Campus, Aurora, CO 80045, USA
6 Division of Pulmonary Medicine and Critical Care, School of Medicine, University of Colorado Anschutz Medical Campus, Aurora, CO 80045, USA
7 Department of Environmental Health Sciences, Mailman School of Public Health, Columbia University, New York, NY 10032, USA
8 Division of Biostatistics, National Jewish Health, Denver, CO 80206, USA
9 Department of Community and Behavioral Health, Colorado School of Public Health, University of Colorado Anschutz Medical Campus, Aurora, CO 80045, USA; megan.cherewick@cuanschutz.edu
* Correspondence: miranda.dally@cuanschutz.edu

Abstract: Vietnam is a country most at risk for experiencing climate change effects, especially increasing temperatures. Agricultural production is one of the biggest contributors to Vietnam's economy. Recent research has explored how climate change will impact agriculture in Vietnam. However, the impact of climate change to the health, safety, and wellbeing of Vietnamese farmers is often overlooked. In this study, we conducted five focus groups with 46 farmers representing three provinces of Vietnam. We used a convergent mixed-methods design and a *Total Worker Health*® framework to assess how farmers in Vietnam experience climate-change-related hazards and describe how famers associate these hazards with impacts to their health, safety, and wellbeing. Multi-dimensional scaling suggests farmers conceptualize hazards separately from health, safety, and wellbeing outcomes, but a thematic analysis of our data indicated that farmers perceive both direct and indirect impacts of climate change to their health, safety, and wellbeing. Direct impacts of climate change described included physical health effects such as heat stress. Indirect impacts included mental health stressors due to productivity demands. Gaps in available health and safety trainings for farmers were also identified. This project demonstrates the need to co-develop safety and health trainings with farmers. System-level approaches both at the societal and community levels are needed. The local governments, cooperatives, Women's Unions, and Farmers' Unions are trusted sources of information that could implement and disseminate these trainings.

Keywords: occupational health; climate adaptation; agricultural workforce

1. Introduction

Vietnam is ranked in the top 10 countries most vulnerable to climate change and climate-related events [1]. Currently, agriculture accounts for 20% of the nation's GDP [2]. The World Bank projects that by 2050, the total GDP could be reduced by 0.7%–2.4% as a result of climate-change-related reductions in agriculture, including the distribution of agricultural production, food supply, and global markets [3]. To date, the focus on climate adaptation strategies within the agriculture sector have primarily focused on technical solutions to plant adaptations [4]. While it is understood that changing weather patterns and extreme climate events will have a great economic impact on the agriculture sector as a whole [5], the effect on the physical and mental wellbeing of those who rely on agriculture as their main source of livelihood is often overlooked.

Vietnam's geography makes it especially susceptible to many climate-related hazards. These include drought, flooding, salinity intrusion, extreme weather events, vector-borne diseases and other biological hazards, and increasing temperatures [3]. In Vietnam, annual average temperatures have increased nearly 0.2 °C per decade since 1980 [6]. Looking forward, annual maximum and minimum temperatures are expected to increase at an even greater rate than annual average temperatures, which will likely amplify the consequences to human health, livelihoods, and ecosystems [3].

Climate change is expected to have negative health consequences worldwide [7]. This includes both physical and mental health effects. Climate change has been linked to increased mortality due to heatwaves [8–10], cardiovascular and respiratory disease exacerbated by air pollution [11], and vector-borne disease transmission [12]. Mental health effects linked to climate change include increased anxiety, depression, and post-traumatic stress disorder [13]. There are individual-, community-, and societal-level factors that affect how vulnerable an individual is to the health consequences of climate change [14,15]. One such vulnerability is work.

Workers, especially those who work outdoors, are extremely vulnerable to the negative health consequences of climate change [16]. Research has emerged showing the link between climate-change-related increases in temperature and the occupational safety, health, and wellbeing of agricultural workers. Climate-change-related increases in temperatures have been linked to an epidemic of chronic kidney disease in agricultural communities clustered around the equator [17]. Associations between increased temperatures and higher rates of traumatic occupational injuries in agricultural workers have also been shown [18].

Climate adaptation studies for agricultural workers are lacking, especially within low- and middle-income countries (LMICs) [19]. For example, while effective interventions to mitigate the effects of heat on physical performance have been developed, often within lab settings [20], these strategies are often not efficacious when applied to the work setting [21]. One potential reason for this discordance is the difficulty in implementing these strategies with consistency and ensuring adherence to the recommendations [22]. This is especially true for smallholder farmers. Smallholder farmers are a subpopulation of agricultural workers who are especially vulnerable to climate change due to their low capacity to adapt to change and their reliance on crops for subsistence and income [23]. There is a need to develop adaptation strategies specifically for smallholder farmers that address not only the health of their corps but their own health, safety, and wellbeing.

Recently, the *Total Worker Health®* approach has emerged as a way to understand the complex interactions between individual-, workplace-, community-, and societal-level factors that influence the health, safety, and wellbeing of workers [24]. Similar to the social–ecological model [25], Total Worker Health (TWH) can guide adaptive strategies, recognizing that workers are nestled within their workplaces that operate within communities and societies. There is a need to understand the vulnerability factors that shape each individual farmer's ability to adapt [26]. Using TWH as a framework, we are provided a holistic view of how work and nonwork factors influence smallholder farmers' abilities to adapt to climate change. Local coping strategies and adaptive capacity can also be influenced by inequalities [5]. Inequalities make it difficult to develop one-size-fits-all approaches to addressing work in hot environments. The ability to adapt also depends on having the relevant skills and training to utilize adaptation strategies [27].

Including farmers' opinions in the development of climate change adaptation strategies will better meet their health, safety, and wellbeing needs [4]. The goal of this study was to understand how farmers in Vietnam view climate-change-related hazards in terms of the impact to their health, safety, and wellbeing. Additionally, we wanted to understand how farmers are currently adapting to climate change and identify additional adaptive capacity needs.

2. Materials and Methods

2.1. Study Population

This study took place in three provinces in Vietnam from August to October 2023 (Figure 1). Convenience sampling was used for province selection with the goal of representation from each of the three main agricultural regions in Vietnam. The first province was Can Tho Province in the Mekong River Delta region, located in the southern part of the country. The primary crop grown in the Mekong River Delta region is rice, which comprises approximately 75% of the total agricultural production value in the country [28]. The second province was Ninh Binh Province, which is in the Red River Delta region and is in the northern part of the country. While the staple crop of the Red River Delta has historically been rice, as a climate-related adaptive strategy there have been recent efforts to increase fruit cultivation [29]. The final province was Lam Dong Province in the Central Highlands of Vietnam, which is located in the middle of the country. The primary crop of the Central Highlands is coffee, which is an important part of the rural economy of Vietnam [30].

Figure 1. Location of the study sites: Can Tho Province in the Mekong River Delta region of Vietnam, Ninh Binh Province in the Red River Delta region of Vietnam, and Lam Dong in the Central Highlands of Vietnam.

Farmers in this study were all from rural areas and not considered part of the formal agricultural sector. All study participants were members of their local cooperatives. Farmers work within a cooperative structure to pool their harvested crop with other farmers to process and sell their product. Should the cooperative initiate it, the farmers within the cooperative will receive both technical and safety training from local organizations and authorities such as the Farmers' Union, the Women's Union, or the Department of Labor, Invalids, and Social Affairs (DOLISA). The Farmers' Union, Women's Union, and the cooperative are all considered important social organizations that provide livelihood support for farmers [1].

2.2. Methodological Approach

In this study, we used a convergent mixed-methods design [31] and a TWH framework. Focus group discussions were held with members of farming cooperatives in the study provinces. Local authorities provided written approval and facilitated the organization of the focus group discussions. Participants in the focus groups were recruited through coordination between the Vietnamese Chamber of Commerce and Industry (VCCI) and the Farmers' Union, with the goal of two focus groups per province and 10 participants per focus group. Eligibility criteria for participants included (1) over the age of 18 and (2) primary occupation being agricultural work. This convenience sample resulted in five total focus groups: a Can Tho male focus group, a Can Tho female focus group, a mixed-gender Ninh Binh focus group, and separate male and female Lam Dong focus groups. At the start of each focus group, participants were provided with informed consent documents. All participants then provided verbal consent to participate. Verbal consent was also obtained to audio record and photograph the focus groups. Upon participation, participants were provided with a VND 155,000 (USD 7) incentive for their participation in the study. Ethics approval was provided by the Colorado Multiple Institutional Review Board (#23-1235) and the Institutional Review Board of the Hanoi University of Public Health (#023-371/DD-YTCC).

Focus groups were conducted in Vietnamese by native speakers. An interpreter was present and provided consecutive interpretation (Can Tho) and simultaneous interpretation (Ninh Binh) to the non-Vietnamese speaker of the research team. Detailed field notes were taken to capture non-verbal cues and other observations. Only native Vietnamese speakers were present at the Lam Dong focus groups. After spending time welcoming the participants, the focus groups were conducted in three main sections: (1) open discussion about climate change, in which the facilitator opened the focus group by asking, "if we talk about climate change now, what do you think about?", (2) pile sorting, and (3) qualitative vignettes. Each focus group took between 60 and 90 min. In Can Tho, male and female focus group discussions were held in the local villages at the cooperatives' headquarters. In Ninh Binh, the combined male and female focus group discussion was held in the city of Ninh Binh at the local Farmers' Union office. In Lam Dong, the focus group with females took place at a local coffee shop in their commune. The focus group with males took place at the commune's Farmers' Union office. Representatives from the Farmers' Union were present at all focus groups. The head of the cooperatives were present and participated in the Can Tho male focus group and the Ninh Binh focus group.

2.2.1. Pile Sorting

The goal of the pile-sorting activity was to understand how participants link climate-change-related hazards to impacts on occupational safety and health. In the past, this activity has been successfully employed to understand how individuals detect and attribute climate change impacts [32]. To conduct the pile-sorting activity, each participant was provided with a set of 26 cards contained in a small plastic bag during the focus group. Each card represented one item, with the item written in Vietnamese on one side and English on the other (Table 1). The items were broadly classified by the research team based on environmental conditions and occupational safety and health conditions. The included items were selected based on background literature review of the most common climate change concerns in Vietnam [1] and published theoretical frameworks for the impact of climate change on occupational health and safety [33–36].

Table 1. List of items provided to participants for the pile-sorting activity.

Environmental Conditions Relating to Climate Change	Theorized Occupational Safety and Health Conditions Relating to Climate Change
Salt intrusion	Injury to people
Drought	Personal illness
Flood	Income
Extreme weather phenomena	Human health
Rising temperatures	Animal health
Heatwaves	Stress
Pollution	Depression
Climate change	Concern
Season change	Job needs
Water pollution	Labor productivity
Pesticide leaks	Type of work
Amount of rain	Diseases in animals
Sea level rise	Disease outbreak in humans

At the start of the activity, participants were instructed to individually group their pile of cards together based on similarities or differences. It was emphasized that there was no right or wrong answer. There was no limit on the number of piles a participant could create. Once participants finished grouping their cards, they were asked to label each pile with a short name to describe the group. The facilitators then invited each participant to explain a little about their reasoning underlying their groupings. At the end of the activity, participants were instructed to rubber band together each of their labeled piles and to place them back in the plastic bag. Each pile label was translated to English after the conclusion of the focus group.

2.2.2. Qualitative Vignettes

The goal of the qualitative vignettes was to assess participants' knowledge and attitudes towards adaptation strategies and to identify potential barriers. We created qualitative vignettes [37–39] to present a hypothetical story to participants. Participants were then asked a series of questions regarding the fictional characters' behaviors. The story was provided to participants in written form and also read aloud by the facilitator. The choice to use fictional characters rather than ask participants to share their own stories was made to alleviate any potential stigma participants may have felt by sharing their own stories. The vignettes presented were context-specific based on the participants' home province but had an overarching theme of heat-related illness (Table 2). Prior research on the impact of climate change on safety and health in Vietnam has shown heat-related illness to be a common theme, and we assumed it would be the climate hazard participants were most concerned with [40]. Initial drafts of the vignette were developed in English and later edited to the specific contexts and translated by Vietnamese members of the research team. Once the vignette was presented, a discussion was facilitated by asking the following: (1) In this situation, what should Mai/Mr. Tuan/Vinh do to ensure safety and health?; (2) Why do you think Mai/Mr. Tuan/Vinh does not do these things?

At the conclusion of the focus group, any follow-up questions from the non-Vietnamese-speaking researcher were asked through an interpreter. Follow-up questions were guided by the researcher's previous findings and experiences. At the end of the focus groups, participants were thanked for their time and asked to complete a brief survey to collect information on demographics (Vietnam Demographic and Health Survey [41]), work information (Agricultural Health Study Farmer Applicator Questionnaire [42]), general health (RAND SF 36 [43]), mental wellbeing (WHO-5 [44]), occupational health (Workplace Health and Safety Survey [45]), and climate change knowledge and beliefs [46].

Table 2. English translation of vignettes presented to focus group participants.

Province	Vignette
Can Tho	In recent years, due to the effects of drought and saltwater intrusion, hundreds of thousands of hectares of rice, crops and fruit trees in the Mekong Delta are at risk of being lost. Mai's house is also in a similar situation, Mai is extremely worried and tries to pump water day and night to save her rice field. Her husband works away from home, so in addition to farming, Mai also has to take care of the housework and take care of the children. Mai has a small child of kindergarten age, so every morning she has to get up early to prepare breakfast for her family, take her child to school, and then go to the fields to work. Mai tries not to drink too much water while working because she is afraid that it will affect her work. Her co-worker saw that Mai looked tired and told her to rest and then continue working. Mai explained that she tried to finish her work, then she also had to buy fresh water to take home to use for eating and living. Because this whole week, tap water has been contaminated with salt, so it cannot be used. After finishing pumping water to irrigate, she went to buy fresh water for cooking and then had to pick up her child. The road was far and it was starting to get dark. She was worried that she would not be able to come back in time to pick up her child. She felt tired and worried. Mai started to cry. She felt lonely and confused, not knowing how to cope without her husband's support.
Ninh Binh	In recent years, many households in the commune are gradually escaping poverty by converting traditional rice cultivation to high-value fruit trees. Mr. Tuan also learned to transfer 3 sao of his family's field to the new Le Guava variety. In the past two years, the weather has been erratic, rainy, and pestilent, so there is no experience in the guava garden of Mr. Tuan's house. He is looking forward to this year's crop to rotate capital. However, yesterday the weather forecast announced that a strong 11-level storm was about to make landfall in the Gulf of Tonkin. This year, the weather is also erratic, with storms coming early. Mr. Tuan was very worried, so early in the morning, he mobilized the whole family to wrap the guavas before the storm came. Before the big storm, it was hot and humid with sweat, but Mr. Tuan did not dare to rest his hands because by noon there was still more than a third of the garden unfinished. His wife and children were worried that he would have a high blood pressure, and advised him to take a break for lunch, but he did not listen. After working for about an hour, Mr. Tuan felt dizzy, could not stand and fainted.
Lam Dong	Vinh has been growing coffee for 5 years, each year Vinh finds that the number of workers who come to apply for the job of harvesting coffee beans is less and less. Coffee is ripe, lack of collectors, so he has to work hard from early morning until late at night hoping to harvest. This morning he went to the field to harvest coffee, it was unusually hot in the morning. He felt more tired than usual while working, but he also got a short break before lunch. During the break, Vinh sat outside in the sun. He brought a bottle of water from home to work and drank it all, but when it ran out, he didn't take any more. After that, Vinh had lunch in the sun. After lunch, Vinh started to have a headache and felt weaker but he still managed to keep working. Vinh just worked and talked with Huy. Huy is not aware of Vinh's bad health, so the two still try to harvest the coffee fields. About twenty minutes later Vinh fainted.

2.3. Data Analysis

First, we analyzed the quantitative data from the survey. Summary statistics were stratified by province. Numerical responses were summarized as mean and standard deviation (SD), and categorical responses were categorized as count (N) and percent (%). Fisher's exact test and ANOVA were used to statistically compare responses.

To analyze the quantitative data generated by the pile-sorting activity, we quantified the number of piles that were created by each participant. The number of piles reflects the number of concepts that participants grouped the items into. Multi-dimensional scaling (MDS) was used to compare the "sameness" of the pile-sorted items [47]. In this study, sameness can be thought of as how often two items were grouped in the same pile. A distance matrix was created in Excel to facilitate the analysis. The distance matrix is analogous to a correlation matrix, with each cell populated by the count of the co-occurrence of two items in a pile. The R functions "dist", "cmdscale", and "seqmds.stress" were used to calculate the underlying data for the MDS plots and fit statistics. The results from the MDS analysis were presented as a 2-dimensional graph. The 2-dimensional graph was examined to identify conceptual clusters with common themes. The stress measure, how well the MDS analysis fits the data, was calculated.

The goal of the qualitative data analysis was to further contextualize the findings of the MDS. The qualitative data analysis followed best practices, including a deductive, theory-driven approach and an inductive, data-driven approach [48,49]. First, the audio recordings from each of the focus groups was transcribed verbatim by a native Vietnamese speaker (T.T.). Then, the transcriber translated the transcripts into English. Once the English transcripts were received by the coder (M.D.), they were read into MAXQDA

2022 (VERBI Software, Berlin, Germany) to facilitate coding. Each transcript was read in three stages. A priori, we identified 26 codes corresponding to the items provided in the pile-sorting activity. Then, during the first reading of the full transcripts, we used inductive coding to identify and add to the original 26 codes in the code book. This resulted in an additional 19 codes. During the second full reading, we applied the codes to each of the relevant passages. To identify thematic areas, in the third stage, we examined similarities and differences between passages with a given code applied. We also examined the co-occurrence of codes within and between the transcripts. This third stage of analysis lead to the development of the identified themes. A code for "quote" was applied to any passage that exemplified an identified theme. Finally, we examined how the identified themes corresponded to the conceptual clusters identified in the MDS analysis.

2.4. Data Quality and Analysis Safeguards

We used detailed written notes taken at the time of the focus group discussion to supplement our analyses. Upon completion of the analysis, we debriefed with the team who conducted the focus groups. Finally, preliminary results were presented to a group of representatives from the Ministry of Labor, Invalids, and Social Assistance, the Ministry of Health, the Farmers' Union, and the Women's Union to assess the cultural appropriateness of our interpretations.

3. Results

There were 46 participants across the five focus groups. Focus groups 1 and 2 were rice farmers from Can Tho (7 males, 10 females); focus group 3 comprised 5 male and 5 female guava farmers from Ninh Binh; and focus groups 4 and 5 were coffee farmers from Lam Dong (10 males, 9 females). A total of 41 participants (89%) completed the demographic survey.

Table 3 provides a summary of participant demographics by province. We observed a statistically significant difference in the age of study participants by province (p-value: <0.001). While 100% of participants from Can Tho and Ninh Binh were Vietnamese, participants from Lam Dong were more ethnically (p-value: 0.04) and religiously diverse (p-value: <0.001). There was a significant difference in the average farm size. Participants in Can Tho on average worked on the largest farms, while workers in Ninh Binh worked on significantly smaller farms (p-value: 0.002). While not statistically different, 50% of participants from Can Tho and Lam Dong had worked in agriculture for more than 20 years, while 50% of participants from Ninh Binh had worked in agriculture for 5–10 years (p-value: 0.36).

3.1. Pile-Sort Clusters

There were 44 participants who provided labeled piles. The two missing piles were from Lam Dong. There were 171 pile-sort piles created across the five focus groups. The median count of piles created per participant was four, with the number of piles ranging from two to seven. There was no observable difference in the number of piles created by province (p-value: 0.06). The median number of items per pile was 6, with the number of items per pile ranging from 2 to 20.

Figure **??** shows the 2-dimensional model resulting from the MDS analysis of the 171 piles from all participants. There were five primary conceptual clusters identified. These included clusters related broadly to (1) climate, weather, and the environment; (2) pollution and pesticides; (3) the physical health of humans and animals; (4) psychological effects; and (5) work demands and livelihood. A third (37%; n = 64) of the sorted piles were related to the cluster of climate, weather, and the environment. Example pile names for this cluster included "weather conditions" and "climate change". The physical health of humans and animals cluster was applied to 28% (n = 48) of piles, and work demands and livelihood was applied to 16% (n = 27) of piles (Table 4).

Table 3. Demographics for participants in 5 focus groups from Can Tho, Ninh Binh, and Lam Dong Provinces, Vietnam.

	Can Tho (N = 17) Mean (SD) Or N (%)	Ninh Binh (N = 10) Mean (SD) Or N (%)	Lam Dong (N = 14) Mean (SD) Or N (%)	*p*-Value
Gender				0.675
Male	7 (41%)	5 (50%)	8 (57%)	
Female	10 (59%)	5 (50%)	6 (43%)	
Age	43 (10)	62 (7)	50 (12)	<0.001
Married	16 (94%)	12 (100%)	10 (86%)	0.758
Head of household	11 (65%)	3 (38%)	10 (77%)	0.216
Vietnamese	15 (100%)	10 (100%)	10 (71%)	0.036
Religion				<0.001
No religion	0	6 (67%)	5 (36%)	
Buddhist	3 (19%)	2 (22%)	5 (36%)	
Catholic	9 (56%)	0	3 (21%)	
Protestant	0	0	1 (7%)	
Hoa Hao	4 (25%)	0	0	
Other	0	1 (1%)	0	
Education				0.944
Primary (passed grade 5)	1 (6%)	0	0	
Secondary (grades 6–9)	9 (53%)	7 (70%)	7 (54%)	
High school (grades 10–12)	5 (29%)	2 (20%)	3 (23%)	
Vocational/college/university	2 (12%)	1 (10%)	3 (23%)	
Weekly sources of news				
Newspaper	13 (81%)	7 (78%)	10 (71%)	0.888
Radio	11 (79%)	7 (88%)	6 (55%)	0.329
TV	16 (100%)	8 (100%)	13 (100%)	1
Farming as main source of income	16 (100%)	9 (90%)	14 (100%)	0.25
Works on their family farm	14 (88%)	7 (100%)	12 (86%)	0.822
Average size of farm (hectares)	186 (131)	29 (40)	109 (53)	0.002
Number of years in agriculture				0.36
5–10 years	4 (25%)	4 (50%)	1 (8%)	
11–20 years	4 (25%)	1 (13%)	5 (39%)	
21–30 years	1 (6%)	1 (13%)	3 (23%)	
Over 30 years	7 (44%)	2 (25%)	4 (31%)	

Note: 5 participants from Lam Dong did not complete a survey. Not all participants responded to all questions.

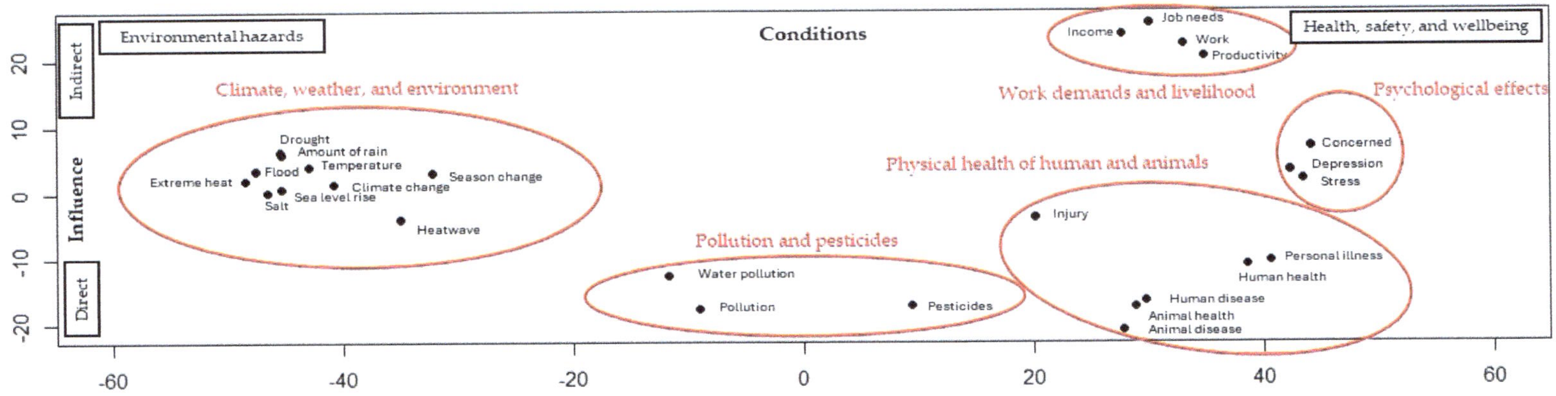

Figure 2. Two-dimensional model solution of the multi-dimensional scaling based on the pile-sorting activity relating environmental hazard conditions that will be influenced by climate change and the direct and indirect occupational health, safety, and wellbeing conditions that are theorized to be impacted by climate change.

Table 4. Concepts identified in the 2-dimensional multi-dimensional scaling (MDS) based on the 171 piles created by the full participant sample and corresponding pile name examples. Note: 4 piles were not given a label by participants.

Theme	Number of Piles (%) N = 171	Pile Name Examples
Climate, weather, and environment	64 (37%)	"Weather conditions", "Climate change", and "Extreme weather events"
Pollution and pesticides	18 (11%)	"Water pollution", "Environment hygiene", and "Climate change impacts, plant protection, chemicals and pollution"
Physical health of humans and animals	48 (28%)	"Effects on human health", "Effects on animal health", and "Feeling unwell leading to being tired"
Psychological effects	10 (6%)	"Mental stress", "Human psychological needs", "Climate affects psychology", and "Anxiety income"
Work demands and livelihood	27 (16%)	"Weather effects to labor, workers' daily life", "Decreasing work hours leads to decreasing income", and "What farmers need"

Figure **??** suggests that farmers were conceptualizing climate-change-related factors distinctly from health and wellbeing factors. However, discussions with farmers highlighted an integrated approach to thinking about climate change and its impact on work and health. As one female worker from Lam Dong noted, "*weather changes affect every aspect of human life and health.*" A male participant from Lam Dong expanded by suggesting implications for mental wellbeing, stating that "*the issues of worry and anxiety are related to the psychological well-being of farmers when their crop yields are not met*". This was echoed throughout each of the focus groups. As a female participant from Ninh Binh stated, "*First, our health, second our crop, are all affected by weather, so it is not very good*".

3.2. Climate Change Impact on TWH

Smallholder farmers in Vietnam acutely recognize the influence of climate change on their health. As women in Can Tho described, "*our health has been negatively impacted since the weather is extremely hot and we suffer from severe dehydration*". Women in Lam Dong suggested chronic effects of working in the heat, noting that "*if the weather is very hot, the elderly are more likely to experience an increase in high blood pressure. Many people have high blood pressure*". Infectious diseases were also of concern, with one participant noting that "*heavy rainfall also brings about many diseases, such as dengue fever. With the increased rainfall, mosquitoes thrive, and this year, dengue fever has been spreading rapidly*".

Notably, the effect of climate on health and safety is not just at work, but within the community as well. Workers who live in low-lying areas and near riverbanks are especially concerned about the risk of landslides. As one participant in Lam Dong noted, "*Heavy rain, slippery conditions, landslides. The terrain here is steep, making transportation difficult. Safety issues arise such as slipping, falling, traffic accidents, etc.*".

While we speculated heat would be the hazard of greatest concern, workers in Lam Dong stated they have not yet experienced the effects of heat as drastically as other regions. Changing rainfall patterns were their biggest concern. Across all five focus groups, changes in temperature were mentioned 53 times, 18 of which came from farmers in Lam Dong (34%, often regarding how they have not yet experienced it). Comparatively, of the 39 total mentions of changing rainfall patterns, 23 came from farmers in Lam Dong (59%). Of these, over half (n = 12) were in relationship to productivity, focusing on "*the consequence is a reduction in crop yields"*. This is an important indication of how climate influences work productivity and what that means for the wellbeing of the worker, a concept that is explored further in the next section.

Only females from Lam Dong mentioned the relationship between unpredictable rain and human health. The major disease of concern was dengue fever. Secondary health effects due to the increased use of pesticides were described: *"Excessive rainfall leads to more plant diseases, which means that farmers have to use more pesticides. This requires more effort and increases exposure to farmers, especially women, who contact with harmful chemicals, affecting their health, particularly their reproductive health"*. Further, the runoff of pesticides into streams and rivers has adverse consequences on those community members who rely on those water sources.

3.3. Climate Change Impact on Livelihood

The impact of climate change on the livelihood of farmers was a common theme throughout the focus groups. Male farmers in Can Tho spent a lot of time discussing how climate change has impacted their productivity (33 of 101 coded segments; 33%), highlighting that *"farmers struggle to meet their needs when working"*. This concept was expanded in Ninh Binh where a farmer explained that *"[changing weather] affects all stages: planting, caring for the crops, harvesting, and even transporting guavas for sale"*.

Reduction in crop yields as it relates to income was a specific concern of men in Can Tho. Income concerns were mentioned 14% of the time in their focus group compared to 8% of the time in the mixed-sex Ninh Binh focus group and 3% of the time in the remaining focus groups. In speaking about the market value of rice, one male in Can Tho described that *"the prices are also unstable and fluctuating. Sometimes, more loss than gain"*, which ultimately leads to a loss of livelihood and *"farmers [having] to abandon farming"*. In Can Tho, both men and women noted the government's role in assisting farmers with their economic standing. While discussions with the males focused on market stabilization, females suggested that *"the government could provide support and arrange additional jobs for women to earn extra income and reduce free time"*.

3.4. Adaptive Strategies, Farmers' Needs, and Trusted Sources of Information

During the focus groups, workers noted adaptive strategies that they have used to address the impacts they are currently experiencing due to climate change. The most described adaptive strategy was based on the principle *"mặtmate trời lên thì mình lặn, mặt trời lặn thì mình lên"*, which is the concept of setting when the sun rises and rising when the sun sets. As workers described, it simply becomes too hot to work. In multiple instances (10 of 28 segments; 36%), farmers reflected on the fact that they now must leave the fields earlier in the day because it is too hot. As a female participant in Can Tho reflected, *"in the past, I left my house at 7 am and went back home at 10 to 11 am, now we have to start at 6 am and stop at 10 am because of the sun, it is too hot, I can't stand it"*, which results in a reduction in the total workable hours in a day. In this case, removing the hazard is the adaptive solution; as one farmer noted, *"there aren't many other options available"*.

Workers in every focus group mentioned hydration as a type of adaptive strategy. A female in Ninh Binh described the importance of hydration by explaining that *"the hot weather still has an impact, and the heat is still very tiring. . .when we're done, we go back to rest, drink water to recover our health"*. However, farmers also emphasized the importance of electrolyte rebalancing. Of the nine passages mentioning rehydration, 56% (n = 5) noted the use of sugared lemonade, tea, or milk instead of water. Responses to the vignettes demonstrated that farmers are aware of the signs of heat stress. In Lam Dong, a female participant explained that *"when during harvesting, if farmers feel tired or have headaches, normally they would try to keep working, it, hence, affects their health, we should change this habit, should think about long-term goals"*. Males in Lam Dong and Ninh Binh both demonstrated knowledge of how to address heat stroke, explaining that *"in the above situation, there is a phenomenon of workers suffering from sunstroke, heat exhaustion, and fatigue. The handling method is to immediately stop working, move the affected person to a cool area, provide light fan ventilation, and offer them green tea with sugar. Loosen clothing if needed. If the person recovers within 5–15 min, that's normal. However, if they remain unresponsive, it is advisable to call a doctor for hospitalization*". When asked if they had ever received formal training on handling these situations, they responded "*we listen to the radio and have witnessed it a few times, on TV and the radio, they also provide instructions on how to handle it. I've also observed a few times when someone fainted, and they provided first aid like that, and the person recovered afterward*".

A common theme that emerged among all focus groups is that occupational safety and health knowledge was commonly gained through personal experience. As a female in Lam Dong summarized, "*[formal] knowledge about occupational safety is often lacking among farmers, leading to their negligence*". As a female in Can Tho noted, while "*there are also occupational safety classes, when applying them in practice, it's quite different from the theory*". This was highlighted when discussing the proper use of personal protective equipment (PPE) when applying pesticides. While workers understand the need to use PPE, two issues were mentioned. The first is that the cost of PPE is prohibitive for farmers. The second is that the use of PPE negatively affects the ability to work. As a female in Can Tho described, "*in outdoor conditions, it can get very hot, and it becomes exhausting and difficult to breathe with a mask on. We wear long-sleeved shirts and gloves, but we can only wear them for a while, then it becomes too hot, and we can't breathe, so we have to take them off.*" Farmers identified a "*need [for] more suitable personal protective clothing*".

When discussing where farmers obtain their information, radio, TV, and social media were mentioned 33% of the time (10 of 30 coded segments). Other important sources of information include the government, specifically the Department of Agriculture, cooperatives, and communal farmers' associations. While farmers feel that "*we cannot address this on an individual scale*", it was well established that information sharing between those with experience is one of the best sources of information, as "*experience and knowledge can [only] be gained from working process*".

4. Discussion

In this study, we explored how agricultural workers in Vietnam experience climate change in relation to their health, safety, and wellbeing. These links were described as both direct and indirect associations. Direct impacts of climate change included physical health effects such as fatigue and dehydration. Indirect impacts included mental health stressors due to productivity demands. While occupational safety and health training has been developed for farmers in Vietnam, this training may not be distributed equitably. For example, it is unclear who is responsible for informal farmers' occupational health and safety, making this group more vulnerable to climate-change-related hazards.

4.1. Health, Safety, and Wellbeing Impacts

Heat stress is the most uniformly reported health concern related to climate hazards between provinces in Vietnam [50]. While it is estimated that nearly 85% of farmers have experienced symptoms of heat-related illness, only 22% sought healthcare services related to the issue [40]. This is consistent with our findings, where fatigue and dehydration were commonly noted in our focus group discussions. Additionally, we found that workers seemed acutely aware of other potential health outcomes related to climate change, including the increase in chronic and infectious diseases. This is consistent with the World Health Organization's statements that climate change and the epidemic of non-communicable diseases are intertwined and that climate change has led to increases in vector-borne diseases [51].

The WHO also notes that mental health issues have been a climate-change-related health outcome [51]. Farmers in Vietnam described anxiety due to income uncertainty that has been amplified by climate change. Research into climate change's impact on the productivity of agricultural plants is extensive [52]. One of the most common adaptive strategies to address this within Vietnam has been agricultural transition, where farmers move from their traditional crop of harvest to a new type of crop. This was demonstrated in our study by the participants in Ninh Binh who participated in a transition from rice to guava over the last seven years. Farmers are often provided technical assistance through their provincial government during such a transition. What our study found is that training on how to address new occupational safety and health hazards resulting from the transition is lacking. This supports other findings in Vietnam that suggest access to safety training is limited [1]. Coupled with our findings that farmers are experiencing negative health and safety consequences of climate change, there is a need to work with farmers, their unions, and government agencies to develop and implement job-related health and safety protective measures.

4.2. Adaptation Needs

Participants in our study suggested that existing occupational safety and health trainings often do not provide practical advice. This can be said of current heat adaptation recommendations as well. While many solutions are well intended and grounded in science, they often fail to consider how they will be implemented in practice [53]. In the recommendations provided by Morrissey et al. in 2021, they suggest that medical examination results should be used to educate workers about how their health conditions may affect their heat tolerance, that health status checks should be conducted before work-shifts, that physiological monitoring should be conducted throughout the workday, and that an emergency action plan should be in place [53]. The differential access to medical care in Vietnam may affect the efficacy of such solutions, highlighting within the TWH framework how societal-level factors can influence individual health outcomes [54]. While it is estimated that 83% of farmers in Vietnam have experienced heat-related illness symptoms, only 22% sought medical care [40]. There is a need to further engage with farmers in Vietnam to identify their needs and to co-develop trainings and interventions to address occupational safety and health issues. By taking a community-based participatory approach [55], situation-specific limitations, such as housing conditions and access to medical care, are taken into account.

Farmers in Vietnam have used their experiences to develop ways to cope with changing temperatures. Workers commonly mentioned a reduction in work hours due to the temperature being too hot. This supports both modelling [56–59] and empirical studies [60,61] that have suggested that increased temperatures will reduce the ability to work. This loss of productivity was an additional stressor our participants described. There is a need to develop solutions to ensure decent work and economic growth for all [62]. To do so, a system-level approach, targeting interventions at the societal and community level, is needed. As suggested by our participants, government intervention in market stabilization and job transition training are two such approaches.

4.3. Limitations

Our study was limited by the number of focus groups we were able to conduct. While the three main agricultural areas of Vietnam were represented, we were only able to conduct a total of five focus groups. As this was a convenience sample, we were limited in our ability to generalize more broadly, even within the participating provinces. Our limited sample size also had implications for the fit of our MDS analysis. Generally, a stress statistic of 0.15 indicates a good fit. Our analysis reached that threshold around the six-dimension solution. This suggests a much more complex association of concepts than we presented in the two dimensions shown. We attempted to address this by providing an in-depth analysis of the focus group discussion to explain the results. While the Farmers' Unions and Cooperatives support the farmers' interests, the presence of these representatives may have influenced responses in our focus group discussions.

5. Conclusions

Climate-change-related hazards are associated with direct and indirect effects on the health, safety, and wellbeing of farmers in Vietnam. As a result of this study, we observed a need to co-develop trainings with agricultural workers in Vietnam to address occupational safety and health needs, ensure they are reflective of the needs and vulnerabilities of each province, and ensure they are delivered equitably. While farmers in Vietnam face a diversity of climate-related hazards, the results from our focus group discussions suggest that heat-stress-related trainings should be prioritized, including how to prevent and respond to heat stress. As farmers in our study noted, they will not be able to solve this problem alone, suggesting that a system-level approach is needed to address the impacts of climate change felt by farmers in Vietnam. The local governments, Women's Unions, and Farmers' Unions are trusted sources of information that could lead these efforts.

Author Contributions: Conceptualization, M.D., T.T.T.T., Q.N., L.K., L.S.N., and M.C.; methodology, M.D., T.T.T.T., T.L.N.N., Q.N., and M.C.; formal analysis, M.D.; investigation, M.D., T.T.T.T., T.L.N.N., and Q.N.; resources, M.D., T.T.T.T., T.L.N.N., Q.N., L.S.N., and M.C.; data curation, M.D. and T.T.T.T.; writing—original draft preparation, M.D.; writing—review and editing, M.D., T.T.T.T., T.L.N.N., Q.N., L.S.N., M.V.D., M.T.-O., J.C., L.K., and M.C.; visualization, M.D.; supervision, M.D. and M.C.; project administration, M.D. and M.C.; funding acquisition, M.D., M.C., L.K., and L.S.N. All authors have read and agreed to the published version of the manuscript.

Funding: The research on which this article is based was originally commissioned by International Labour Organization's Vision Zero Fund with additional support from Training Grant T42 OH00929 funded by the Centers for Disease Control and Prevention. The financial support of Vision Zero Fund is gratefully acknowledged. However, this article has not been prepared, reviewed or endorsed by the ILO and the ILO assumes no responsibility for its content and accuracy. Responsibility rests solely with the author or authors.

Data Availability Statement: The data presented in this study are available on request from the corresponding author due to privacy concerns.

Conflicts of Interest: The authors declare no conflicts of interest.

References

1. Tran, V.T.; An-Vo, D.A.; Cockfield, G.; Mushtaq, S. Assessing Livelihood Vulnerability of Minority Ethnic Groups to Climate Change: A Case Study from the Northwest Mountainous Regions of Vietnam. *Sustainability* **2021**, *13*, 7106. [CrossRef]
2. Mishra, A.K.; Pede, V.O. Perception of climate change and adaptation strategies in Vietnam Are there intra-household gender differences? *Int. J. Clim. Chang. Strateg. Manag.* **2017**, *9*, 501–516. [CrossRef]
3. The World Bank. *Vietnam: The Economics of Adaptation to Climate Change Study*; The World Bank: Washington, DC, USA, 2010.
4. Ylipaa, J.; Gabrielsson, S.; Jerneck, A. Climate Change Adaptation and Gender Inequality: Insights from Rural Vietnam. *Sustainability* **2019**, *11*, 2805. [CrossRef]
5. IMHEN; UNDP. Sumary for Policy Makers. In *Vietnam Special Report on Managing the Risks of Extreme Events and Disasters to Advance Climate Change Adaptation*; Vietnam Natural Resources and Environment Publishing House: Hanoi, Vietnam, 2015; pp. 2–27.
6. Otto, M. A Tool for the Estimation and Analysis of Local Climate and Population Heat Exposure. Available online: www.climatechip.org (accessed on 15 August 2019).
7. Watts, N.; Amann, M.; Ayeb-Karlsson, S.; Belesova, K.; Bouley, T.; Boykoff, M.; Byass, P.; Cai, W.; Campbell-Lendrum, D.; Chambers, J.; et al. The Lancet Countdown on health and climate change: From 25 years of inaction to a global transformation for public health. *Lancet* **2018**, *391*, 581–630. [CrossRef]
8. Bobb, J.F.; Peng, R.D.; Bell, M.L.; Dominici, F. Heat-related mortality and adaptation to heat in the United States. *Environ. Health Perspect.* **2014**, *122*, 811–816. [CrossRef]
9. Curriero, F.C.; Heiner, K.S.; Samet, J.M.; Zeger, S.L.; Strug, L.; Patz, J.A. Temperature and mortality in 11 cities of the eastern United States. *Am. J. Epidemiol.* **2002**, *155*, 80–87. [CrossRef]
10. Schifano, P.; Leone, M.; De Sario, M.; De'donato, F.; Bargagli, A.M.; D'Ippoliti, D.; Marino, C.; Michelozzi, P. Changes in the effects of heat on mortality among the elderly from 1998–2010: Results from a multicenter time series study in Italy. *Environ. Health* **2012**, *11*, 58. [CrossRef] [PubMed]
11. Sacks, J.D.; Stanek, L.W.; Luben, T.J.; Johns, D.O.; Buckley, B.J.; Brown, J.S.; Ross, M. Particulate matter-induced health effects: Who is susceptible? *Environ. Health Perspect.* **2011**, *119*, 446–454. [CrossRef]

12. Rocklov, J.; Dubrow, R. Climate change: An enduring challenge for vector-borne disease prevention and control. *Nat. Immunol.* **2020**, *21*, 479–483. [CrossRef]
13. Dodgen, D.; Donato, D.; Kelly, N.; La Greca, A.; Morganstein, J.; Reser, J.; Ruzek, J.; Schweitzer, S.; Shimamoto, M.M.; Tart, K.T.; et al. Chapter 8: Mental Health and Well-Being. In *The Impacts of Climate Change on Human Health in the United States: A Scientific Assessment*; U.S. Global Change Research Program: Washington, DC, USA, 2016; pp. 217–246.
14. Gronlund, C.J. Racial and socioeconomic disparities in heat-related health effects and their mechanisms: A review. *Curr. Epidemiol. Rep.* **2014**, *1*, 165–173. [CrossRef]
15. Keim, M.E. Building human resilience: The role of public health preparedness and response as an adaptation to climate change. *Am. J. Prev. Med.* **2008**, *35*, 508–516. [CrossRef] [PubMed]
16. Roelofs, C.; Wegman, D. Workers: The climate canaries. *Am. J. Public Health* **2014**, *104*, 1799–1801. [CrossRef]
17. Glaser, J.; Lemery, J.; Rajagopalan, B.; Diaz, H.F.; García-Trabanino, R.; Taduri, G.; Madero, M.; Amarasinghe, M.; Abraham, G.; Anutrakulchai, S.; et al. Climate Change and the Emergent Epidemic of CKD from Heat Stress in Rural Communities: The Case for Heat Stress Nephropathy. *Clin. J. Am. Soc. Nephrol.* **2016**, *11*, 1472–1483. [CrossRef] [PubMed]
18. Dally, M.; Butler-Dawson, J.; Sorensen, C.J.; Van Dyke, M.; James, K.A.; Krisher, L.; Jaramillo, D.; Newman, L.S. Wet Bulb Globe Temperature and Recorded Occupational Injury Rates among Sugarcane Harvesters in Southwest Guatemala. *Int. J. Environ. Res. Public Health* **2020**, *17*, 8195. [CrossRef] [PubMed]
19. Yang, Y.; Chang, A.P. Heat Stress Intervention Research in Construction: Gaps and Recommendations. *Ind. Health* **2017**, *55*, 201–209. [CrossRef]
20. Zhao, Y.; Yi, W.; Chan, A.P.; Wong, F.K.; Yam, M.C. Evaluating the Physiological and Perceptual Responses of Wearing a Newly Designed Cooling Vest for Construction Workers. *Ann. Work. Expo. Health* **2017**, *61*, 883–901. [CrossRef]
21. Schall, M.C.; Sesek, C.R.F.; Pascoe, C.D. Aluminet: An Intervention for Heat-Related Illness Among Construction Workers. *Int. J. Environ. Res. Public Health* **2018**, *15*, 247.
22. Glaser, J.; Wegman, D.H.; Arias-Monge, E.; Pacheco-Zenteno, F.; Prince, H.; Chavarria, D.; Martinez-Cuadra, W.J.; Jakobsson, K.; Hansson, E.; Lucas, R.A.; et al. Workplace Intervention for Heat Stress: Essential Elements of Design, Implementation, and Assessment. *Int. J. Environ. Res. Public Health* **2022**, *19*, 3779. [CrossRef]
23. Donatti, C.I.; Harvey, C.A.; Martinez-Rodriguez, M.R.; Vignola, R.; Rodriguez, C.M. Vulnerability of smallholder farmers to climate change in Central America and Mexico current knowledge and research gaps. *Clim. Dev.* **2019**, *11*, 264–286. [CrossRef]
24. National Institute for Occupational Safety and Health (NIOSH). What Is Total Worker Health? 2020. Available online: https://www.cdc.gov/niosh/twh/totalhealth.html (accessed on 18 January 2023).
25. Centers for Disease Control and Prevention. The Social-Ecological Model: A Framework for Prevention. 2022. Available online: https://www.cdc.gov/violenceprevention/about/social-ecologicalmodel.html (accessed on 10 January 2024).
26. Huynh, P.T.A.; Resurreccion, B.P. Women's differentiated vulnerability and adaptations to climate-related agricultural water scarcity in rural Central Vietnam. *Clim. Dev.* **2014**, *6*, 226–237. [CrossRef]
27. Mitchell, T.; Tanner, T.; Lussier, K. 'We Know What We Need!' South Asian Women Speak Out on Climate Change Adapation, Action Aid International and IDS. 2007. Available online: https://actionaid.org/publications/2007/we-know-what-we-need-south-asian-women-speak-out-climate-change-adaptation#downloads (accessed on 15 September 2023).
28. Asian Development Bank. *Vietnam Environment and Climate Change Assessment*; Asian Development Bank: Mandaluyong, Philippines, 2013.
29. Binh, L.N.; Duy, L.N.D.; Dung, N.T.P.; Lien, D.T.P.; Minh, N.N.; Phuong, D.D.C. *Overview and Situation of Vegetable Production in Vietnam-Case Study of Sweet Potato and Purple Onion*; Michigan State University: East Lansing, MI, USA, 2017.
30. Giovannucci, D.; Lewin, B.; Swinkels, R.; Varangis, P. *Socialist Republic of Vietnam Coffee Sector Report*; World Bank: Washington, DC, USA, 2004; SSRN 996116.
31. Creswell, J.W.; Clark, V.L.P. *Designing and Conducting Mixed Methods Research*, 3rd ed.; SAGE Publications: Thousand Oaks, CA, USA, 2017.
32. Horne, L.R.; De Urioste-Stone, S.; Daigle, J.; Noblet, C. *Using Pile Sorts to Understand Perceptions of Climate Change, In Sustainable Tourism VIII*; Complutense University of Madrid: Madrid, Spain, 2018; pp. 175–183.
33. Dally, M.; Newman, L.S. Worker Health. In *Global Climate Change and Human Health*; Lemery, J., Knowlton, K., Sorenson, C.J., Eds.; John and Wiley and Sons: Hoboken, NJ, USA, 2021.
34. Schulte, P.A.; Bhattacharya, A.; Butler, C.R.; Chun, H.K.; Jacklitsch, B.; Jacobs, T.; Kiefer, M.; Lincoln, J.; Pendergrass, S.; Shire, J.; et al. Advancing the framework for considering the effects of climate change on worker safety and health. *J. Occup. Environ. Hyg.* **2016**, *13*, 847–865. [CrossRef] [PubMed]
35. Schulte, P.A.; Chun, H. Climate change and occupational safety and health: Establishing a preliminary framework. *J. Occup. Environ. Hyg.* **2009**, *6*, 542–554. [CrossRef] [PubMed]
36. Schulte, P.A.; Jacklitsch, B.L.; Bhattacharya, A.; Chun, H.; Edwards, N.; Elliott, K.C.; Flynn, M.A.; Guerin, R.; Hodson, L.; Lincoln, J.M.; et al. Updated assessment of occupational safety and health hazards of climate change. *J. Occup. Environ. Hyg.* **2023**, *20*, 183–206. [CrossRef] [PubMed]
37. Finch, J. The Vignette Technique in Survey Research. *Sociology* **1987**, *21*, 105–114. [CrossRef]
38. Harrits, G.S.; Møller, M.Ø. Qualitative Vignette Experiments: A Mixed Methods Design. *J. Mix. Methods Res.* **2020**, *15*, 526–545. [CrossRef]

39. Jackson, M.; Harrison, P.; Swinburn, B.; Lawrence, M. Using a Qualitative Vignette to Explore a Complex Public Health Issue. *Qual. Health Res.* **2015**, *25*, 1395–1409. [CrossRef]
40. Tran, N.Q.L.; Nghiem, S.; Chu, C.; Luong, M.A.; Ho, T.T.; Phung, D. The Prevalence of Heat-related Illnesses and Associated Factors among Rice Farmers in Vietnam. *J. Agromedicine* **2023**, *28*, 486–496. [CrossRef]
41. Committee for Population Family and Children. *Vietnam Demographic and Health Survey 2002*; Committee for Population, Family and Children: Hanoi, Vietnam, 2003. Available online: https://dhsprogram.com/pubs/pdf/fr139/fr139.pdf (accessed on 15 September 2023).
42. National Institutes of Health. *Agricultural Health Study Farmer Applicator Questionnaire*; Agricultural Health Study: Iowa, NC, USA, 1996. Available online: https://aghealth.nih.gov/collaboration/qx/farmer.pdf (accessed on 15 September 2023).
43. Ngo-Metzger, Q.; Sorkin, D.H.; Mangione, C.M.; Gandek, B.; Hays, R.D. Evaluating the SF-36 Health Survey (Version 2) in Older Vietnamese Americans. *J. Aging Health* **2008**, *20*, 420–436. [CrossRef]
44. Topp, C.W.; Østergaard, S.D.; Søndergaard, S.; Bech, P. The WHO-5 Well-Being Index: A systematic review of the literature. *Psychother. Psychosom.* **2015**, *84*, 167–176. [CrossRef]
45. Institute for Work & Health. *Workplace Health and Safety Survey*; Workplace Safety and Health Institute: Washington, DC, USA, 2016. Available online: https://www.iwh.on.ca/sites/iwh/files/iwh/tools/ohs_vulnerability_measure_questionnaire_2016.pdf (accessed on 15 September 2023).
46. Kabir, M.I.; Rahman, M.B.; Smith, W.; Lusha, M.A.F.; Azim, S.; Milton, A.H. Knowledge and perception about climate change and human health: Findings from a baseline survey among vulnerable communities in Bangladesh. *BMC Public Health* **2016**, *16*, 266. [CrossRef]
47. Hout, M.C.; Papesh, M.H.; Goldinger, S.D. Multidimensional scaling. *Wiley Interdiscip. Rev. Cogn. Sci.* **2013**, *4*, 93–103. [CrossRef] [PubMed]
48. Maxwell, J.A. The SAGE Handbook of Applied Social Research Methods. In *Designing a Qualitative Study*; Sage Publishing: London, UK, 2008; Volume 2.
49. Charmaz, K. *Constructing Grounded Theory*; Sage Publishing: London, UK, 2014.
50. McKinley, J.D.; LaFrance, J.T.; Pede, V.O. Climate Change Adaptation Strategies Vary with Climatic Stress: Evidence from Three Regions of Vietnam. *Front. Sustain. Food Syst.* **2021**, *5*, 762650. [CrossRef]
51. World Health Organization. *Cliamte Change and Noncommunicable Diseases: Connections*; World Health Organization: Geneva, Switzerland, 2023. Available online: https://www.who.int/news/item/02-11-2023-climate-change-and-noncommunicable-diseases-connections#:~:text=Climate%20change%20is%20already%20impacting,diseases,%20and%20mental%20health%20issues (accessed on 5 June 2024).
52. Shrestha, S.; Mahat, J.; Shrestha, J.; Madhav, K.C.; Paudel, K. Influence of high-temperature stress on rice growth and development. A review. *Heliyon* **2022**, *8*, e12651. [CrossRef]
53. Morrissey, M.C.; Casa, D.J.; Brewer, G.J.; Adams, W.M.; Hosokawa, Y.; Benjamin, C.L.; Grundstein, A.J.; Hostler, D.; McDermott, B.P.; McQuerry, M.L.; et al. Heat Safety in the Workplace: Modified Delphi Consensus to Establish Strategies and Resources to Protect the US Workers. *Geohealth* **2021**, *5*, e2021GH000443. [CrossRef] [PubMed]
54. Nguyen, D.N.; Nguyen, L.H.; Nguyen, C.T.; Pham, H.Q.; Hwang, J.; Vu, G.T.; Tran, B.X.; Latkin, C.A.; Ho, C.S.; Ho, R.C. Health Status and Health Service Utilization among Vietnamese Farmers in a Mountainous Province. *Int. J. Enviriron. Res. Public Health* **2019**, *16*, 4768. [CrossRef]
55. Eriksen, M.; Rothenberg, R. Community-based participatory research (CBPR). Editorial. *Health Educ. Res.* **2012**, *27*, 553–554. [CrossRef]
56. Andrews, O.; Le Quéré, C.; Kjellstrom, T.; Lemke, B.; Haines, A. Implications for workability and survivability in populations exposed to extreme heat under climate change: A modelling study. *Lancet Planet. Health* **2018**, *2*, e540–e547. [CrossRef]
57. Kjellstrom, T.; Freyberg, C.; Lemke, B.; Otto, M.; Briggs, D. Estimating population heat exposure and impacts on working people in conjunction with climate change. *Int. J. Biometeorol.* **2018**, *62*, 291–306. [CrossRef] [PubMed]
58. Kjellstrom, T.; Holmer, I.; Lemke, B. Workplace heat stress, health and productivity—An increasing challenge for low and middle-income countries during climate change. *Glob. Health Action* **2009**, *2*, 291–306. [CrossRef]
59. Kjellstrom, T.; Kovats, R.S.; Lloyd, S.J.; Holt, T.; Tol, R.S. The direct impact of climate change on regional labor productivity. *Arch Environ. Occup Health* **2009**, *64*, 217–227. [CrossRef]
60. Dally, M.; Butler-Dawson, J.; Krisher, L.; Monaghan, A.; Weitzenkamp, D.; Sorensen, C.; Johnson, R.J.; Carlton, E.J.; Asensio, C.; Tenney, L.; et al. The impact of heat and impaired kidney function on productivity of Guatemalan sugarcane workers. *PLoS ONE* **2018**, *13*, e0205181. [CrossRef] [PubMed]
61. Sahu, S.; MSett Kjellstrom, T. Heat exposure, cardiovascular stress and work productivity in rice harvesters in India: Implications for a climate change future. *Ind. Health* **2013**, *51*, 424–431. [CrossRef] [PubMed]
62. United Nations. *The 17 Sustainable Development Goals*; United Nations Publications: New York, NY, USA. Available online: https://sdgs.un.org/goals (accessed on 8 November 2023).

climate

MDPI

Article

Assessing the Value of a Human Life in Heat-Related Mortality: Lessons from COVID-19 in Belgium

Koen De Ridder

VITO—Flemish Institute for Technological Research, 2400 Mol, Belgium; koen.deridder@vito.be

Abstract: This study evaluates the cost of heat-related mortality using economic impacts and mortality data from the COVID-19 pandemic in Belgium as a proxy. By examining the economic loss measured by gross domestic product (GDP) decline and excess mortality during the first COVID-19 wave (March–June 2020), a new estimate for avoided heat-related mortality is derived. The results show that the cost per avoided death is EUR 377,000 ± EUR 222,000, significantly lower than numerical values of the commonly used Value of a Statistical Life (VSL). However, when this cost is divided by the expected remaining (eight) life years at the age of death, the resulting monetary value for a saved life year, in a EUR 47,000 ± EUR 28,000 range, aligns well with commonly used values for the Value of a Life Year (VOLY). Thus, the present study contributes to the ongoing debate on the most appropriate methods for valuing human life in the context of heat-related mortality. By comparing our results with both VSL and VOLY, we underscore the limitations of VSL in the context of heat-related mortality and advocate for VOLY as a more accurate and contextually relevant metric. These findings may offer useful insights for policymakers in evaluating and prioritizing investments in heat-related mortality-prevention strategies.

Keywords: extreme heat; mortality; COVID-19

1. Introduction

Extreme heat induces an increased incidence of premature deaths, especially in vulnerable population groups such as the elderly and persons suffering from cardiovascular disease [1]. This is a cause of concern, given that climate change projections indicate a significant rise in the frequency, duration, and intensity of extreme heat episodes. Even if the goals of the Paris Agreement to avoid dangerous climate change are achieved, a substantial rise in the frequency of deadly heat episodes is still expected [2].

In temperate climate zones, heatwaves cause more fatalities than any other weather-related disaster [3]. A notable example stems from the comparison between Hurricane Katrina in 2005, which resulted in approximately 1500 deaths [4], and the 2003 European heatwave, which led to 70,000 reported deaths [5]. Additionally, the combined death toll from the 2003 European heatwave and the 2010 Russian mega-heatwave accounted for over 80% of all deaths caused by natural disasters in Europe between 1970 and 2012 [6]. The heat-related mortality burden during the summer of 2022, the hottest season on record in Europe, has been estimated at 62,000 [7].

In July 2024, the UN Secretary General issued a call to action on extreme heat [8], underlining the urgency of the problem, among other posing severe risks to health and economic stability worldwide. The statement highlighted the need to protect those most affected by extreme heat, such as children, the urban poor, and outdoor workers; and called to boosting resilience using data and science.

City dwellers face a heightened risk of mortality from extreme heat [9–11]. Cities experience air temperatures in excess of rural values, especially at night, temperatures being higher by up to 7–8 °C and more. In an analysis covering Belgian cities [12], it has been found that urban areas experience twice as many heatwave days as their rural surroundings. In addition, there is evidence [13] that the urban heat island intensity itself may increase during heatwaves. While some of the increased urban mortality can be attributed to the vulnerability of urban populations, such as a higher proportion of isolated elderly individuals, it is also directly linked to the urban heat island (UHI) effect itself [14].

The mortality impact of heatwaves is often downplayed by citing the 'harvesting' phenomenon, where frail individuals die a few days or weeks earlier than they would have otherwise. However, this effect diminishes with stronger heatwaves. For instance, during the 2003 European heatwave, the harvesting effect was modest [15]. Overall, the number of life years remaining at death from heat-related causes has been estimated at approximately eight years [16,17].

The monetary cost of heat-related mortality is typically calculated by invoking the Value of a Statistical Life (VSL), which is an individual willingness to pay for a reduced mortality risk [18]. For example, if a survey reveals that individuals are willing to pay EUR 30 to reduce their risk of dying by one in 100,000, then the total willingness to pay would be EUR 3,000,000 (EUR 30 × 100,000). This amount represents the VSL, which in this case is EUR 3,000,000. Hence, the VSL is not the value of a specific individual's life, but rather an aggregation of individual values for small changes in the risk of death.

From this definition, it should be clear that the VSL constitutes a highly subjective metric, being fraught with substantial uncertainty. For instance, VSL estimates across the US and Europe range from EUR 1 million to EUR 10 million [19]. Use of the VSL in the context of heat-related mortality has raised concern, considering that, often, this quantity has been established in the context of, e.g., transportation accident risks, which is very different from the mortality risks associated with heat exposure [20]. In particular, economic assessments using the Value of a Statistical Life (VSL) do not explicitly account for age, thus assuming that life expectancy at the time of death is similar across all mortality causes. However, temperature–mortality relationships indicate that heat exposure disproportionately affects older populations [11]. Some have questioned the validity of using VSL in heat-related mortality, stating that "the literature offers no heatwave-specific willingness to pay" [21], or that "the use of the VSL can be controversial" [22].

Hence, while a deeper understanding of the economic impacts of climate change on health is necessary to alert decision makers to the urgency of mitigation and to support concrete adaptation actions [21], and while the cost of heat-related mortality often dominates economic assessments of climate change impacts [23], there remains a significant level of uncertainty in determining the monetary value to be assigned to a human life.

In the present paper, we attempt to estimate the monetary cost attributed to heat-related mortality in an objective manner, by leveraging the reported economic impact of measures put in place to contain mortality during the first wave of the COVID-19 pandemic in Belgium. Concretely, the cost of avoided mortality is estimated by equating the following: (1) the welfare loss incurred during the first wave of the pandemic (March–June 2020), using gross domestic product (GDP) as a metric; (2) the estimated number of averted deaths during the same period, owing to the measures that were implemented by the government to contain the pandemic.

2. Data and Methods

2.1. COVID-19 Mortality as a Proxy for Heat-Related Mortality

Before estimating GDP loss and the associated avoided mortality, the validity of employing COVID-19 mortality as a basis for establishing heat-related mortality costs is assessed, questioning whether COVID-19 mortality serves as an adequate proxy for heat-related mortality.

The available evidence seems to indicate that it does. The health profile of people vulnerable to severe COVID-19 is very similar to that of people who are sensitive to (and die from) excessive heat [24], including the following populations:

- Elderly individuals, especially those with multiple chronic conditions and those living in nursing homes;

- People with underlying medical conditions such as cardiovascular and cerebrovascular disease, hypertension, chronic pulmonary disease, kidney disease, diabetes, obesity, Alzheimer's disease, and dementia;
- Socially isolated persons (homeless people, migrants, older people living alone).

In addition, the expected remaining lifetime at the time of death is highly similar for heat- and COVID-19-related mortality, amounting to 7.7 years for COVID-19 [25] versus 8 years for heat-related mortality [16,17].

Considering the above elements, it is fair to say that many of the risk factors for severe COVID-19 overlap with those for heat-related mortality. Therefore, COVID-19 mortality appears to constitute a reasonable proxy for heat-related mortality.

2.2. Gross Domestic Product (GDP)

In Belgium, the first wave of the pandemic lasted from March to June 2020, leading to the largest decline in economic output since World War II. This decline was mainly related to the containment measures put in place by the government, drastically reducing the transmission of infections within the population. Starting on 14 March 2020, Belgium enforced a nationwide lockdown, closing schools, universities, restaurants, cafes, gyms, and non-essential businesses. Mobility was heavily restricted, and public gatherings were banned. These measures were gradually eased beginning 4 May 2020, allowing a phased return to social and economic activities. Note that only non-pharmaceutical measures were involved, as vaccination campaigns only began in 2021.

Gross domestic product (GDP) measures the value of all final goods and services produced within a country in a given year. Alternatively, it can be considered as the total income earned by households and businesses in a year [26]. A decline in GDP indicates a reduction in economic activity and is associated with job losses, reduced working hours, lower wages, and overall lower income levels for businesses and individuals. All these factors constitute a direct cost to the community.

In Belgium, quarterly GDP is published by the National Bank of Belgium [27]. Their estimates are based on national and regional accounts, which are compiled according to the requirements of the European System of Accounts (ESA 2010) [28], which is an internationally compatible EU accounting framework for a systematic and detailed description of an economy.

Table 1 presents reported GDP values for quarters 1 and 2, as well as their sum, for recent years. The consideration of only the first two quarters relates to the fact that the paper focuses on the first COVID-19 wave, which occurred between March and June 2020, i.e., within quarters 1 and 2 of that year.

Table 1. Quarterly values (in billion EUR) of Belgium's gross domestic product for 2014–2020, as reported by the National Bank of Belgium [27].

	2014	2015	2016	2017	2018	2019	2020
Q1	97.1	100.1	102.4	107.7	110.3	114.7	116.3
Q2	101.0	104.9	108.9	112.5	116.0	119.9	105.7
Q1 + Q2	198.2	205.0	211.3	220.2	226.3	234.6	222.0

2.3. COVID-19 Mortality

In early March 2020, the Belgian health institute Sciensano was mandated by the regional and federal authorities with the task to gather daily nationwide COVID-19 mortality figures [29]. Considering that the standard registration procedure of the specific causes of death is a rather slow process, an ad hoc surveillance was established to allow the near-real-time monitoring of COVID-19 mortality. A centralized data registration was established, gathering data from hospitals, long-term health care facilities, and general practitioners.

Death reporting was exhaustive, including all deaths potentially attributable to the pandemic, irrespective of the diagnostic method and setting. Subsequent studies showed that the ratio between observed overall excess mortality and reported COVID-19 mortality was remarkably close to unity in Belgium [30–33]. This consistent reporting of comprehensive and reliable mortality figures makes the Belgian data particularly suitable for use in quantitative studies as the one presented here.

Daily mortality data of the first COVID-19 wave in Belgium (March–June 2020) were acquired from the dashboard managed by Sciensano [34]. The total number of reported COVID-19 deaths over this period amounted to 9721 (Figure 1).

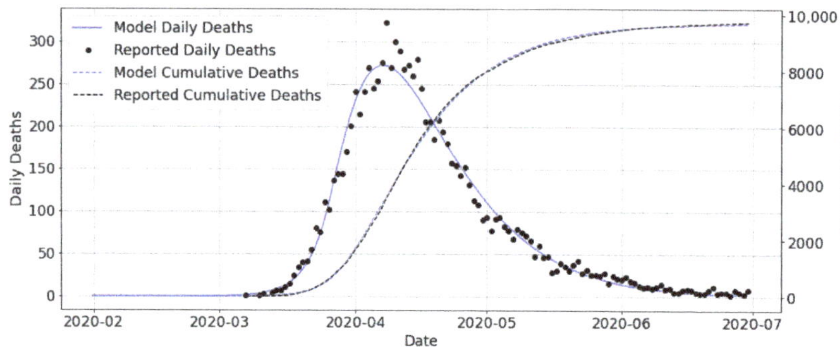

Figure 1. Daily number of COVID-19 deaths in Belgium in the first half of 2020. The dots represent the reported values, and the solid blue line represents values simulated with the model by De Visscher [35]; see Section 2.4 for details. The dashed lines show the cumulative number of deaths based on the reported (black) and modelled (blue) daily values.

2.4. Epidemiological Modelling

One of the main challenges of the approach presented here is the estimation of the hypothetical number of avoided deaths that would have occurred in the absence of any containment measures. Apart from adopting insights from the literature, we will apply an epidemiological model to evaluate scenarios of pandemic containment, in particular scenarios based on measures taken by individuals at a modest or no economic cost, as opposed to the nationwide lockdown imposed by the government.

The epidemiological model used here to simulate the evolution of the pandemic and to assess the potential number of averted deaths is the compartmental model developed by De Visscher [35]. This model simulates the time evolution of the number of uninfected (U), infected (I), sick (S), very sick (V), better (B), recovered (R), and deceased (D) individuals. In the remainder of this paper the 'better' and 'recovered' categories will be disregarded as our focus is on mortality only and considering that the omission of these categories has no bearing on mortality outcomes.

The model is based on the following system of coupled ordinary differential equations (an overhead dot denoting a time derivative):

$$\dot{U} = -k_1(I + S/2 + V/3)(U/P) \qquad (\equiv -r_1) \qquad (1)$$

$$\dot{I} = k_1(I + S/2 + V/3)(U/P) - k_2 I \qquad (\equiv r_1 - r_2) \qquad (2)$$

$$\dot{S} = k_2 I - k_3 S - k_5 S \qquad (\equiv r_2 - r_3 - r_5) \qquad (3)$$

$$\dot{V} = k_3 S - k_4 V - k_6 V \qquad (\equiv r_3 - r_4 - r_6) \qquad (4)$$

$$\dot{D} = k_4 V \qquad (\equiv r_4) \tag{5}$$

In these equations, the total population (P) is set at 11.5 million, the total Belgian population in early 2020. The symbols $r_1, \ldots, r_6$ shown between brackets refer to the transfer rates between compartments, as graphically shown in Figure 2. The coefficients $k_2, \cdots, k_6$ are fixed to values that represent COVID-19 characteristics [35], such as the incubation time. The coefficient k_1 holds a central position, being related to the reproduction number (see above) through $R_0 \approx ak_{11}$, with $a \approx 10$ days (the units of k_{11} being days^{-1}). Unlike the other coefficients, it is an adjustable parameter in the model.

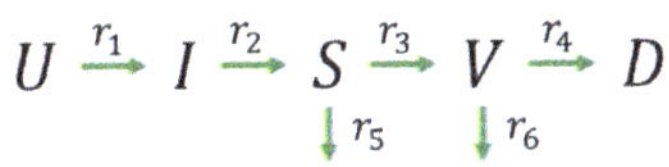

Figure 2. Pathway from infection to death across the various model compartments, involving different transition rates ($r_1, \ldots, r_6$) incorporated in Equations (1)–(5).

The model contains a mechanism to simulate the impact of measures aiming to reduce R_0, by introducing an effectiveness E, with $R_{0,\text{ after}} = (1 - E)R_{0,\text{ before}}$; the subscripts refer to the reproduction number before and after the introduction of a measure, such as a lockdown. A measure that is 100% effective ($E = 1$) would bring R_0 to zero, effectively terminating the spread of the virus. A measure with an effectiveness of 0% corresponds to a situation without any mitigating measures at all.

The system of ordinary differential Equations (1)–(5) is solved using Runge–Kutta–Fehlberg (RKF45) integration over the period from 1 February to 30 June 2020. The coefficients R_0 and E were adjusted until the model yielded a satisfactory match with daily number of deaths observed in Belgium (Figure 1), resulting in $R_0 = 3$ and $E = 0.91$. The value obtained for the reproduction number R_0 matches nicely with independent estimates of $R_0 \approx 2.9$ [36]. The infection fatality rate (IFR), having a default value in the model of 1.5%, was kept as this value coincides well with reported values for Belgium of 1.47% [32] and 1.251–1.672% [37].

With these parameter values, the modelled cumulative number of deaths matches the reported number within 0.5% (Figure 1). The time tendencies simulated for the various model compartments using these calibrated parameters are shown for Belgium for February–June 2020 in Figure 3.

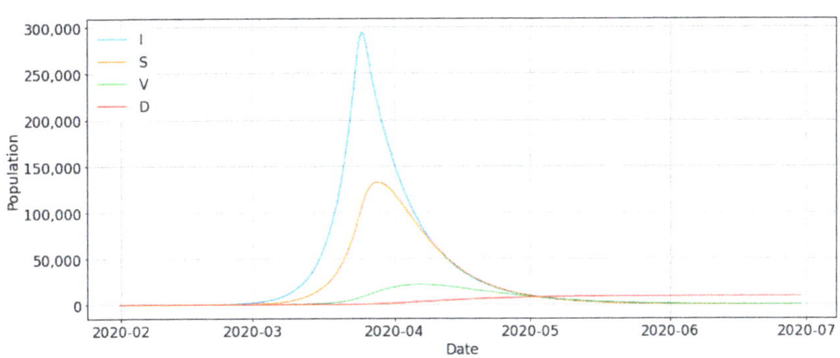

Figure 3. Simulated evolution of the COVID-19 pandemic in Belgium for the compartments of infected (I), sick (S), very sick (V), and deceased (D) individuals. Notice the shift over time of the peaks from the infected towards the sick, very sick, and deceased compartments.

The conclusion of this subsection is that the model is capable of correctly reproducing daily and accumulated deaths in the period February–June 2020, including the effect of the lockdown (through the parameter E). More details regarding the model and its applications are available in the paper by De Visscher [35].

3. Results

3.1. Net Loss of Gross Domestic Product (GDP)

The economic cost resulting from the containment measures is assessed by examining the net GDP decrease occurring during quarters 1 and 2 of 2020 (i.e., covering the first COVID-19 wave) in comparison to what would have occurred in a hypothetical non-COVID-19 scenario.

According to data from the National Bank of Belgium (Table 1), in the years leading up to 2020, the GDP value of the combined first and second quarters consistently rose by approximately EUR 7.3 billion compared to the same period of the previous year. Extrapolating this tendency (Figure 4) through linear regression yields, we obtain a predicted GDP value for the first two quarters of the year 2020 (in the absence of COVID-19) of EUR 241.4 billion $\pm$ EUR 2.7 billion (95% confidence interval).

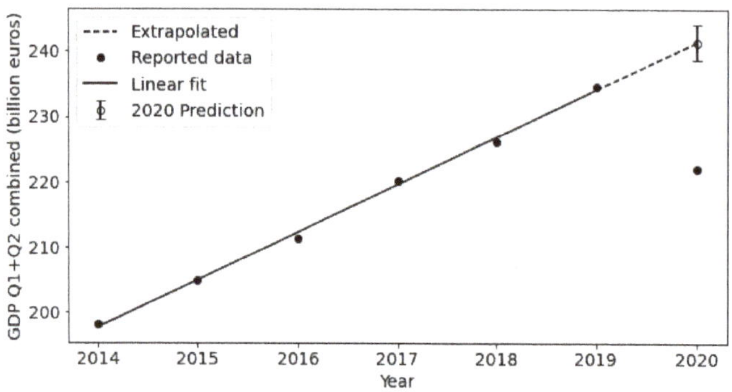

Figure 4. Gross domestic product (GDP) for Belgium (quarters 1 and 2 combined), for the years 2014–2020. The reported values are denoted by dots, and the red line corresponds to a linear fit; the dashed portion represents an extrapolation of the fitting curve to the extrapolated (counterfactual) value for 2020 (open circle). The error bar represents the 95% confidence interval on this extrapolated value.

Conversely, during the first half of 2020, the reported actual GDP reached a value of only EUR 222.0 billion, implying a net decrease of EUR 19.4 billion $\pm$ EUR 2.7 billion during the first wave of COVID-19 in Belgium in comparison to the hypothetical situation in which no pandemic had arisen.

3.2. Number of Lives Saved Due to the Lockdown

The number of lives saved due to the lockdown was estimated by comparing the number of actual deaths (approximately 10,000, see above) with the number of deaths that would have occurred in the absence of the lockdown.

The latter was assessed by means of the compartmental model described in Section 2.4, setting the efficiency of the imposed measures at 0% ($E = 0$), thus removing any mitigating measures. Figure 5 shows that, over the concerned period February—June 2020, this would have led to 161,000 deaths, or approximately 151,000 extra deaths compared to the actual reported number.

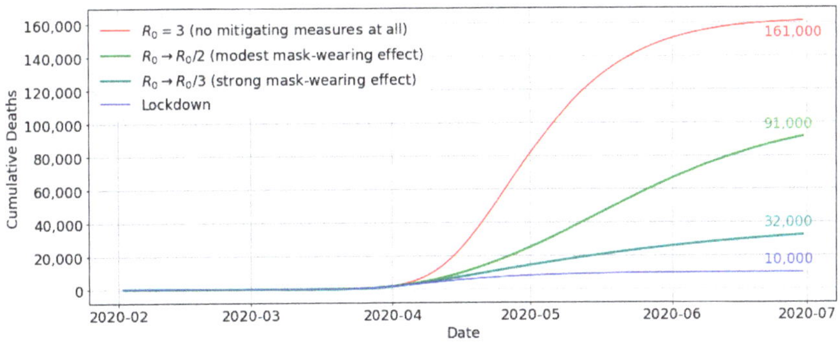

Figure 5. Cumulative number of deaths simulated by means of the De Vischer model [35] for the following scenarios: no mitigation (red), modest mask-wearing effect (green), strong mask-wearing effect (turquoise), lockdown (blue). The numbers on the right correspond to the simulated total number of deaths for each scenario at the end of June 2020.

Other studies have reached similar numbers. For instance, a study on European COVID-19 mortality [38] found that, for Belgium, the absence of mitigating measures would have led to 120,000 deaths by May 4. (The model of Section 2.4 used in the present analysis is a bit more conservative, predicting 93,000 deaths for that date.)

Yet, these high estimates do not account for individual changes in behaviour. They ignore the fact that, in the presence of a raging pandemic, people would have applied self-imposed measures, thus reducing the reproduction number [38]. Several studies ascertain that such measures, such as wearing a mask or working remotely, have an effect of the same magnitude as the official, population-level, public health intervention measures [39,40].

Masks are the most cost-effective of non-pharmaceutical interventions, having relatively few adverse effects on the economy [41]. Analysis has shown the strong potential effect of mask wearing on reducing the reproduction number, with reductions by a factor in the range of 2−3 for plausible ranges of public uptake (75−90%) and mask effectiveness (0.5−0.7) [42]. When implementing this range of R_0, reductions in the model described above (Section 2.4), the number of deaths is reduced from 161,000 (no mitigation) to 91,000 (R_0 reduced by a factor of 2) resp. 33,000 (R_0 reduced by a factor of 2) (Figure 5).

This means that, in the absence of any government interventions, a simple and cost-effective measure such as self-imposed mask-wearing would have resulted in a number of deaths between 33,000 and 91,000, or 62,000 ± 29,000. In other words, compared to the 10,000 deaths that effectively occurred during this period, 52,000 ± 29,000 deaths may have been avoided due to the measures imposed by the government. The large uncertainty in this estimate reflects the hypothetical nature of estimating the potentially avoided (counterfactual) number of deaths. Nevertheless, our estimate is consistent with an independent assessment [43], suggesting that 50,000 deaths were avoided due to the lockdown during the first COVID-19 wave in Belgium.

3.3. Cost per Avoided Death

In the previous subsections, the following findings were presented:

- The net GDP loss caused by the lockdown amounts to EUR 19.4 billion ± EUR 2.7 billion;
- The number of avoided deaths due to the lockdown is estimated at 52,000 ± 29,000.

Putting these two results together, the cost per avoided death is obtained by dividing the net GDP loss by the number of avoided deaths. Incorporating error propagation, a value of EUR 377,000 ± EUR 222,000 is obtained as the cost per prevented fatality, i.e., the value that has implicitly been assigned to a human life through the containment measures.

4. Discussion

The monetary value assigned to a human life in this study, estimated at EUR 377,000 ± EUR 222,000, stands in stark contrast to the Value of a Statistical Life (VSL) typically cited in the literature, which ranges between EUR 1 million and EUR 10 million [19] (see the Introduction Section).

This discrepancy is not unexpected given the fundamentally different constructs these values represent. While the VSL is derived from willingness-to-pay (WTP) studies that assess how much individuals are willing to pay to reduce their risk of dying, our estimate is based on actual economic losses incurred during the COVID-19 pandemic. The VSL method aggregates individual preferences across small changes in mortality risk, often in contexts like transportation accidents, which involve different demographic profiles and risk perceptions than those associated with heat-related mortality [22,44].

Moreover, the VSL does not explicitly account for age, instead assuming a uniform life expectancy across all causes of death. However, empirical evidence consistently shows that heat exposure disproportionately affects older populations, who have fewer remaining life years at the time of death [11]. This assumption introduces a significant bias when applying VSL to heat-related mortality, where the affected demographic group primarily consists of the elderly. As such, using VSL in this context can lead to overestimation of the economic value of avoided deaths.

Our study's estimate aligns more closely with the Value of a Life Year (VOLY), which is designed to account for the remaining life expectancy at the time of death. When we divide our estimate of EUR 377,000 ± EUR 222,000 by the average remaining life years (approximately eight years for heat-related deaths), the resulting VOLY of EUR 47,000 ± EUR 28,000 falls within the range recommended for European countries (EUR 25,000–EUR 100,000) [44]. This suggests that VOLY may be a more appropriate metric for valuing lives in the context of climate-related mortality, particularly for older populations.

Furthermore, the significant disparity between VSL and VOLY observed here echoes findings in the broader literature, where differences of up to an order of magnitude have been reported [45,46]. Such disparities underscore the inherent uncertainties in these metrics and the importance of context-specific valuation methods. One study noted that relying on VSL can lead to substantially higher estimates compared to VOLY, especially when applied to scenarios involving predominantly older individuals [17], concluding that "this is a major and far from resolved source of uncertainty in economic terms". Our analysis, therefore, suggests that VOLY should be the preferred approach for economic evaluations of heat-related mortality, given its closer alignment with the actual demographic characteristics of the affected population.

It is insightful to apply the monetary value assigned to human life to estimates of projected future heat-related mortality. Currently, average heat-related mortality in Belgium stands at approximately 70 excess deaths per year (reference period 1980–2010). Future projections estimate that the annual average of heat-related deaths could increase to around 2800 by the end of the century under a high climate scenario (2837 annual deaths under RCP8.5 are cited in [23] and 2777 annual deaths under SSP5-8.5 in [47]). Multiplying this projected figure of 2800 annual deaths by the previously calculated value of human life results in an estimated economic burden of approximately EUR 1 billion per year, though this figure comes with a substantial uncertainty range (EUR 0.43 billion to EUR 1.67 billion).

Applying this methodology to other types of mortality, such as that associated with respiratory infections, cardiac arrest, or lung cancer, appears feasible as long as the VOLY figure (EUR 43,000) is used. The primary challenge lies in accurately estimating the expected remaining life years at the age of death, which may vary significantly across different ailments compared to the eight years assumed for COVID-19 and heat-related mortality.

5. Conclusions

In this paper, we made an attempt to assess the cost of heat-related mortality by using COVID-19 mortality as a proxy. Our approach focused on Belgium during the first COVID-19 wave (March–June 2020), since this period was exceptionally well documented, and highly accurate daily mortality figures were available. The underlying idea is that the economic losses caused by to the lockdown can be equated to the numerous lives saved due to the lockdown, leading to a unit cost per life saved.

First, it was argued that COVID-19 mortality is indeed a suitable proxy for heat-related mortality. Not only are vulnerable population groups for both COVID-19 and heat nearly identical (elderly people and people with pre-existing disease), the potential number of life years remaining at the age of death is also very similar, being approximately eight years for either.

The monetary cost that can be assigned to a human life under COVID-19 (and by extension, to heat-related mortality) was established by linking the following factors:

- The net loss in the gross domestic product (GDP) occurring in the first half of the year 2020 (i.e., the period encompassing the first COVID-19 wave) due to the measures taken by the government, which was estimated to represent a loss of 19.4 billion ± EUR 2.7 billion;
- The number of avoided deaths due to the lockdown, estimated using evidence from the literature and through epidemiological modelling, considering the hypothetical number of people that were saved due to the measures taken by the government, their number amounting to 52,000 ± 29,000.

The resulting monetary cost of avoiding one death was found to be EUR 377,000 ± EUR 222,000. This amount turns out to be an order of magnitude lower than commonly used values for the Value of a Statistical Life (VSL). Among other things, this was attributed to the relatively short stretch of expected life years remaining at the age of death for extreme heat as compared to all causes of death. As a consequence, when reducing our estimate for the monetary cost of a human life to an annual cost, by dividing by the presumed eight life years remaining at the age of death, the resulting range of EUR 19,000–EUR 75,000 is much closer to the generally accepted values for the Value of a Life Year (VOLY).

The novelty of the approach presented here is that, rather than assessing the cost of a human life based on an individual willingness to pay, it is assessed as an actual *collective* willingness to pay. This may even be the more appropriate way to go, considering that the cost of climate action and adaptive interventions is borne by collectivity. The main new insight is that our analysis identifies the VOLY, together with a life expectance at the age of death of 8 years, as the preferred approach for quantifying heat-related mortality cost. As mentioned in the Introduction Section, quite some uncertainty regarding the VSL vs. VOLY approach has existed so far; our findings clearly identify the latter as the most appropriate metric.

The relevance and application of these results lie in their utility for supporting cost–benefit assessments of heat-related health climate actions. Mortality cost figures provide critical evidence for decision making, bolstering the case for investments in health protection and serving as a baseline against which the effectiveness of protective measures can be monitored [48]. The significance of these protective measures cannot be overstated. For instance, it has been demonstrated that the heat-related health action plans implemented after the deadly 2003 heatwave in France resulted in a nearly 70% reduction in mortality during the extreme heatwave of 2006 [49]. More recently, an analysis of heat-related mortality in Europe concluded that the death toll during the summer of 2023, which reached nearly 50,000, would have been 80% higher without adaptive actions [50]. Many of these adaptive actions focus on early warning and behavioural changes, which have been found to be highly cost effective; in fact, estimates for France show that the annual operating costs for a heat-related health action plan involves an annual operating cost of EUR 455,000 [21], which is at the same level of cost as that estimated here for a single human life.

This shows that there is ample room to consider additional measures, in particular those targeting infrastructure and urban planning. Urban interventions are not only relevant because of the elevated temperatures occurring in cities, such interventions may also be more efficient considering the high density of people that may benefit from them [51]. One measure that is considered particularly useful in an urban context consists of increasing the abundance of green urban infrastructure [47,52,53]. A study involving 100 European cities [54] has shown that a large potential for additional cooling by greening remains in many cities. While such infrastructural and planning measures are more costly than, e.g., behavioural changes, they bring benefits that last for years or decades. Moreover, urban greening measures bring many other benefits, including a positive impact on the physical and mental health of urban dwellers.

Our analysis has several limitations. First, the number of averted fatalities only accounts for mortality directly associated with COVID-19 infections. It does not account for mortality that might otherwise have been affected by the pandemic, owing to the reduced health care for non-COVID-19 disease, including cancers going undetected and urgent care being postponed [31]. At the same time, the lockdowns have also led to a reduction in certain types of fatalities, such as those related to traffic accidents. Still, balancing all these effects is far from straightforward, which is why they were left out.

Moreover, our approach is limited with respect to the monetary metric used, i.e., net GDP loss, which was used to estimate the amount of wealth a country is prepared to forego to save people from dying. Indeed, other ways to quantify the willingness to pay for a human life (or life year) are available. For instance, the WHO considers a cost per QALY (Quality Adjusted Life Year) of one–three times the GDP per capita as cost-effective [55]. Interestingly, with a Belgian GDP per capita of EUR 43,000 (2019 value), we arrive at a range of EUR 43,000–EUR 129,000 per QALY, suggesting slightly lower values for the VOLY, i.e., not significantly different from our VOLY estimate of EUR 19,000–EUR 75,000.

Finally, the results obtained here cannot easily be transposed to other regions, especially low- or middle-income countries, considering that any pandemic containment measures would probably induce GDP shocks of a different intensity. Also, such countries generally feature a different population age distribution, with a lower share of elderly people, which would affect epidemiological aspects such as the infection fatality rate (IFR). Still, with suitable data regarding actual mortality and net GDP loss, the methodology described in this paper could be replicated to other regions, which will be the subject of future research.

Funding: This research received no external funding.

Data Availability Statement: Belgian COVID-19 mortality data are available from the 'Belgium COVID-19 Dashboard, Sciensano' Belgium COVID-19 Epidemiological Situation—Deaths. Available online: https://lookerstudio.google.com/embed/reporting/c14a5cfc-cab7-4812-848c-0369173148ab/page/QTSKB (accessed on 1 August 2024). (Right-Click on the Graph 'Daily New Deaths' and Select 'Export' to Download a Csv File Containing the Data). Belgian quarterly gross domestic product (GDP) data are available from the National Bank of Belgium. National Accounts—Quarterly and Annual Aggregates. 2023. Available online: https://stat.nbb.be/ (accessed on 27 June 2024). (On the left-hand menu, select 'National accounts', then 'Quarterly and annual aggregates', again select 'Quarterly and annual aggregates', then 'Gross value added by branch of activity'; adjust the 'Time' range and consider the 'Total economy' figures.)

Conflicts of Interest: Author Koen De Ridder was employed by the company VITO—Flemish Institute for Technological Research. The author declare that the research was conducted in the absence of any commercial or financial relationships that could be construed as a potential conflict of interest.

References

1. van Daalen, K.R.; Tonne, C.; Semenza, J.C.; Rocklöv, J.; Markandya, A.; Dasandi, N.; Jankin, S.; Achebak, H.; Ballester, J.; Bechara, H.; et al. The 2024 Europe report of the Lancet Countdown on health and climate change: Unprecedented warming demands unprecedented action. *Lancet Public Health* **2024**, *9*, e495–e522. [CrossRef] [PubMed]

2. Matthews, T.K.R.; Wilby, R.L.; Murphy, C. Communicating the deadly consequences of global warming for human heat stress. *Proc. Natl. Acad. Sci. USA* **2017**, *114*, 3861–3866. [CrossRef]
3. Borden, K.A.; Cutter, S.L. Spatial patterns of natural hazards mortality in the United States. *Int. J. Health Geograph.* **2008**, *7*, 64. [CrossRef] [PubMed]
4. Beven, J.L., II; Avila, L.A.; Blake, E.S.; Brown, D.P.; Franklin, J.L.; Knabb, R.D.; Pasch, R.J.; Rhome, J.R.; Stewart, S.R. Atlantic Hurricane Season of 2005. *Mon. Weather Rev.* **2008**, *136*, 1109–1173. [CrossRef]
5. Robine, J.-M.; Cheung, S.L.K.; le Roy, S.; van Oyen, H.; Griffiths, C.; Michel, J.-P.; Herrmann, F.R. Death toll exceeded 70,000 in Europe during the summer of 2003. *Comptes. Rendus. Biol.* **2008**, *331*, 171–178. [CrossRef]
6. Golnaraghi, M.; Etienne, C.; Guha-Sapir, D.; Below, R. *Atlas of Mortality and Economic Losses from Weather, Climate and Water Extremes (1970–2012)*; World Meteorological Organization: Geneva, Switzerland, 2014; ISBN 978-92-63-11123-4. Available online: https://library.wmo.int/idurl/4/57564 (accessed on 22 August 2024).
7. Ballester, J.; Quijal-Zamorano, M.; Méndez Turrubiates, R.F.; Pegenaute, F.; Herrmann, F.R.; Robine, J.M.; Basagaña, X.; Tonne, C.; Antó, J.M.; Achebak, H. Heat-related mortality in Europe during the summer of 2022. *Nat. Med.* **2023**, *29*, 1857–1866. [CrossRef] [PubMed]
8. United Nations Secretary-General's Call to Action on Extreme Heat. 25 July 2024. Available online: https://www.un.org/sites/un2.un.org/files/unsg_call_to_action_on_extreme_heat_for_release.pdf (accessed on 2 August 2024).
9. Gabriel, K.M.A.; Endlicher, W.R. Urban and rural mortality rates during heatwaves in Berlin and Brandenburg, Germany. *Environ. Poll.* **2011**, *159*, 2044–2050. [CrossRef] [PubMed]
10. Dousset, B.; Gourmelon, F.; Laaidi, K.; Zeghnoun, A.; Giraudet, E.; Bretin, P.; Maurid, E.; Vandentorren, S. Satellite monitoring of summer heatwaves in the Paris metropolitan area. *Int. J. Climatol.* **2011**, *31*, 313–323. [CrossRef]
11. Huang, W.T.K.; Masselot, P.; Bou-Zeid, E.; Fatichi, S.; Paschalis, A.; Sun, T.; Gasparrini, A.; Manoli, G. Economic valuation of temperature-related mortality attributed to urban heat islands in European cities. *Nat. Commun.* **2023**, *14*, 7438. [CrossRef]
12. Wouters, H.; De Ridder, K.; Poelmans, L.; Willems, P.; Brouwers, J.; Hosseinzadehtalaei, P.; Tabari, H.; Vanden Broucke, S.; van Lipzig, N.P.M.; Demuzere, M. Heat stress increase under climate change twice as large in cities as in rural areas: A study for a densely populated midlatitude maritime region. *Geophys. Res. Lett.* **2017**, *44*, 8997–9007. [CrossRef]
13. De Ridder, K.; Maiheu, B.; Lauwaet, D.; Daglis, I.A.; Keramitsoglou, I.; Kourtidis, K.; Manunta, P.; Paganini, M. Urban Heat Island Intensification during Hot Spells—The Case of Paris during the Summer of 2003. *Urban Sci.* **2017**, *1*, 3. [CrossRef]
14. Keller, R.C. *Fatal Isolation: The Devastating Paris Heatwave of 2003*; University of Chicago Press: Chicago, IL, USA, 2015; 240p.
15. Toulemon, L.; Barbieri, M. The mortality impact of the August 2003 heatwave in France: Investigating the 'harvesting' effect and other long-term consequences. *Popul. Stud.* **2008**, *62*, 39–53. [CrossRef]
16. Bosello, F.; Schechter, M. Integrated socio-economic assessment. In *Regional Assessment of Climate Change in the Medi-Terranean. Vol. 2: Agriculture, Forests, Ecosystem Services and People*; Navarra, A., Tubiana, L., Eds.; Springer: Berlin/Heidelberg, Germany, 2013; 402p.
17. Sánchez-Martínez, G.; Williams, E.; Yu, S.S. The economics of health damage and adaptation to climate change in Europe: A review of the conventional and grey literature. *Climate* **2015**, *3*, 522–541. [CrossRef]
18. OECD. *Mortality Risk Valuation in Environment, Health and Transport Policies*; OECD Publishing: Paris, France, 2012. [CrossRef]
19. Ciscar, J.C.; Rising, J.; Kopp, R.E.; Feyen, L. Assessing future climate change impacts in the EU and the USA: Insights and lessons from two continental-scale projects. *Environ. Res. Lett.* **2019**, *14*, 084010. [CrossRef]
20. Ščasný, M.; Alberini, A. Valuation of Mortality Risk Attributable to Climate Change: Investigating the Effect of Survey Administration Modes on a VSL. *Int. J. Environ. Res. Public Health* **2012**, *9*, 4760–4781. [CrossRef] [PubMed]
21. Adélaïde, L.; Chanel, O.; Pascal, M. Health effects from heat waves in France: An economic evaluation. *Eur. J. Health Econ.* **2022**, *23*, 119–131. [CrossRef] [PubMed]
22. Szewczyk, W.; Feyen, L.; Matei, A.; Ciscar, J.C.; Mulholland, E.; Soria, A. *Economic Analysis of Selected Climate Impacts*; EUR 30199 EN; Publications Office of the European Union: Luxembourg, 2020. Available online: https://data.europa.eu/doi/10.2760/845605 (accessed on 22 August 2024).
23. Forzieri, G.; Cescatti, A.; Batista e Silva, F.; Feyen, L. Increasing risk over time of weather-related hazards to the European population: A data-driven prognostic study. *Lancet Planet. Health* **2017**, *1*, e200–e208. [CrossRef]
24. World Health Organization. *Care for Vulnerable Population Groups: Updated Evidence on Risk Factors and Vulnerability in Heat and Health in the WHO European Region: Updated Evidence for Effective Prevention*; Sánchez-Martínez, G., de'Donato, F., Kendrovski, V., Eds.; World Health Organization: Copenhagen, Denmark, 2021; pp. 97–120. Available online: https://iris.who.int/bitstream/handle/10665/339462/9789289055406-eng.pdf (accessed on 22 August 2024).
25. Vandresse, M. Belgium—Excess Mortality in 2020: 124,000 Years of Life Lost (Federal Planning Bureau Fact Sheet No. 5). 2021. Available online: https://www.plan.be/uploaded/documents/202105100808300.FACTSHEET_005_DEMO_12407_E.pdf (accessed on 27 June 2024).
26. Shapiro, D.; MacDonald, D.; Greenlaw, S.A. *Principles of Economics*, 3rd ed.; OpenStax Rice University, 2022. Available online: https://openstax.org/details/books/principles-economics-3e (accessed on 22 August 2024).
27. National Bank of Belgium. National Accounts—Quarterly and Annual Aggregates. 2023. Available online: https://stat.nbb.be/ (accessed on 27 June 2024).

28. European Union. European System of Accounts (ESA 2010). Eurostat, Manual and Guidelines. 2013. Available online: https://doi.org/10.2785/16644 (accessed on 30 July 2024).
29. Renard, F.; Scohy, A.; Van der Heyden, J.; Peeters, I.; Dequeker, S.; Vandael, E.; Van Goethem, N.; Dubourg, D.; De Viron, L.; Kongs, A.; et al. Establishing an ad hoc COVID-19 mortality surveillance during the first epidemic wave in Belgium, 1 March to 21 June 2020. *Eurosurveillance* **2021**, *26*, 2001402. [CrossRef]
30. Bustos Sierra, N.; Bossuyt, N.; Braeye, T.; Leroy, M.; Moyersoen, I.; Peeters, I.; Scohy, A.; Van der Heyden, J.; Van Oyen, H. All-cause mortality supports the COVID-19 mortality in Belgium and comparison with major fatal events of the last century. *Arch. Public Health* **2020**, *78*, 117. [CrossRef]
31. Luyten, J.; Schokkaert, E. Belgium's response to the COVID-19 pandemic. *Health Econ. Policy Law* **2021**, *17*, 37–47. [CrossRef]
32. Molenberghs, G.; Faes, C.; Verbeeck, J.; Deboosere, P.; Abrams, S.; Willem, L.; Aerts, J.; Theeten, H.; Devleesschauwer, B.; Bustos Sierra, N.; et al. COVID-19 mortality, excess mortality, deaths per million and infection fatality ratio, Belgium, 9 March 2020 to 28 June 2020. *Eurosurveillance* **2022**, *27*, 2002060. [CrossRef] [PubMed]
33. COVID-19 Excess Mortality Collaborators. Estimating excess mortality due to the COVID-19 pandemic: A systematic analysis of COVID-19-related mortality, 2020–2021. *Lancet* **2022**, *399*, 1513–1536. [CrossRef]
34. Belgium COVID-19 Epidemiological Situation—Deaths. *Belgium COVID-19 Dashboard, Sciensano*. (Right-Click on the Graph 'Daily New Deaths' and Select 'Export' to Download a Csv File Containing the Data). Available online: https://lookerstudio.google.com/embed/reporting/c14a5cfc-cab7-4812-848c-0369173148ab/page/QTSKB (accessed on 1 August 2024).
35. De Visscher, A. The COVID-19 pandemic: Model-based evaluation of non-pharmaceutical interventions and prognoses. *Nonlinear Dyn.* **2020**, *101*, 1871–1887. [CrossRef]
36. Abrams, S.; Wambua, J.; Santermans, E.; Willem, L.; Kuylen, E.; Coletti, P.; Libin, P.; Faes, C.; Petrof, O.; Herzog, S.A.; et al. Modelling the early phase of the Belgian COVID-19 epidemic using a stochastic compartmental model and studying its implied future trajectories. *Epidemics* **2021**, *35*, 100449. [CrossRef]
37. COVID-19 Forecasting Team. Variation in the COVID-19 infection-fatality ratio by age, time, and geography during the pre-vaccine era: A systematic analysis. *Lancet* **2022**, *399*, 1469–1488. [CrossRef] [PubMed]
38. Flaxman, S.; Mishra, S.; Gandy, A.; Unwin, H.J.T.; Mellan, T.A.; Coupland, H.; Whittaker, C.; Zhu, H.; Berah, T.; Eaton, J.W.; et al. Estimating the effects of non-pharmaceutical interventions on COVID-19 in Europe. *Nature* **2020**, *584*, 257–261. [CrossRef]
39. Jamison, J.C.; Bundy, D.; Jamison, D.T.; Spitz, J.; Verguet, S. Comparing the impact on COVID-19 mortality of self-imposed behavior change and of government regulations across 13 countries. *Health Serv. Res.* **2021**, *56*, 874–884. [CrossRef]
40. Lison, A.; Banholzer, N.; Sharma, M.; Mindermann, S.; Unwin, H.J.T.; Mishra, S.; Stadler, T.; Bhatt, S.; Ferguson, N.M.; Brauner, J.; et al. Effectiveness assessment of non-pharmaceutical interventions: Lessons learned from the COVID-19 pandemic. *Lancet Public Health* **2023**, *8*, e311–e317. [CrossRef]
41. Agyapon-Ntra, K.; McSharry, P.E. A global analysis of the effectiveness of policy re-sponses to COVID-19. *Sci. Rep.* **2023**, *13*, 5629. [CrossRef] [PubMed]
42. Fisman, D.N.; Greer, A.L.; Tuite, A.R. Bidirectional impact of imperfect mask use on reproduction number of COVID-19: A next generation matrix approach. *Infect. Dis. Model.* **2020**, *5*, 405–408. [CrossRef]
43. Molenberghs, G.; Buyse, M.; Abrams, S.; Hens, N.; Beutels, P.; Faes, C.; Verbeke, G.; Van Damme, P.; Goossens, H.; Neyens, T.; et al. Infectious diseases epidemiology, quantitative methodology, and clinical research in the midst of the COVID-19 pandemic: Perspective from a European country. *Contemp. Clin. Trials* **2020**, *99*, 106189. [CrossRef]
44. Desaigues, B.; Ami, D.; Bartczak, A.; Braun-Kohlová, M.; Chilton, S.; Czajkowski, M.; Farreras, V.; Hunt, A.; Hutchison, M.; Jeanrenaud, C.; et al. Economic valuation of air pollution mortality: A 9-country contingent valuation survey of value of a life year (VOLY). *Ecol. Indic.* **2011**, *11*, 902–910. [CrossRef]
45. Botzen, W.J.W.; Martinius, M.L.; Bröde, P.; Folkerts, M.A.; Ignjacevic, P.; Estrada, F.; Harmsen, C.N.; Daanen, H.A.M. Economic valuation of climate change–induced mortality: Age dependent cold and heat mortality in the Netherlands. *Clim. Chang.* **2020**, *162*, 545–562. [CrossRef]
46. Chiabai, A.; Spadaro, J.V.; Neumann, M.B. Valuing deaths or years of life lost? Economic benefits of avoided mortality from early heat warning systems. *Mitig. Adapt. Strateg. Glob. Chang.* **2018**, *23*, 1159–1176. [CrossRef]
47. Ignjačević, P.; Botzen, W.; Estrada, F.; Daanen, H.; Lupi, V. Climate-induced mortality projections in Europe: Estimation and valuation of heat-related deaths. *Int. J. Disaster Risk Reduct.* **2024**, *111*, 104692. [CrossRef]
48. WHO. *2021 WHO Health and Climate Change Global Survey Report*; World Health Organization: Geneva, Switzerland, 2021; 92p. Available online: https://www.who.int/publications/i/item/9789240038509 (accessed on 22 August 2024).
49. Fouillet, A.; Rey, G.; Wagner, V.; Laaidi, K.; Empereur-Bissonnet, P.; Le Tertre, A.; Frayssinet, P.; Bessemoulin, P.; Laurent, F.; De Crouy-Chanel, P.; et al. Has the impact of heat waves on mortality changed in France since the European heat wave of summer 2003? A study of the 2006 heat wave. *Int. J. Epidemiol.* **2008**, *37*, 309–317. [CrossRef] [PubMed]
50. Gallo, E.; Quijal-Zamorano, M.; Méndez Turrubiates, R.F.; Tonne, C.; Basagaña, X.; Achebak, H.; Ballester, J. Heat-related mortality in Europe during 2023 and the role of adaptation in protecting health. *Nat. Med.* **2024**. [CrossRef]
51. Vargo, J.; Stone, B.; Habeeb, D.; Liu, P.; Russell, A. The social and spatial distribution of temperature-related health impacts from urban heat island reduction policies. *Environ. Sci. Policy* **2016**, *66*, 366–374. [CrossRef]
52. Johnson, D.; See, L.; Oswald, S.M.; Prokop, G.; Krisztin, T. A cost–benefit analysis of implementing urban heat island adaptation measures in small- and medium-sized cities in Austria. *Environ. Plan. B Urban Anal. City Sci.* **2021**, *48*, 2326–2345. [CrossRef]

53. Iungman, T.; Cirach, M.; Marando, F.; Barboza, E.P.; Khomenko, S.; Masselot, P.; Quijal-Zamorano, M.; Mueller, N.; Gasparrini, A.; Urquiza, J.; et al. Cooling cities through urban green infrastructure: A health impact assessment of European cities. *Lancet* **2023**, *401*, 577–589. [CrossRef]
54. Lauwaet, D.; Berckmans, J.; Hooyberghs, H.; Wouters, H.; Driesen, G.; Lefebre, F.; De Ridder, K. High resolution modelling of the urban heat island of 100 European cities. *Urban Clim.* **2024**, *54*, 101850. [CrossRef]
55. Robinson, L.A.; Hammitt, J.K.; Chang, A.Y.; Resch, S. Understanding and improving the one-and-three- times GDP per capita cost-effectiveness thresholds. *Health Policy Plan* **2017**, *32*, 141–145. [CrossRef] [PubMed]

climate

Article

Tackling Complexity: Integrating Responses to Internal Displacements, Extreme Climate Events, and Pandemics

Roberto Ariel Abeldaño Zuñiga [1,2,*], Gabriela Narcizo de Lima [3,4] and José Carlos Suarez-Herrera [5]

1 Yhteiskuntadatatieteen Keskus, Valtiotieteellinen Tiedekunta, Helsingin Yliopisto, 00150 Helsinki, Finland
2 Helsinki Institute of Urban and Regional Studies (URBARIA), University of Helsinki, 00150 Helsinki, Finland
3 Geography Department, Faculty of Arts and Humanities, Porto University, 4150-564 Porto, Portugal; gabrielalima@letras.up.pt
4 Centre of Studies in Geography and Spatial Planning, Porto University, 4150-564 Porto, Portugal
5 Office of Research and Knowledge Transfer, Mid-Atlantic University, 35017 Las Palmas de Gran Canaria, Spain; josecarlos.suarez@pdi.atlanticomedio.es
* Correspondence: ariel.abeldanozuniga@helsinki.fi

Abstract: Background: During 2020 and 2021, over 50.2 million individuals were forced to leave their homes to escape the impacts of climate-related disasters, unable to practice social isolation or self-quarantine. A considerable proportion of them reside in densely populated areas with a lack of basic services such as water and sanitation and limited access to essential healthcare. This study aimed to estimate the internal displacements during 2020 and 2021 due to climate-related events, and review the evidence for proposing policy recommendations. Methods: Data from the Internal Displacement Monitoring Centre were used for assessing internal displacement by disasters during 2020 and 2021. In addition, the authors conducted a bibliographic review to analyse the responses to internal displacements in climate-related disasters. Results: There were 883 severe storms and 1567 flood events resulting in 50.2 million internal displacements globally. Through the documents reviewed, the legal framework, the vulnerabilities and current challenges of internally displaced persons, and the response policy recommendations were analysed. Conclusions: The increased awareness of displacement and migration, particularly driven by climate-related factors, aligns with international agreements emphasising coordinated action. This recognition becomes even more critical in the context of the convergence of climate-related displacements and the potential for future pandemics.

Keywords: internal displacements; climate-related disasters; COVID-19; pandemics; disaster risk reduction

1. Introduction

Published reports have detailed the increased risk faced by internally displaced people during the global coronavirus (COVID-19) pandemic [1,2]. Over 50.2 million individuals were forced to leave their homes to escape the impacts of climate-related disasters, unable to practice social isolation or self-quarantine. A significant proportion of them reside in densely populated areas with a lack of basic services such as water and sanitation and limited access to essential healthcare.

The extreme climate events that led to forced displacement directly impacted regions already affected by the pandemic, exacerbating health risks and challenging response capabilities. The massive mobility of people in disaster situations heightened the vulnerability of these individuals and overwhelmed healthcare systems. This connection underscores the need for integrated approaches and global resilience to address the complex interactions between climate-induced displacement and the management of health crises such as the COVID-19 pandemic. Although the COVID-19 pandemic may have ended in 2023, the relevance of the issue persists due to its lasting impacts and the ongoing need to address the intricate interconnections between extreme climate events and health crises. The climate emergency continues to trigger massive displacements, affecting vulnerable communities worldwide. The lessons learned during the pandemic emphasise the importance of a global and coordinated response to crises. Moreover, the post-pandemic world requires resilient strategies to simultaneously address multiple challenges, such as climate-induced displacements, emphasising the need to sustain attention and action on this relevant issue.

The humanitarian impact of internal displacements caused by extreme climate events is significant and manifests in many ways [3]. Affected communities not only experience the loss of homes and belongings but also the disruption of social networks and support systems. This situation heightened the vulnerability of displaced individuals as access to basic services, such as clean water and healthcare, was severely disrupted or limited. The precarious living conditions in temporary shelters increased susceptibility to other diseases, not only to COVID-19 but also to the continuity of care for other non-communicable chronic diseases. Moreover, communities already facing economic and social challenges were further adversely affected, creating a cycle of humanitarian challenges. Highlighting these humanitarian impacts in past years is crucial to raise awareness about the need for preparedness to provide timely humanitarian responses to future challenges. This study aimed to (1) estimate the internal displacements during 2020 and 2021 due to climate-related events, and (2) review the evidence for proposing policy recommendations. So, this study underscores how the convergence of these events directly affected individuals, their families, and communities. It emphasises the importance of addressing not only immediate needs but also working towards long-term solutions that strengthen community resilience and mitigate future risks.

2. Materials and Methods

Data from the Internal Displacement Monitoring Centre (IDMC) based in Geneva, Switzerland, were used for assessing internal displacement by disasters during 2020 and 2021. Together with the Norwegian Refugee Council, this body monitors the situation of internal displacement due to circumstances of violence and disaster [4]. For this study, the authors retrieved all internal displacements caused by disaster situations recorded in the IDMC database.

The IDMC is the leading NGO for collecting, processing, curating, and publishing information on internal displacement due to disasters and conflicts globally. Their methods and terminology have been clearly described on their website [4]. The IDMC prioritises its data sources, with the first level comprising data primarily produced by government agencies, UN organisations, the International Organization for Migration (IOM), humanitarian clusters, and local authorities. The second level includes data from international and local NGOs, civil society, human rights organisations, and academia. Finally, the third level contains data from international and local media, affected populations, and non-state armed groups [4]. The first-level sources are considered the most reliable and are used as the primary basis for producing estimates. The second-level sources are consulted only when information from first-level sources is unavailable. The third-level sources are only used if their reliability can be rigorously assessed and verified [4].

According to the definition of internal displacement by the IDMC, this term refers to the "estimated number of internal displacement movements to have taken place during the year. Figures include individuals who have been displaced more than once. In this sense, the number of internal displacements does not equal the number of people displaced during the year" [4].

In addition to the database consulted for conducting the analysis of internal displacement by disasters mentioned above, an extensive bibliographic review was also carried out. Due to the specificity of the topic of responses to internal displacement in climate-related disasters, the selection of documents for the corpus was based on intentional criteria. The papers, book chapters, reports, and institutional reports were traced using various sources. Websites of governmental agencies and international organisations (WHO/OMS,

PAHO/OPS, UNISDR/EIRD) were searched to gather data on documents related to the search terms "internal displacements" and "disaster situations". The period selected was 2020–2021, and the languages of the documents searched are Spanish, English, Portuguese, and French.

The resulting documents were subjected to a screening process to review the relevance of each article (and associated keywords) to the object of study. Documents were included if they represented original research, government documents, or literature reviews on the topics cited and explicitly documented these concepts in the body of the manuscript.

To assess the articles that met the inclusion requirements, the research objectives and results were summarized. The argumentation of this article was structured in light of the results provided by the selected articles and related to our categories of analysis. To organise the report of the results, the authors created categories related to the responses: (1) internal displacements during 2020 and 2021 due to climate-related events, (2) legal framework, (3) vulnerabilities and current challenges of internally displaced persons, and (4) response policy recommendations.

3. Results

3.1. Internal Displacement during 2020 and 2021 Due to Climate-Related Events

The analysis of data from the Internal Displacement Monitoring Centre [4] and the bibliographic review addressing responses to internal displacement in climate-related disasters has yielded relevant results. However, it is important, initially, to define some of the concepts discussed for a comprehensive understanding of the results.

Climate change refers to long-term alterations in global climate patterns, primarily resulting from human activities releasing greenhouse gases into the atmosphere. These changes have significant impacts on climate, such as rising average temperatures, alterations in rainfall patterns, and more frequent extreme weather events, directly influencing ecosystems, biodiversity, and the daily lives of communities. Extreme climate events, in turn, are intensified and are uncommon episodes within the climate system. They encompass phenomena like severe storms, heatwaves, floods, wildfires, and hurricanes, causing significant impacts on the environment, infrastructure, and society. These events often lead to disasters, population displacements, loss of human lives, and substantial damage to ecosystems. Understanding these concepts is crucial for addressing the complex and interconnected challenges related to the environment and climate, seeking adaptation and mitigation strategies to ensure the resilience of communities in the face of these adverse scenarios [5,6].

According to data recorded by the Internal Displacement Monitoring Centre [4], during the years 2020 and 2021, there were 883 severe storm events worldwide (including cyclones, hurricanes, and typhoons) and 1567 flood events resulting in forced displacements of people. These events impacted 160 countries, with a more significant impact in regions of the Global South, such as Central America, Sub-Saharan Africa, and East and Southeast Asia. In total, these events led to 50.2 million internal displacements globally (Figure 1), against the backdrop of the health crisis stemming from the COVID-19 pandemic. Table 1 displays the top twenty events with the highest records of internal displacements during 2020 and 2021 [4]. It can be observed that these events cumulatively accounted for over 28.3 million displacements, impacting countries in Asia, Central America, and Africa most significantly. The interplay of climate-related disasters and the recurrent waves of the pandemic during those years has exacerbated the challenges faced by displaced populations and required comprehensive policy responses.

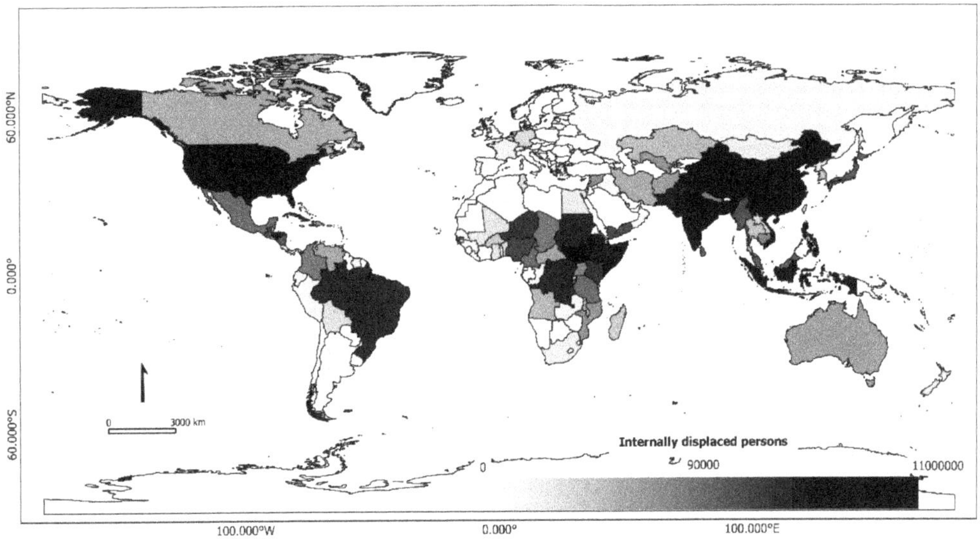

Figure 1. Internal displacements resulting from climatic events—2020 to 2021.

Table 1. Top twenty climate-related events with the highest records of internal displacements during 2020 and 2021.

N	Affected Countries and Territories and Event Name	Date of Event (Start)	Internal Displacements
1	India, Bangladesh, Myanmar, Bhutan: Cyclone Amphan	16 May 2020	4,950,000
2	The Philippines, Palau, Viet Nam: Typhoon Rai (known as Odette in the Philippines)	11 Dec 2021	3,915,000
3	China: Flood Season	1 Jun 2020	3,760,000
4	Bangladesh: Flood	26 Jun 2020	1,921,000
5	The Philippines, Viet Nam: Typhoon Vamco (known as Ulysses in the Philippines)	8 Nov 2020	1,560,000
6	India, Bangladesh: Tropical Cyclone Yaas	23 May 2021	1,505,000
7	China: Flood	16 Jul 2021	1,490,000
8	China, The Philippines, Taiwan: Typhoon "In-Fa"/"Fabian"	14 Jul 2021	1,430,000
9	The Philippines, Vietnam: Typhoon Goni (known as Rolly in the Philippines)	26 Oct 2020	1,250,000
10	Pakistan: Flood	15 Jun 2020	810,000
11	Colombia, Nicaragua, Honduras, El Salvador, Guatemala, Belize: Hurricane Iota	14 Nov 2020	743,000
12	The Philippines, Viet Nam: Tropical Cyclone Conson (known as Jolina in the Philippines)	6 Sep 2021	727,000
13	India, Bangladesh: Tropical Cyclone Yaas	23 May 2021	703,000
14	The Philippines, Viet Nam: Typhoon Molave (known as Quinta in the Philippines)	23 Oct 2020	623,000
15	Dominican Republic, Haiti, Cuba, Puerto Rico, USA: Hurricane Laura	21 Aug 2020	585,000
16	Somalia: Gu rains	31 Mar 2020	505,000
17	India: Flood	1 Jun 2021	500,000
18	China: Floods	10 Aug 2020	490,000
19	China: Flood	1 Sep 2021	467,000
20	Sudan: Flood	1 Jul 2020	454,000
	Total Internal displacements		**28,388,000**

3.2. Legal Framework

In the existing legal framework for the protection of internally displaced persons, international instruments such as the Guiding Principles on Internal Displacement can be found, developed by the UN Secretary General's Representative on Internal Displacement [7]. These principles underscore the responsibility of states to provide protection and assistance to internally displaced persons within their borders.

The awareness of displacement and migration resulting from disasters, climate change, and environmental degradation has grown significantly since 2010, leading to discussions at various levels. This recognition is reflected in international legally binding agreements that emphasise the necessity for coordinated, cross-sectoral action [8].

One crucial agreement is within the United Nations Framework Convention on Climate Change. Migration issues were initially incorporated into the UNFCCC in the 2010 Cancun Adaptation Framework, gaining further anchoring with the 2015 Paris Agreement [9]. The Paris Agreement set up a task force on Displacement under the Executive Committee of the Warsaw International Mechanism for Loss and Damage Associated with Climate Change [8].

In addition to legally binding agreements, several international policy frameworks integrate references to displacement and migration prompted by disasters, climate change, and environmental degradation. The Sendai Framework for Disaster Risk Reduction 2015–2030 [10] seeks to minimise the impact of disasters on people, the environment, and economies, emphasising action in economic, social, and environmental policy areas.

The Global Compact for Safe Orderly and Regular Migration of 2018 [11] represents a critical advancement in global governance on environmental migration. It includes specific actions to address the adverse drivers and structural factors compelling people to move due to natural disasters, climate change, and environmental degradation.

The 2030 Agenda for Sustainable Development recognises the serious threat of climate change and environmental degradation to the achievement of Sustainable Development Goals [12]. Leveraging investments in activities addressing both sustainable development and climate action is highlighted as an opportunity.

Initiatives like the Nansen Initiative's Agenda for the Protection of Cross-Border Displaced Persons [13] and the Platform on Disaster Displacement [14] provide a coherent approach to the protection of people displaced across borders in the context of disasters and climate change.

Recent reports, including the United Nations Secretary General's High-Level Panel on Internal Displacement [15] and the Action Agenda on Internal Displacement [16], emphasise proactive and systematic approaches to addressing internal displacement as part of the UN's work on climate change.

Regionally, specific agreements are observed, such as the African Union's Kampala Convention on Internal Displacement in Africa [17] and the American Convention on Human Rights in the Americas [18]. These treaties aim to ensure fundamental rights for internally displaced persons, including protection against inhumane living conditions, but they are not the only ones, as many countries have developed national policies and legislation to address issues related to internal displacement.

Collectively, these agreements and frameworks underscore the urgency of addressing displacement caused by climate change and disasters. They emphasise the need for coordinated efforts, legal preparedness, and global cooperation to protect affected populations and build resilience in the face of environmental challenges. However, the applicability of these legal frameworks in the context of pandemics or epidemics is a challenge. The response to internal displacement must be adjusted to address the complexity of multi-threat scenarios. It is essential to consider how public health measures impact the specific rights and needs of internally displaced persons, such as access to safe shelter, continuous medical care, and basic services. The assessment of current protection should analyse the flexibility of existing legal frameworks to adapt to health crises, ensuring that internally displaced persons are not only protected from disease but also from the additional vulnerability associated with displacement. Collaboration between government entities, humanitarian organisations, and civil society is crucial to ensure a comprehensive and legally robust response in these complex scenarios.

Legal preparedness plays a crucial yet often overlooked role in effective disaster risk management. In the aftermath of a disaster, time becomes a critical factor, and having well-established laws in place is essential. Laws can be instrumental in addressing the complex challenges arising from climate-related displacement, particularly when contemplating planned relocation as a proactive measure to mitigate the risk of further displacement due to climate-related events or as a durable solution for existing displacement. Nevertheless, some legal gaps have been identified that should be addressed to strengthen the protection of internally displaced persons in the context of pandemics and epidemics, such as the lack of specific provisions considering the unique needs of this population during health crises.

Adaptation of Guiding Principles to Health Crises: The Guiding Principles on Internal Displacement may not adequately address circumstances arising from a pandemic. Therefore, it is necessary to review and adapt these principles to include specific provisions ensuring the protection and healthcare of internally displaced persons in public health emergencies.

Specific Regional Legal Framework: Some regions may lack a specific legal framework addressing the complexities of protecting internally displaced persons in the context of pandemics. Therefore, it is necessary to develop or strengthen regional treaties that consider the intersections between displacement and health crises and adapt existing regulations.

Intensifying actions to prevent and minimise: Governments need to broaden their efforts to prevent, minimise, and address climate-related displacement. However, this necessitates a review and strengthening of existing laws to ensure that such actions are grounded in solid legal foundations.

Promoting early action: The promotion of actions before displacement, including climate change adaptation and early measures, requires governments to incorporate clear and specific legal provisions to guide such initiatives.

Protection against Discrimination: Current regulatory frameworks may not offer sufficient protection against health-related discrimination of internally displaced persons. A potential solution would be to incorporate explicit provisions ensuring equal access to medical care and other services.

Addressing protection needs: Legal frameworks should explicitly address the protection and assistance needs of people displaced due to the adverse effects of climate change. The quest for durable solutions must be legally strengthened, especially for populations experiencing protracted displacement.

Intersectoral Coordination Mechanisms: A coordinated approach between sectors, such as health and disaster management, may be lacking in the legal framework. To address this gap, provisions need to be established that encourage collaboration and coordination between various entities to ensure holistic and effective responses [19].

Increasing Finance for Local Adaptation: Increasing climate adaptation finance for local organisations requires legal frameworks that promote local engagement and leadership, ensuring that laws support and facilitate the effective use of funds by local entities [20].

Integration of Climate Finance: Legal coherence is crucial to ensure the integration of climate finance with humanitarian, disaster risk reduction, and development finance. Donors should operate within established legal frameworks to optimise synergies among these financial streams.

The International Health Regulations [21] play a crucial role in managing global health emergencies. These regulations are an international legal instrument aimed at preventing the international spread of diseases and contributing to their control. Although the IHR is not specifically designed to address internal displacement, it can have relevant implications and connections in the context under consideration:

Global Coordination: The IHR establishes a framework for the notification and response to events of international public health concern. This promotes global collaboration and coordination of efforts to address pandemics and epidemics, which is essential when dealing with population displacement associated with health crises.

Integration with Humanitarian Responses: In situations where internal displacement is affected by extreme climate events and pandemics, the IHR can provide a framework for integrating humanitarian responses with public health measures. This ensures coordinated and efficient action across different sectors.

Protection of Public Health: The regulations aim to protect public health, including the care and protection of people affected by internal displacement during health emergencies. They can influence the formulation of policies and protocols to ensure adequate medical care for this population.

Adaptation to Local and Mobile Contexts: These regulations recognise the need to adapt to specific contexts and the mobility of populations. This flexibility may be relevant when considering protocols for internally displaced persons in emergencies. We promote a balance between, on the one hand, a legal framework based on the principles of participatory inclusive governance, and on the other hand, strategic alliances that take shape in the territories between the social actors involved in the structuring of a coordinated global agenda.

In summary, the International Health Regulations provide an international framework that can influence how countries manage situations of internal displacement during health crises. Their effective implementation can contribute to a more coordinated and efficient response at the intersection of displacement, extreme climate events, and pandemics.

3.3. Identified Vulnerabilities and Current Challenges

Climate change amplifies existing challenges and vulnerabilities, leading to compounding crises for communities worldwide. Notable instances include Angolan citizens crossing into Namibia in December 2020 due to drought-induced shortages of food, water, healthcare, and employment. In November 2020, Honduras faced consecutive hurricanes (Eta and Iota), causing massive displacement, damaging crops, and worsening the plight of families already suffering from COVID-19 and poverty [22]. Iraq struggles with water shortages, drought, geopolitical tensions, and water mismanagement, placing local communities at risk of displacement and insecurity. In Yemen, a prolonged armed conflict intensified in 2020, coupled with extreme flooding that devastated communities and exacerbated health crises. Over 300,000 people, primarily internally displaced persons, were affected, leading to secondary displacement [20]. These interconnected challenges underscore the urgent need for comprehensive and sustainable climate and humanitarian interventions.

In Central America, specific vulnerabilities are evident in the high exposure to extreme climate events, such as hurricanes and storms, impacting communities in the Caribbean region already prone to social and economic instability [23]. During the pandemic, mobility restrictions limited access to shelters and essential services, exacerbating the vulnerability of these communities. In sub-Saharan Africa, rural populations dependent on agriculture face climate vulnerabilities, while the lack of infrastructure and basic services increases their exposure. The pandemic amplified disparities in healthcare access and challenged response capacities [24]. In Southeast Asia, urban vulnerabilities are accentuated by climate events and high population density. During the pandemic, urban communities experienced disruptions in essential services and faced economic hardships [25]. In all regions, the interaction between contextual vulnerabilities and the pandemic underscored the need to address structural issues to build resilience and prepare for future displacement.

The gaps in responding to internal displacement caused by extreme climate events during the pandemic were conspicuous in various areas. Firstly, mobility restrictions imposed to contain the spread of COVID-19 affected the timely delivery of humanitarian assistance, leaving vulnerable communities without access to shelters, essential supplies, and adequate healthcare. Additionally, the lack of coordination between humanitarian actors and government authorities contributed to inefficiencies in resource distribution. Economic difficulties exacerbated gaps as many displaced individuals lost their jobs and faced challenges covering basic needs. As we move into the future, gaps persist in terms of preparedness and response capacity, with additional challenges related to forecasting extreme climate events and global coordination. Addressing these gaps is crucial to enhance the resilience of affected communities and ensure more effective responses in similar situations in the future.

3.4. Response Policy Recommendations

Integrated Response Centres: Establish integrated response centres that operate jointly to address both displacement and health crises. These centres should bring together experts in health, disaster management, coordination, and social services to facilitate a coordinated and efficient response. In 2021, a comprehensive needs analysis was conducted in response to the Rohingya refugee situation in Bangladesh. The purpose was to provide an analysis of how needs have changed in 2020, with an emphasis on the impact of the COVID-19 pandemic on multisectoral needs. The report highlights the need to increase funding for the integrated humanitarian response [26,27].

Integrated Contingency Plans: Develop contingency plans that incorporate specific protocols to address situations where displacement and pandemics coexist. These plans should outline specific actions, roles, and responsibilities of the various involved stakeholders. In cases of emergencies, external humanitarian assistance may take days, and in certain instances, weeks, to become available. Hence, it is crucial to establish a preparedness plan grounded in the existing capacity of the country to effectively address the initial stages of an emergency, particularly in regions prone to seasonal weather events like Central America or Southeast Asia. In 2015, the Inter-Agency Standing Committee formulated guidelines for crafting Advanced Preparedness Actions and Contingency Planning, considering the interplay of intricate events involving displaced populations in epidemic or pandemic scenarios [28]. Typically, the fundamental components of a contingency plan—coordinative measures, the composition of aid packages, procurement, and logistics arrangements—are applicable across various potential emergencies within a country. Specialised contingency plans should only be devised when a distinctly different humanitarian response is warranted, for instance, due to significant geographical differences in the affected areas.

Integrated Surveillance Systems: Implement integrated surveillance systems that monitor both the spread of diseases and population displacement caused by extreme climatic events. These systems would collect relevant data on disease incidence and population movement patterns. The gathered information would provide a robust foundation for more informed and adaptive decision-making in real time. Furthermore, it would facilitate a more agile and coordinated response, enabling authorities and humanitarian organisations to anticipate potential crises and allocate resources efficiently. This comprehensive approach not only strengthens the ability to manage emergencies more effectively but also contributes to identifying underlying links and patterns between climate-induced displacement and disease spread, informing long-term preventive strategies. Two projects starting in Burkina Faso and Ethiopia [29,30] are examples of these integrated approaches, where they plan to implement surveillance methods in line with the International Health Regulations.

Coordinated International Collaboration: Promotes a more integrated international coordination, where international public health and disaster management agencies work hand in hand to share information, resources, and experiences in real time. We are aware that cyclones, hurricanes, and typhoons simultaneously impact multiple countries in Central America and Southeast Asia each season. We also know that this will continue to occur every year. In an increasingly interconnected world, fostering collaborative efforts on an international (regional) scale is essential to address complex challenges effectively. By encouraging collaboration between global health and disaster management organisations, we can establish a more comprehensive approach to crisis response. This involves the sharing of critical information about disease outbreaks and climatic events, pooling resources to ensure a more efficient and targeted response, and exchanging best practices and lessons learned. Integrated international coordination not only enhances the ability to respond promptly to emergencies but also builds a collective reservoir of knowledge and expertise. This collaborative synergy allows for a more cohesive global response, minimising the duplication of efforts and maximising the impact of interventions. By breaking down silos and fostering a culture of shared responsibility, we can collectively navigate the intricate landscape of health and disaster management, enhancing our collective resilience to unforeseen challenges. This approach ensures that countries and communities are better equipped to face the dual challenges of climate-induced displacement and disease outbreaks in a more synchronised and effective manner. In this regard, the Sendai Framework for Disaster Risk Reduction 2015–2030 is the key international framework to drive international cooperation, as reported by the mid-term evaluation conducted by Australia [31].

Anticipatory humanitarian funding options and strengthening protection and human rights: A component of debt relief is crucial, particularly as the costs of mitigating disaster displacement during the pandemic prove inevitably higher [32]. Simultaneously, reinforcing the protection and human rights of people on the move is essential. Governments must uphold obligations under international law, safeguarding the rights and welfare of internally displaced people and internal migrants [33–36]. The goal is to establish a form of legal protection similar to that for refugees but applicable to individuals moving within their own country [37].

Adaptive Quarantine Protocols: Establishing quarantine and isolation protocols that are adaptable to displacement contexts, recognising the mobility of affected populations and ensuring continuous medical care. The proposal for Adaptive Quarantine Protocols aims to address the complexity of displacement situations by establishing flexible guidelines that cater to the specific needs of affected populations. Recognising the inherent mobility of displacement, these protocols should be designed to be applicable in dynamic contexts, where communities may be constantly on the move due to extreme climate events or other emergencies. Adaptability involves considering the availability of resources, healthcare infrastructure, and the continuity of medical care, as implemented by Ukraine in 2021 [38]. Additionally, sociocultural characteristics of displaced populations must be considered to ensure that protocols are culturally sensitive. This measure not only aims to manage the spread of diseases but also to ensure the overall well-being of displaced communities by providing continuous medical care and addressing specific needs that arise in displacement contexts.

Integrated drills: Conduct integrated drills that replicate realistic scenarios where displacement and pandemics coexist. This involves creating simulated situations that mirror the challenges posed by both displacement events and pandemics simultaneously. The purpose of these integrated drills is to assess the effectiveness of integrated responses and make necessary adjustments to plans. Furthermore, these simulations can be implemented by using artificial intelligence tools [39]. By incorporating elements of both displacement and pandemic response into these simulations, emergency responders and relevant authorities can assess the coordination, communication, and overall efficacy of their strategies. This approach not only helps identify potential gaps or weaknesses in the existing response frameworks but also facilitates the refinement of protocols to enhance overall preparedness. Furthermore, integrated drills provide an invaluable opportunity for different agencies, including health departments, disaster management teams, and humanitarian organisations, to collaborate and streamline their efforts. This collaborative approach can foster a better understanding of how various sectors can work together seamlessly during complex emergencies, improving the overall resilience of communities facing the dual challenges of displacement and pandemics. The insights gained from these simulations contribute to continuous learning, ensuring that response strategies remain adaptive and robust in the face of evolving threats and crises.

4. Discussion

The global impact of 50.2 million internal displacements during 2020–2021 due to severe climate events [4] highlights the pressing need for a legal framework. International agreements such as the Paris Agreement, Sendai Framework, and Global Compact for Migration provide a foundation for coordinated global responses [10,11], yet adaptation to health crises is a challenge.

The existing legal frameworks, including the Guiding Principles on Internal Displacement [7], must be tailored to health crisis contexts. The emphasis on intensifying actions, promoting early intervention, and protection against discrimination aligns with the evolving nature of displacement crises.

Legal preparedness should facilitate intersectoral coordination, addressing gaps between health and disaster management sectors. Ensuring financial support for local adaptation and integrating climate finance within legal frameworks is imperative for effective resource utilisation and synergies between humanitarian, development, and climate funds.

In response to these challenges, the proposed response policies provide a roadmap for mitigating the impact of dual challenges. Integrated response centres, contingency plans, and surveillance systems address immediate needs, emphasising the importance of collaborative efforts. Anticipatory humanitarian funding, adaptive quarantine protocols, and integrated drills contribute to long-term resilience.

The nexus of climate-induced displacement, pandemics, and legal preparedness demands a nuanced approach. Strengthening legal frameworks, fostering international collaboration, and integrating health and displacement responses are critical. The identified vulnerabilities underscore the urgency of tailored policies, ensuring a holistic and adaptive strategy to safeguard displaced populations amidst evolving challenges.

The comprehensive analysis of the intersection between internal displacement, extreme climate events, and pandemics underscores the complexity of managing emergencies impacting displaced populations. The assessment of existing legal frameworks reveals that, while there are robust international and regional instruments for the protection of internally displaced persons, there are significant gaps in their applicability during health crises.

Policy recommendations emphasise the need to adapt and strengthen legal frameworks, considering specific protocols for displacement scenarios during pandemics. The importance of integrated international coordination, where public health and disaster management sectors collaborate effectively, is highlighted. Practical strategies, such as the implementation of integrated drills and the establishment of adaptable quarantine protocols, are proposed.

The connection between climate-related displacement and disease spread, as evidenced by the COVID-19 pandemic, underscores the urgency of integrated approaches. The adaptation of the Guiding Principles on Internal Displacement and the consideration of region-specific legal frameworks are essential. The implementation of integrated surveillance systems provides an effective tool for monitoring and responding to the dynamics of these interconnected phenomena.

Among the limitations of this study, we must mention the lack of variables related to the duration of displacement, which hinders analysis of displacement periods and potential exposure to COVID-19 contagions during emergencies [22,24]. Another limitation is related to the lack of other sources of data for producing accurate estimations on internal displacement or for triangulation purposes with existing data sources. We are aware of the limitations of a single discourse, considering the multiplicity of interests involved in the management of extreme weather events and pandemic outbreaks, which makes it difficult to structure a coordinated global agenda among the different actors from both the public and private sectors. In this study, we were not able to analyse the effects of policy stringency on COVID-19 spread and the impact of these policies on internally displaced populations. On the one hand, the impacts of the policies have been so varied that no single article comprehensively analyses all findings. There are separate epidemiological, economic, and mental health studies, for example. Notably, the impact of policies varied significantly depending on the cultural context of each country or subnational unit, as illustrated by a previous study [40].

5. Conclusions

We can conclude that addressing the complexity of the intersection between displacements, extreme climate events, and pandemics requires a global, coordinated, and adaptable response. Preparedness, adaptation of legal frameworks, and collaboration between sectors and countries are crucial for effectively managing emergencies and ensuring the protection and care of internally displaced persons in health crises.

The analyses and proposals presented in this work serve not only as an endpoint but as a precursor for further, more targeted inquiries. The identified complexities, gaps, and recommendations function as guiding signposts, directing attention toward the necessity of ongoing research endeavours. The call for adaptability, collaboration, and preparedness necessitates a continual exploration of these themes to foster a deeper comprehension of the global, coordinated, and adaptable response required in navigating the intricate landscape of health crises.

Author Contributions: R.A.A.Z., G.N.d.L. and J.C.S.-H. contributed to the conception of the work, the analysis, and the interpretation of data. All authors have drafted the work and approved the submitted version. R.A.A.Z., G.N.d.L. and J.C.S.-H. agree to be personally accountable for the author's contributions and for ensuring that questions related to the accuracy or integrity of any part of the work, even ones in which the author was not personally involved, are appropriately investigated, resolved, and documented in the literature. All authors have read and agreed to the published version of the manuscript.

Funding: This research received no external funding. The APC was funded by the University of Helsinki Library. Open access funding provided by University of Helsinki.

Data Availability Statement: Publicly available datasets were analysed in this study. This data can be found here: https://gidd.idmcdb.org/ (accessed on 1 December 2023).

Conflicts of Interest: The authors declare no conflicts of interest.

References

1. Ag Ahmed, M.A.; Ly, B.A.; Diarra, N.H.; Traore, F.B.; Diarra, D.; Kande, I.F.; Dembele, M.; Doumbia, S.; Alami, H. Challenges to the Implementation and Adoption of Physical Distancing Measures against COVID-19 by Internally Displaced People in Mali: A Qualitative Study. *Confl. Health* **2021**, *15*, 88. [CrossRef] [PubMed]
2. de Ruiter, M.C.; van Loon, A.F. The Challenges of Dynamic Vulnerability and How to Assess It. *iScience* **2022**, *25*, 104720. [CrossRef] [PubMed]
3. Abeldaño Zuñiga, R.A.; González Villoria, A.M. Still Ignored and Still Invisible: The Situation of Displaced People and People Affected by Disasters in the COVID-19 Pandemic. *Sustain. Sci.* **2021**, *16*, 1749–1752. [CrossRef] [PubMed]
4. Internal Displacements Monitoring Centre. Available online: https://www.internal-displacement.org/ (accessed on 4 March 2023).
5. Intergovernmental Panel on Climate Change. Climate Change 2021. The Physical Science Basis. 2021. Available online: https://www.ipcc.ch/report/ar6/wg1/downloads/report/IPCC_AR6_WGI_Full_Report.pdf (accessed on 1 December 2023).
6. Intergovernmental Panel on Climate Change. Anexo I: Glosario. In *Calentamiento global de 1,5 °C, Informe Especial del IPCC Sobre los Impactos del Calentamiento Global de 1,5 °C con Respecto a los Niveles Preindustriales y las Trayectorias Correspondientes que Deberían Seguir las Emisiones Mundiales de Gases de Efecto i*; IPCC: Geneve, Switzerland, 2018. Available online: https://www.ipcc.ch/site/assets/uploads/sites/2/2019/10/SR15_Glossary_spanish.pdf (accessed on 1 December 2023).
7. United Nations High Commissioner for Human Rights Guiding Principles on Internal Displacement. Available online: https://www.unhcr.org/media/guiding-principles-internal-displacement (accessed on 15 December 2023).
8. Directorate-General for European Civil Protection and Humanitarian Aid Operations (ECHO). *Addressing Displacement and Migration Related to Disasters, Climate Change and Environmental Degradation: Commission Staff Working Document: July 2022*; ECHO: Luxembourg, 2022.
9. United Nations Framework Convention on Climate Change. *Paris Agreement of the United Nations Framework Convention on Climate Change*; United Nations Framework Convention on Climate Change: Paris, France, 2015.
10. United Nations Office for Disaster Risk Reduction. *Sendai Framework for Disaster Risk Reduction 2015–2030*, 1st ed.; UNDRR, Ed.; United Nations Office for Disaster Risk Reduction: Geneve, Switzerland, 2015.
11. United Nations—General Assembly. *Global Compact for Safe, Orderly and Regular Migration*; United Nations—General Assembly: New York, NY, USA, 2019; Volume 30, pp. 774–816.
12. United Nations Sustainable Development Goals. Available online: https://sdgs.un.org/es/goals (accessed on 9 January 2022).
13. Nansen Initiative. *Agenda for the Protection of Cross-Border Displaced Persons in the Context of Disasters and Climate Change*; Nansen Initiative: Grand-Saconnex, Switzerland, 2015.

14. The Platform on Internal Displacements. Available online: https://disasterdisplacement.org/ (accessed on 1 January 2024).
15. United Nations United Nations Secretary-General's High-Level Panel on Internal Displacement. Available online: https://www.un.org/internal-displacement-panel/ (accessed on 1 January 2024).
16. United Nations High Commissioner for Refugees. *UNHCR Institutional Plan on Solutions to Internal Displacement*; UNHCR: Geneva, Switzerland, 2023.
17. African Union Kampala Convention. *African Union for the Protection and Assistance of Internally Displaced Persons in Africa*; African Union Kampala Convention: Addis Ababa, Ethiopia, 2009.
18. Organization of the American States The American Convention on Human Rights. Available online: https://humanrightscommitments.ca/wp-content/uploads/2015/11/American-Convention-on-Human-Rights.pdf (accessed on 3 January 2024).
19. Suarez-Herrera, J.C.; de A. Hartz, Z.M.; da Cruz, M.M. Alianzas Estratégicas En Salud Global: Perspectivas Innovadoras Para Una Epidemiología Del Desarrollo Sostenible. In *Epidemiología en la era del Desarrollo Sustentable*; Cometto, M.C., Abeldaño Zuñiga, R.A., Moyado Flores, S., Eds.; Universidad Nacional de Córdoba—Editorial Analéctica: Córdoba, Argentina, 2021; pp. 179–214.
20. International Federation of Red Cross and Red Crescent Societies Displacement in a Changing Climate. *Localized Humanitarian Action at the Forefront of the Climate Crisis*; International Federation of Red Cross and Red Crescent Societies: Geneva, Switzerland, 2021.
21. World Health Organization. *The International Health Regulations (2005)*; WHO: Geneva, Switzerland, 2005.
22. Abeldaño Zuñiga, R.A.; Narcizo de Lima, G. The 2020 Hurricanes, Internal Displacements, and COVID-19 in Latin America and the Caribbean Countries. In *The Role of GIS in COVID-19 Management and Control*; CRC Press: Boca Raton, FL, USA, 2023; pp. 111–132, ISBN 9781000876314.
23. Abeldaño Zuñiga, R.A.; Osorio Guzmán, E.U.; González Villoria, R.A.M. Gendered Impacts of COVID-19 on People Internally Displaced Due to Disasters in Latin America and the Caribbean. In *The Impacts of COVID-19 on Migration and Migrants from a Gender Perspective*; IOM, Ed.; IOM: Geneva, Switzerland, 2022; ISBN 9789292682842.
24. Abeldaño Zuñiga, R.A.; de Lima, G.N.; Quiroga, D.E. Climate Change and Population Displacement in Africa. In *The Routledge Handbook of African Demography*; Odimegwu, C., Adewoyin, Y., Eds.; Routledge: London, UK, 2022; pp. 510–532, ISBN 9780429508561.
25. Abeldaño Zuñiga, R.A. Socio-Economic Impact of Disasters in the Philippines: The First Two Decades of the 21st Century. In *Disasters in the Philippines*; Bristol University Press: Bristol, UK, 2023; pp. 25–46.
26. Inter Sector Coordination Group Joint Multi-Sector Needs Assessment (J-MNSA). *Refugee and Host Communities*; J-MNSA: Dhaka, Bangladesh, 2020.
27. Inter Sector Coordination Group Secretariat. *2021 Joint Response Plan: Rohingya Humanitarian Crisis*; Inter Sector Coordination Group Secretariat: Cox's Bazar, Bangladesh, 2021.
28. IASC Task Team on Preparedness and Resilience. *Guidance: Emergency Response Preparedness*; IASC Task Team on Preparedness and Resilience: Geneva, Switzerland, 2015.
29. The World Bank. The Pandemic Fund: Strengthening Disease Surveillance Systems, Laboratory Capacity, and Staff Skills in Public Health Emergency Preparedness and Coordination in Burkina Faso. Available online: https://www.worldbank.org/en/programs/financial-intermediary-fund-for-pandemic-prevention-preparedness-and-response-ppr-fif/brief/strengthening-disease-surveillance-systems-laboratory-capacity-and-staff-skills-in-public-health-emergency-preparedness-and-coordination-in-burkina-faso (accessed on 6 June 2023).
30. The World Bank Pandemic Fund Grant: Democratic Republic of Ethiopia. Available online: https://www.worldbank.org/en/programs/financial-intermediary-fund-for-pandemic-prevention-preparedness-and-response-ppr-fif/brief/pandemic-multi-sectoral-prevention-preparedness-and-response-in-ethiopia (accessed on 6 June 2023).
31. Commonwealth of Australia. *Australia's National Midterm Review of the Sendai Framework for Disaster Risk Reduction 2015–2030 Report*; Commonwealth of Australia: Canberra, Australia, 2022.
32. Thalheimer, L.; Jjemba, E.; Simperingham, E. The Role of Forecast-Based Financing. *Forced Migr. Rev.* **2022**, 34–36.
33. Intergovernmental Authority on Development. *Protocol on Free Movement of Persons in the IGAD Region*; Intergovernmental Authority on Development: Djibouti, Republic of Djibouti, 2020.
34. Internal Displacements Monitoring Centre. *Internal Displacement and Food Security*; Internal Displacements Monitoring Centre: Geneva, Switzerland, 2023.
35. United Nations Office for Disaster Risk Reduction El Impacto de Los Huracanes Obliga a Miles de Desplazados a Volver a Empezar. Available online: https://www.undrr.org/es/news/los-efectos-de-iota-y-eta-aun-reverberan-el-impacto-de-los-huracanes-obliga-miles-de (accessed on 24 February 2022).
36. United Nations Office for Disaster Risk Reduction (UNDRR). *Global Assessment Report on Disaster Risk Reduction 2019*; United Nations Office for Disaster Risk Reduction (UNDRR): Geneva, Switzerland, 2019.
37. Oelgemöller, C. The Global Compacts, Mixed Migration and the Transformation of Protection. *Interventions* **2021**, *23*, 183–190. [CrossRef]
38. International Monetary Fund Policy Responses to COVID-19. Available online: https://www.imf.org/en/Topics/imf-and-covid19/Policy-Responses-to-COVID-19 (accessed on 6 June 2023).

39. Piccialli, F.; di Cola, V.S.; Giampaolo, F.; Cuomo, S. The Role of Artificial Intelligence in Fighting the COVID-19 Pandemic. *Inf. Syst. Front.* **2021**, *23*, 1467–1497. [CrossRef] [PubMed]
40. Tu, K.; Chen, S.; Mesler, R.M. Policy stringency and the spread of COVID-19: The moderating role of culture and its implications on first responses. *Health Policy* **2023**, *137*, 104896. [CrossRef] [PubMed]

MDPI AG
Grosspeteranlage 5
4052 Basel
Switzerland
Tel.: +41 61 683 77 34

Climate Editorial Office
E-mail: climate@mdpi.com
www.mdpi.com/journal/climate

www.ingramcontent.com/pod-product-compliance
Lightning Source LLC
Chambersburg PA
CBHW061517210926
51850CB00008B/1948